Mitochondria

a practical approach

TITLES PUBLISHED IN THE PRACTICAL APPROACH SERIES

Affinity chromatography
Animal cell culture
Biochemical toxicology
Biological membranes
Carbohydrate analysis
Centrifugation (2nd Edition)
DNA cloning
Drosophila
Electron microscopy in molecular biology
Gel electrophoresis of nucleic acids
Gel electrophoresis of proteins
H.p.l.c. of small molecules
Human cytogenetics
Human genetic diseases
Immobilised cells and enzymes
Iodinated density-gradient media
Microcomputers in biology
Mutagenicity testing
Neurochemistry
Nucleic acid and protein sequence analysis
Nucleic acid hybridisation
Oligonucleotide synthesis
Photosynthesis: energy transduction
Plant cell culture
Spectrophotometry and spectrofluorimetry
Steroid hormones
Teratocarcinomas and embryonic stem cells
Transcription and translation
Virology

Mitochondria

a practical approach

Edited by

V M Darley-Usmar

Department of Biochemistry, Wellcome Research Laboratories, Langley Court, Beckenham, Kent BR3 3BS, UK

D Rickwood

Department of Biology, University of Essex, Wivenhoe Park, Colchester, Essex CO4 3SQ, UK

M T Wilson

Department of Chemistry, University of Essex, Wivenhoe Park, Colchester, Essex CO4 3SQ, UK

OXFORD · WASHINGTON DC

IRL Press Limited,
PO Box 1,
Eynsham,
Oxford OX8 1JJ,
England

British Library Cataloguing in Publication Data

Mitochondria: a practical approach.
1. Mitochondria
I. Darley-Usmar, V.M. II. Rickwood, D.
III. Wilson, M.T.
574.87′342 QH603.M5

ISBN 1-85221-034-6 (hardbound)
ISBN 1-85221-033-8 (softbound)

Printed by Information Printing Ltd, Oxford, England

Preface

Studies of the mitochondrion have always been at the heart of biochemistry and its associated areas of research. This is hardly surprising considering the functions that this organelle performs, functions which are central to the life of most eukaryotic cells. Although we have learned a great deal about mitochondria over the last decades, the field of mitochondrial research remains a fertile one and there is no doubt there is still much to learn using the new techniques that are now available.

This book is intended to help all those people who wish to do experiments using either mitochondria or systems derived from mitochondria. Such research workers may be relative novices to the field, postgraduates or scientists coming into the area of mitochondrial investigations from other disciplines. They may, on the other hand, be researchers ('mitochondriacs' even) who are widely experienced in some aspects of mitochondrial biochemistry but who now wish to broaden the scope of their research interests. Indeed, such shifts of emphasis are quite likely to occur in mitochondrial research as there has grown up, over the years, a dichotomy in the approach to working on this most important of organelles. On the one hand there have been intensive studies of the components of the major metabolic pathways associated with the mitochondrion and on the other a great deal of work has been carried out on the biogenesis of mitochondria in terms of the molecular biology and origin of the various components. The aim of this book is to present in a single volume the most important techniques in both these areas of mitochondrial research.

In the course of compiling these methods some techniques have inevitably been related to mitochondria from a particular species or cell type. However, this should not pose a serious problem in view of the similarity of mitochondrial functions in a wide range of organisms. What we have tried to do is to give a representative method for all the important investigative procedures.

Unfortunately, constraints of space have made it necessary for us to leave out some important topics. Probably the foremost of these is a full discussion of the lipid composition of mitochondria. This we have, with regret, relegated to Appendix II.

We thank the contributors to this volume who have not only written in a way we hope and believe is clear and informative, but who have also made our job as editors so much easier than it could otherwise have been.

V.M.Darley-Usmar
D.Rickwood
M.T.Wilson

Contributors

M.Barat-Gueride
Biologie Génèrale, Bât 400, Université de Paris-Sud, Orsay 91405, France

D.S.Beattie
Department of Biochemistry, School of Medicine, Medical Center, Morgantown, WV 26506, USA

L.G.Brent
Veterans Administration Medical Center, 4500 S Lancaster Road, Dallas, TX 75216, USA

R.A.Capaldi
Institute of Molecular Biology, University of Oregon, Eugene, OR 97401, USA

V.M.Darley-Usmar
Department of Biochemistry, The Wellcome Research Laboratories, Langley Court, Beckenham, Kent BR3 3BS, UK

A.Dawson
University of East Anglia, School of Biological Sciences, Norwich, Norfolk NR4 7TJ, UK

R.Docherty
Department of Biochemistry, UTHSC at Dallas, 5323, Harry Hines Bvd., Dallas, Texas, USA

B.Dujon
Centre de Genetique Moleculaire, Centre National de la Recherche Scientifique, 91190 Gif-sur-Yvette, France

W.W.Hauswirth
Department of Immunology and Medical Microbiology, University of Florida, Gainsville FL 32610, USA

M.Klingenburg
Institute of Physiological Chemistry, University of Munich, Goethestrasse, Munich 2, FRG

R.Krämer
Institute of Physiological Chemistry, University of Munich, Goethestrasse, Munich 2, FRG

R.Lang
Biology Department, Open University, Walton Hall, Milton Keynes MK7 6AA, UK

L.O.Lim
Department of Immunology and Medical Microbiology, University of Florida, Gainsville, FL 32610, USA

P.N.Lowe
Department of Biochemical Microbiology, The Wellcome Research Laboratories, Langley Court, Beckenham, Kent BR5 3BS, UK

F.Malatesta
Department of Experimental Medicine and Biochemical Sciences, University of 'Tor Vergata', Rome, Italy

F.Millett
Department of Chemistry, Chemistry Building, Fayetteville, AR 72701, USA

C.I.Ragan
Department of Biochemistry, School of Biochemical and Physiological Sciences, University of Southampton, Southampton SO9 3TU, UK

D.Rickwood
Department of Biology, University of Essex, Wivenhoe Park, Colchester, Essex CO4 3SQ, UK

J.B.Robinson,Jr
Veterans Administration Medical Center, 4500 S Lancaster Road, Dallas, TX 75216, USA

P.Sarti
Centro di Biologica Moleculaire, Istituto di Chimica Biologia, Citta Universitara, 00185 Roma, Italy

K.Sen
Department of Biochemistry, School of Medicine, Medical Center, Morgantown WV 26506, USA

P.Srere
Veterans Administration Medical Center, 4500 S Lancaster Road, Dallas, TX 75216, USA

B.Sumegi
Veterans Administration Medical Center, 4500 S Lancaster Road, Dallas, TX 75216, USA

S.Takamiya
Institute of Molecular Biology, University of Oregon, Eugene, OR 97401, USA

G.Turner
Microbiology Department, The Medical School, Bristol University, Bristol BS8 1TD, UK

M.T.Wilson
Department of Chemistry, University of Essex, Wivenhoe Park, Colchester, Essex CO4 3SQ, UK

Contents

ABBREVIATIONS **xvi**

1. ISOLATION AND CHARACTERISTICS OF INTACT MITOCHONDRIA **1**
D.Rickwood, M.T.Wilson and V.M.Darley-Usmar

Introduction 1
Isolation of Mitochondria 3
General principles 3
Isolation of mitochondria from mammalian cells 4
Isolation of mitochondria from fungi 6
Isolation of mitochondria from plant tissues 8
Purity and Functional Characterization of Intact Mitochondria 10
The oxygen electrode 11
Spectral characterization 14
Analysis of purity on the basis of DNA 16
References 16

2. ELECTRON MICROSCOPY OF MITOCHONDRIA **17**
R.D.A.Lang

Introduction 17
Fundamental principles 17
Choice of method 17
Thin Sectioning 18
Principles 18
Experimental procedures 18
Negative Staining 19
Principle of the method 19
Experimental procedures 19
Freeze-etching 21
Principles of the technique 21
Experimental procedure 22
Rapid freezing methods 26
Interpretation and analysis of freeze-etch replicas 29
References 33

3. TRANSPORT ACROSS MEMBRANES **35**
A.Dawson, M.Klingenburg and R.Krämer

Cation Transport: Introduction 35
Proton Transport 35
Fundamental concepts 35
Measurement of H^+-pumping and stoichiometries 37

Measurement of proton-motive force 41
Calcium Transport 51
Fundamental concepts 51
Methodology 52
Metallochromic indicators 54
Ca^{2+}-sensitive electrodes 57
Accumulation of Ca^{2+} studied using $^{45}Ca^{2+}$ 61
Methods for Measuring Metabolite Transport in Mitochondria 64
Introduction 64
Pre-treatment of mitochondria 65
Transport measurements 67
Special equipment for higher time resolution in transport studies 71
Measuring Metabolite Transport by Mitochondrial Carrier Proteins in Reconstituted Systems 72
Introduction 72
Reconstituted systems of mitochondrial carriers 73
Basic principles of transport measurements in proteoliposomes 74
Procedures 75
Acknowledgements 76
References 77

4. SUB-FRACTIONATION OF MITOCHONDRIA AND ISOLATION OF THE PROTEINS OF OXIDATIVE PHOSPHORYLATION 79
C.I.Ragan, M.T.Wilson, V.M.Darley-Usmar and P.N.Lowe

Introduction 79
Sub-Fractionation of Mitochondria 79
Preparation of sub-mitochondrial particles 79
Sub-fractionation of mitochondria using detergents 80
Assay of Fractions with Marker Enzymes 81
Assay for malate dehydrogenase—a matrix marker 81
Assay for monoamine oxidase—an outer membrane marker enzyme 82
Calculation of the purity of mitochondrial fractions 82
Preparation of Mitochondria or Mitochondrial Fragments from Beef Heart 83
Mitochondrial preparation suitable for complexes I, II, III and IV 84
Additional mitochondrial preparation suitable for complex IV (cytochrome c oxidase) 84
Preparation of Complex I: General Strategy 85
Purification procedure 85
Properties of complex I 89
Preparation of Complex II: General Strategy 91
Purification procedure for complex II 91
Properties of complex II 94

Preparation of Complex III: General Strategy 96
Purification procedure for complex III 97
Properties of complex III 98
Preparation of Ferrocytochrome c: Oxygen Oxidoreductase EC 1.9.3.1 (Complex IV): General Strategy 101
Assay of purification by absorbance spectroscopy 101
The red/green split 102
Ammonium sulphate fractionation 104
Final steps 105
Yonetani method 106
Determination of concentration and criteria of purity 107
Measurement of enzyme activity 107
Preparation of ATPase (ATP Synthetase) 108
Purification procedure 108
Properties of ATPase 109
References 111

5. RECONSTITUTION AND MOLECULAR ANALYSIS OF THE RESPIRATORY CHAIN 113
V.M.Darley-Usmar, R.A.Capaldi, S.Takamiya, F.Millett, M.T.Wilson, F.Malatesta and P.Sarti

Introduction 113
Polyacrylamide Gel Electrophoresis of Inner Mitochondrial Membrane Proteins 113
Sample preparation 114
Choice of SDS–PAGE system 114
Large-Scale Isolation of Cytochrome c Oxidase Subunits 117
Preparation of subunits MtI, MtII, MtIII and C_{IV} of cytochrome c oxidase 117
Preparation of polypeptides C_V, C_{VI}, ASA and AED of cytochrome c oxidase 118
Preparation of polypeptides $C_{VII}-C_{IX}$ of cytochrome c oxidase 121
Analysis of Molecular Structure by Radiolabelling and Antibody Binding 122
Detection of radiolabelled proteins after SDS–PAGE 122
Western blotting of mitochondrial proteins 123
Preparation of Selectively Modified Analogues of Cytochrome c 129
Modification of cytochrome c with m-trifluoromethylphenyl-isocyanate 130
Chromatographic separation of CF_3PhNHCO-cytochrome c derivatives 130
Peptide mapping of CF_3PhNHCO-cytochrome c derivatives 134
Reaction of Cytochrome c Analogues with the Cytochrome bc_1 Complex and Cytochrome c Oxidase 134

Reaction of cytochrome c derivatives with complex III 134
Reaction of cytochrome c derivatives with complex IV 137
Interpretation of kinetic results 138
The Use of Chemical Modification and Peptide Mapping to Locate the Cytochrome c Binding Site on Subunit II of Cytochrome c Oxidase 139
Reaction of cytochrome c oxidase with ETC 139
Digestion of subunit II and peptide mapping 141
Reconstitution of Mitochondrial Complexes into Artificial Phospholipid Vesicles 143
Equipment and materials 143
Strategies for incorporation of membrane proteins into phospholipid vesicles 144
Functional Assays of Reconstituted Cytochrome Oxidase 147
Proton pumping assay 147
Respiratory control ratio 151
References 152

6. AN ENZYMATIC APPROACH TO THE STUDY OF THE KREBS TRICARBOXYLIC ACID CYCLE 153
J.B.Robinson,Jr, L.G.Brent, B.Sumegi and P.A.Srere

Introduction 153
Disruption of Mitochondria 153
Detergent treatments 153
Sonication 155
Osmotic shock 156
Freeze–thawing and lyophilization 157
Permeabilization of mitochondria using toluene 157
Release of enzyme activity by digitonin 158
Assay of Krebs Tricarboxylic Acid Cycle Enzymes 159
Factors affecting the assay of enzymes 159
General spectrophotometric techniques 160
Assays of individual mitochondrial enzymes 160
Isolation of Mitochondrial Enzymes 164
An overview of protein purification 165
Examples of enzyme isolation: isolation of citrate synthase, malate dehydrogenase and fumarase from yeast mitochondria 168
Acknowlcdgements 170
References 170

7. METHODS FOR STUDYING THE GENETICS OF MITOCHONDRIA 171
W.W.Hauswirth, L.O.Lim, B.Dujon and G.Turner

Introduction to the Genetics of Mitochondria from Animal and Plant Cells 171

Isolation of mtDNA from Mammalian Tissue and Tissue Culture Cells 171
Large-scale preparation of animal mitochondria 172
Small-scale preparation of animal mitochondria 173
Isolation of mtDNA from purified mitochondria 173
Very small-scale preparation of mtDNA 174
Direct mtDNA analysis from animal tissue 175
Isolation of displacement loops of mtDNA 175
Isolation of mtDNA from tissue culture cells 176
Isolation of Mitochondrial RNA from Mammalian Tissue and Tissue Culture Cells 177
Phenol purification of mtRNA 178
CsCl purification of mtRNA 178
Isolation of Mitochondrial DNA and RNA from higher plant tissues 178
Preparation of plant mitochondria 179
Preparation of plant mtDNA 179
Preparation of plant mtRNA 180
Analysis of mtDNA and mtRNA from Animal and Plant Tissues 181
Rapid detection of restriction site polymorphisms in animal tissue 181
Detection of mtDNA by the biotin–avidin method 184
Introduction: Nature of the Mitochondrial Mutations of Saccharomyces cerevisiae 187
Media for Working with Yeast Mitochondrial Mutants 189
Composition and preparation of media 189
Use of media and precautions in growing mitochondrial mutants 191
Basic Genetic Techniques for Yeast 195
Isolation of mitochondrial mutants 195
Segregation tests 203
Tests for recombination 209
Complementation tests 219
Quantitative and qualitative tests for suppressiveness 220
Analysis of Yeast Mitochondrial DNA 223
Purification of mitochondrial DNA 223
Quick, small-scale preparation of yeast mitochondrial DNA (minilysates) 224
Colony hybridization 225
Analysis of Yeast Mitochondrial Translation Products 226
Mitochondrial Genetics of Aspergillus nidulans 227
Introduction 227
Isolation and identification of extranuclear mutants 227
Vegetative segregation of mitochondrial alleles 233
Recombination of mitochondrial markers 233
Uniparental inheritance in the sexual cycle 234
Physical analysis of the mitochondrial genome 236
A nuclear gene encoding a mitochondrial function 242

Acknowledgements 242
References 242

8. **DNA REPLICATION AND TRANSCRIPTION** **245**
M.Barat-Gueride, R.Docherty and D.Rickwood

Introduction 245
Replication of Mitochondrial DNA 245
Introduction 245
Systems for studying DNA replication 246
Methods for studying DNA replication 254
Methods for identifying the origin of replication 262
Transcription in Mitochondria 265
Introduction 265
Systems used to study transcription 265
Characteristics of mitochondrial RNA polymerases 267
Isolation of RNA from mitochondria 268
Methods to study the rates of synthesis and initiation in mitochondria 270
Determination of the accuracy of initiation of transcripts 275
Hybridization analysis of transcription 275
Inhibitors of transcription in mitochondria 279
References 280

9. **SYNTHESIS OF MITOCHONDRIAL PROTEINS** **283**
D.S.Beattie and K.Sen

Introduction 283
Cytoplasmic Synthesis of Mitochondrial Proteins 283
Methods for Studying the Import of Proteins into Mitochondria 284
Antibodies to membrane proteins 284
Import of proteins into mitochondria in vivo 285
In vitro import of polypeptides into mitochondria 290
In vitro transcription–translation systems 297
Import Studies Using Other Mitochondrial Systems 298
Import of proteins into mammalian mitochondria in vitro 298
Import of proteins into mammalian mitochondria in vivo 299
Isolation of Mitochondrial Polysomes 302
Isolation of mitochondria 302
Isolation of polysomes 303
Assay for presence of polysomes 304
In Vitro Synthetic Systems 304
Protein synthesis by isolated mitochondria 304
Protein synthesis on isolated ribosomes 306

Protein synthesis on isolated polysomes 307
Acknowledgements 307
References 308

APPENDICES

I. Single letter code for amino acids 311
II. Phospholipid content of mitochondria 313

INDEX **317**

Abbreviations

ARS	autonomous replication sequence
BCIP	5-bromo-4-chloro-3-indolyl phosphate
BSA	bovine serum albumin
COV	cytochrome oxidase vesicles
DAPI	4,6-diamidine-2-phenylindole hydrochloride
DBM-paper	diazobenzyloxymethyl paper
DEPC	diethylpyrocarbonate
DMO	5,5′-dimethyloxazolidine-2,4-dione
DTT	dithiothreitol
EGTA	ethyleneglycobis(β-aminoethyl)ether tetraacetic acid
ETC	1-ethyl-3(-3-[^{14}C]trimethylaminopropyl) carbodiimide
FAME	fatty acid methyl ester
FITC	fluorescein isothiocyanate
HEDTA	*N*-hydroxyethyl ethylenediamine triacetic acid
HRP	horseradish peroxidase
MM	minimal medium
MNNG	*N*′-methyl-*N*′-nitro-*N*′-nitrosoguanidine
Mops	3-(*N*-morpholino) propanesulphonic acid
mtDNA	mitochondrial DNA
mtRNA	mitochondrial RNA
NBT	nitroblue tetrazolium
NM	nutrient medium
NTA	nitrilotriacetate
PAGE	polyacrylamide gel electrophoresis
PBS	phosphate-buffered saline
PCA	perchloric acid
PEG	polyethylene glycol
PVC	polyvinyl chloride
PVP	polyvinyl pyrrolidone
RC	respiratory competent
RCR	respiratory control ratio
RD	respiratory deficient
SDS	sodium dodecylsulphate
SMP	sub-mitochondrial particle
SSC	standard saline citrate
TAME	*N*α-*p*′-tofyl-L-arginine methyl ester
TPMP	triphenylmethylphosphonium
TPP	tetraphenylphosphonium
TMPD	tetramethyl-*p*-phenylenediamine dihydrochloride
Xgal	5-bromo-4-chloro-3-indolyl β-D-galactopyranoside

CHAPTER 1

Isolation and characteristics of intact mitochondria

D.RICKWOOD, M.T.WILSON and V.M.DARLEY-USMAR

1. INTRODUCTION

Eukaryotic cells are extremely diverse, covering the plant and animal kingdoms as well as the fungi. Whether the eukaryotes are single-celled or complex multi-cellular organisms it is remarkable that the structure and functions of mitochondria are so similar. There are four major structural regions, these are the outer and inner membranes and the two spatial regions delineated by these membranes, namely the intermembrane space and the matrix. These structural regions are broadly associated with specific and different functions. *Table 1* lists the major metabolic functions of the mitochondrion, their associated macromolecules and intra-mitochondrial location. In this book we have concentrated on the practical aspects of those functions shown in *Table 1* which form major areas of current research interest. For this reason some topics have not been covered here; for example rather little is known of outer membrane structure and function. In other cases, for example the proteases and proteins which guide the import and processing of those mitochondrial proteins coded for and synthesized in the cytosol, their presence has been implied but, except in a few cases (e.g. ref. 1), they have not been isolated. For a more detailed theoretical discussion of mitochondrial function the reader is referred to a text that reviews many aspects of mitochondria (2).

Differences between mitochondria from different species and between different tissues in the same organism reflect, in the most part, the relative importance of the different metabolic pathways contained within the mitochondrion to a given cell or organism. Morphologically and genetically all mitochondria appear to be strikingly similar. This fact has allowed researchers to generalize both methods of preparation and analysis for mitochondria from a wide variety of sources. Deeper analysis invariably reveals important and intriguing differences between mitochondria from different sources, some of which are of considerable significance to the metabolism of the organism. A case in point is the way in which mitochondria from brown adipose tissue have been modified specifically to generate heat (3).

In order to assess any of the properties of mitochondria it is important to be able to purify them free of contaminants that may interfere with the activity of interest. However, the 'purity' of mitochondria depends on the area of interest and this also determines the method of assay used to determine the purity. For example, if one is interested in electron transport processes then usually one is not interested in whether or not the mitochondria are contaminated with nuclear DNA, while such contamina-

Table 1. Compartments of the mitochondrion and their associated metabolic functions.

Outer membrane	
Function	*Associated enzyme(s)*
Oxidation of neuroactive aromatic amines	Monoamine oxidase (4)
Cardiolipin biosynthesis	e.g. glycerol phosphate acyl transferase (5)
Transport of nuclear coded and cytoplasmically synthesized proteins	Not yet identified (6)
Electron transfer	NADH cytochrome *c* reductase (7) (rotenone-insensitive). The function of this protein is not yet fully defined.
Intermembrane space	
Function	*Associated enzymes*
Maintenance of adenine nucleotide balance	Adenylate kinase Nucleoside diphosphokinase Nucleoside monophosphokinase (8)
Electron transfer from complex III to complex IV of the respiratory chain	Cytochrome *c* (see Chapter 5)
Processing of proteins imported from cytoplasm	Not yet isolated (see Chapter 9)
Inner membrane	
Function	*Associated macromolecules*
Oxidative phosphorylation	The phospholipid bilayer which is essential to maintain the proton gradient. Four electron transfer complexes, three of which couple electron transfer to formation of a proton gradient, and a proton-driven ATP synthetase (see Chapters 4,5).
Transport of pyridine nucleotides	ADP/ATP translocase (see Chapter 3)
Ca^{2+} ion transport	e.g. Ca^{2+} ATPase (see Chapter 3)
Transport of metabolites	Pyruvate carrier, $H_2PO_4^-/OH^-$ antiport, dicarboxylate carriers, citrate/malate antiport, carnitine shuttle
Matrix	
Function	*Associated macromolecules*
Oxidation of pyruvate to acetyl CoA	The pyruvate dehydrogenase complex (9)
Oxidation of ketone bodies	e.g. 3-ketoacid CoA transferase (10)
Oxidation of amino acids	e.g. glutaminase, glutamate dehydrogenase, aspartate aminotransferase, α-ketoglutarate transaminases (11)
Part of the urea cycle	Carbamylphosphate synthetase, ornithine transcarbamylase (12)
Oxidation of fatty acids to acetyl CoA	Fatty acyl-CoA dehydrogenase, enoyl hydratase, β-hydroxyacyl-CoA dehydrogenase, β-ketoacyl-CoA thiolase (13)
Protection against oxidative stress	Superoxide dismutase, catalase, glutathione peroxidase, glutathione reductase (14)
Processing of proteins imported from the cytoplasm	Proteases for specific signal peptides of imported proteins (see Chapter 9)
Inheritance of genes coding for mitochondrial RNA and some proteins	Mitochondrial DNA, DNA polymerase and primase (see Chapters 7 and 8)
Synthesis of 13 membrane components of the proteins of oxidative phosphorylation	Ribosomes and apparatus for transcription and translation (see Chapters 8 and 9)

Table 2. Purification of mitochondria by isopycnic sucrose gradients.

1.	Prepare continuous sucrose gradients from 1–2 M sucrose containing 1 mM EDTA, 0.1% BSA and 10 mM Tris-HCl, pH 7.5. Prepare gradients either by using a simple gradient maker or, more easily, by allowing four solutions of 1.0, 1.3, 1.6 and 2.0 M sucrose buffered and containing EDTA and BSA to diffuse overnight. Cool the gradients to 5°C before use.
2.	Gently but thoroughly resuspend the pellet of crude mitochondria in 0.8 M sucrose buffered and containing EDTA and BSA; it is often convenient to use a loose-fitting Potter homogenizer for this step.
3.	Centrifuge the gradients for 2 h at 80 000 *g* at 5°C. The intact mitochondria form a brown band at about 1.19 g/ml. On occasions brown bands denser and lighter than the intact mitochondria are also found; these represent damaged mitochondria.
4.	It is possible to unload the whole gradient into fractions but usually one removes the band of intact mitochondria using a Pasteur pipette.
5.	Dilute the gradient solution containing the mitochondria by the addition of 2 vols of 1 mM EDTA, 10 mM Tris-HCl, pH 7.4 and pellet the pure mitochondria by centrifugation at 20 000 *g* for 10 min at 5°C.

tion could be prejudicial for studies of the transcription of mitochondrial DNA.

This chapter describes in detail the isolation of mitochondria from different types of cells and the methods that can be used to assess their integrity and purity.

2. ISOLATION OF MITOCHONDRIA

2.1. General principles

The experimental approach for isolating mitochondria is basically the same irrespective of the source of tissue; plant or animal. The first steps involve rupture of the cell membrane while maintaining the structural integrity of the mitochondria. After breaking open the cells, differential centrifugation is used sometimes together with isopycnic density gradients, to separate the mitochondria from other organelles and cell debris. However, the yield of mitochondria and the ease with which pure preparations may be obtained depends very much on the type of tissue and, of course, the amount of tissue available. The exact details of the method chosen will depend not only on the tissue source but also on the type of experiment for which the mitochondria are to be isolated. For example, in some instances, such as the isolation of electron transport complexes, mitochondrial integrity may be sacrificed to obtain high yield. In this chapter we describe experimental protocols for preparing mitochondria from a wide range of organisms; in addition, in the subsequent chapters, the methods most suitable for a specific application are also described.

2.1.1 *Purification of mitochondria by isopycnic centrifugation*

After lysis of the cells, low-speed centrifugation is used to remove unlysed cells, nuclei and large membrane fragments prior to pelleting the mitochondria by centrifugation at 10 000 *g* for 10 min. The 'mitochondrial' pellet in fact contains a wide range of components including lysosomes, peroxisomes and membrane fragments. Although repeated washing of this crude pellet can be used as a method of purifying mitochondria, it is not very efficient either in terms of the final purity or yield. Usually the preferred method of purifying the mitochondria is to use isopycnic centrifugation. Usually one uses a 1–2 M sucrose gradient centrifuged at 80 000 *g* for 2 h at 5°C (*Table 2*). In some instances it may be advantageous to use one of the newer types of gradient

Table 3. Gradients for the isopycnic purification of mitochondria.

Gradient medium	*Concentration range*	*Centrifugation conditions g/time (min)*	*Density of mitochondria (g/ml)*
Sucrose	1.0–2.0 M	80 000/120	1.19
Percoll[a]	25–60%(v/v)	40 000/20	1.10
Metrizamide	20–50%(w/v)	80 000/60	1.16
Nycodenz	20–50%(w/v)	80 000/60	1.17

[a]The gradient also contains 0.25 M sucrose as the osmotic balancer.

fractionation such as 25–60% Percoll gradients, or 20–50% metrizamide or Nycodenz gradients (*Table 3*) as often gradients of these media give better resolution and more intact mitochondria because they are less viscous and expose the mitochondria to less osmotic stress (18). Whichever type of gradient medium is used, it is advisable to use continuous gradients because, although discontinuous gradients do appear to give sharper bands, artefactual fractionations can occur at the interfaces. The mitochondria form distinct brown bands within the gradient and they can be removed by using a Pasteur pipette. When separating mitochondria on gradients it is important not to expose them to excessive centrifugal force because the combination of high centrifugal force and the hydrostatic force generated in the gradient by the centrifugal force can disrupt the integrity of the mitochondrial membranes.

Mitochondria isolated by isopycnic centrifugation can be considered intact based upon their buoyant density in the gradient medium since damage to the outer or inner membranes alters this characteristic. However, this is only a fairly crude indication of the integrity of mitochondria and it gives no information as to their functionality.

2.2 **Isolation of mitochondria from mammalian cells**

2.2.1 *Isolation of mitochondria from liver*

Rat-liver mitochondria have long been used by investigators of mitochondrial structure and function because of the ease with which it is possible to prepare intact, pure mitochondria in high yields and they remain the preparation of choice for many workers. As a result of the historical pre-eminence of liver mitochondria in studies on mitochondrial functions, a large number of experimental protocols have been published in the literature (e.g. ref. 15). However, all are essentially similar in terms of homogenization method and medium. The protocol that has been used by the authors is as follows.

(i) Starve an adult (250 g) rat overnight to deplete the levels of glycogen and fatty acids in the liver.

(ii) Kill the animal by decapitation, exsanguinate as much as possible, remove the liver and place it into ice-cold homogenization medium [0.3 M sucrose, 1 mM EGTA, 5 mM Mops, 5 mM KH_2PO_4, 0.1% bovine serum albumin (BSA) (fatty acid free)] adjusted to pH 7.4 with KOH. Other sugars can be used in place of sucrose (15).

(iii) With small scissors chop the liver into small (2 mm) cubes and wash in ice-cold homogenizing medium to remove as much blood as possible; the final washing medium should be free of blood.

(iv) Transfer the chopped liver to a pre-cooled glass–Teflon motorized Potter–Elvejhem homogenizer, add 2 ml of cold homogenization medium to each gram of chopped liver and homogenize the tissue using six up and down strokes of the pestle rotating at 500–1000 r.p.m.

(v) Filter the homogenate into a beaker through gauze, rinse the homogenizer with more medium and pool with the rest of the homogenate.

(vi) Transfer the homogenate to centrifuge tubes and centrifuge at 5°C for 10 min at 1000 *g*.

(vii) Carefully decant the supernatant and centrifuge it for 10 min at 10 000 *g* at 5°C; discard the supernatant.

(viii) Resuspend the pellet in 0.5 ml of the homogenizing medium using a 5 mm diameter cooled glass rod. The mitochondria form a soft brown pellet, if a dark brown button is found this should be discarded as it consists of pelleted red blood cells.

The mitochondria obtained are relatively pure and simple repeated washing (two or three times) in homogenizing medium may provide a pure enough preparation, alternatively the mitochondria can be purified by isopycnic centrifugation on sucrose gradients (*Table* 2).

2.2.2 *Isolation of mitochondria from heart muscle*

The difficulty of preparing intact mitochondria from heart muscle arises because of the problem of disrupting this fibrous tissue. It is possible to use a blender with rotating blades to disrupt muscle tissue. Although useful for bulk preparation (see Chapter 4) such treatment can adversely affect the integrity of the mitochondria. An alternative approach is to use proteolytic enzymes so that the cells of the minced and digested muscle tissue can be lysed by homogenizing them using a glass–Teflon Potter–Elvejhem homogenizer. Nagarse, a mixture of bacterial proteolytic enzymes, has been used to digest muscle tissue but it is not possible to inhibit the activity of this mixture of enzymes thus making it impossible to store mitochondria prepared using this method for prolonged periods (16). Alternatively collagenase or trypsin can be used; these have the advantages of specific proteolytic activity which can be inhibited. The protocol used by the authors is as follows.

(i) Obtain fresh heart muscle, if from an abbatoir it should be transported to the laboratory in ice. Remove the fat and pericardium then mince the tissue; it is suggested that only 500 g of tissue is processed at a time.

(ii) Suspend the minced heart in an equal volume of 0.3 M sucrose, 1 mM $CaCl_2$, 5 mM Mops, 5 mM KH_2PO_4, 0.1% BSA adjusted to pH 7.4 with KOH.

(iii) Add collagenase, 7 mg/100 ml of suspension, mix well and incubate at 0°C for 40 min.

(iv) Terminate the incubation by the addition of EGTA to a final concentration of 2 mM.

(v) Drain the digestion medium from the minced tissue and resuspend the tissue in fresh medium containing 0.3 M sucrose, 1 mM EGTA, 5 mM Mops, 5 mM KH_2PO_4, 0.1% BSA adjusted to pH 7.4 with KOH (300 ml of medium for each 100 g of tissue).

(vi) Disaggregate the digested tissue with a glass rod. Then further disaggregate the minced tissue using a ratio of one volume of mince to 8 volumes of buffer with six up and down strokes in a loose-fitting glass–Teflon homogenizer at 100–300 r.p.m. Pool the samples and then homogenize again using a tight-fitting homogenizer. It is important not to process too much tissue at one time.

(vii) Centrifuge the homogenate at 1500 *g* for 10 min at 5°C, carefully decant the supernatant and centrifuge it at 10 000 *g* for 10 min at 5°C. Discard the clear red supernatant.

(viii) A lighter layer of damaged mitochondria is found on top of the dark brown pellet of mitochondria, remove this by gentle washing. This pellet can then be purified by repeated washing and this gives a good yield of intact mitochondria, alternatively the mitochondria can be purified by isopycnic centrifugation on sucrose gradients as described in *Table 2*.

2.2.3 *Isolation of mitochondria from tissue culture cells*

Cultured mammalian cells are usually quite difficult to break open, this, together with the often small amount of cells available and the low activity of some of the enzymes, makes it difficult to isolate pure intact mitochondria from tissue culture cells (17). The following method has been used by the authors.

(i) Harvest 5×10^7 cells, wash them three times with 5 ml of Ca^{2+}, Mg^{2+} free phosphate-buffered saline (PBS) and freeze at −70°C; this helps to weaken the cells prior to homogenization, but unfortunately it also uncouples the mitochondria. Resuspend the cells in 5 ml of 0.25 M sucrose, 1 mM EGTA, 10 mM Hepes–NaOH, pH 7.4 containing 0.5% BSA and centrifuge the suspension at 500 *g* for 2 min at 5°C. Discard the cloudy supernatant and resuspend the cells in 5 ml of the same medium. (An alternative way of facilitating the lysis of cells, which avoids the necessity of freezing, is to use 10 mM triethanolamine–acetate buffer, pH 7.0 in the homogenizing medium instead of Hepes–NaOH as this makes the cells fragile and they lyse readily when homogenized.)

(ii) Homogenize the cells in a tight-fitting glass–Teflon homogenizer using 10 up and down strokes at 500 r.p.m.

(iii) Centrifuge the homogenate at 1500 *g* for 10 min at 5°C. Keep the supernatant and re-extract the nuclear pellet with a further 5 ml of homogenizing medium; combine the two supernatants.

(iv) Centrifuge the combined post-nuclear supernatants at 10 000 *g* for 10 min at 5°C. The crude mitochondrial pellet obtained contains about 5 mg of protein; the mitochondria can be purified by isopycnic centrifugation on sucrose gradients (*Table 2*) if required.

2.3 **Isolation of mitochondria from fungi**

It is more difficult to prepare pure mitochondria from fungal cells than from animal cells because of the presence of a tough cell wall in all but a few mutant strains (e.g. slime mutants of *Neurospora*). Most work has centred on yeast strains especially *Saccharomyces*, as well as the filamentous fungi *Aspergillus* and *Neurospora*. The major problem in this case is to break open the cells. There are two approaches to this, either the cell wall can be digested enzymatically to give protoplasts, or broken physically by using various grinding methods.

The disadvantages of enzymatically protoplasting cells are that it involves prolonged incubation under adverse conditions (e.g. high concentrations of cells and thiols) using a crude extract from snail gut (See Section 3.3.3 of Chapter 9) or mushrooms (19) and the conversion to protoplasts may be incomplete. However, a number of workers do prefer to use the protoplasting technique for preparing mitochondria from yeast for a variety of applications.

In the authors' experience grinding is a rapid method of breaking open cells but usually it is necessary to compromise between maximizing the degree of breakage and avoiding fragmentation of cellular structures (20) which in turn can contaminate the mitochondrial preparation. In the authors' experience of *S. cerevisiae*, grinding with glass beads can give rapid preparations of pure mitochondria, similar grinding protocols have been devised for *Aspergillus* using a carborundum wheel (21) and also *Neurospora*. The method used by the authors is as follows and is based on the method of Lang *et al.* (22).

(i) Harvest the yeast cells by centrifugation at 1000 *g* for 20 min at 5°C, then wash them by resuspension in cold distilled water followed by re-centrifugation as before.

(ii) Suspend each gram of cells in two volumes of cold lysis medium containing 0.6 M sorbitol, 1 mM EDTA, 10 mM Tris-HCl, pH 7.4, put them into a screw-top bottle and then add 3 g of acid-washed glass beads (diameter 0.5–0.75 mm) for each millilitre of cell suspension.

(iii) Shake the cells through a distance of about 80–100 cm twice a second (2 Hz) for 30 sec and then cool in ice. Repeat this shaking procedure until a high proportion of the cells are lysed, usually this needs a total shaking time of 3 min. The degree of lysis can be checked microscopically after staining with Trypan blue.

(iv) Pour off the supernatant from the glass beads and wash the beads with the lysis medium once or twice to recover the remainder of the cells.

(v) Centrifuge the cell lysate at 2000 *g* for 10 min at 5°C to pellet unbroken cells, cell wall debris and nuclei.

(vi) Very carefully pour off the supernatant, avoid disturbing the pelleted material, if you think that there may be some contamination of the supernatant by the pellet then repeat step (v).

(vii) Centrifuge the supernatant at 15 000 *g* for 10 min at 5°C to pellet the mitochondria.

(viii) Gently resuspend the mitochondria in 0.8 M sucrose, 1 mM EDTA, 10 mM Tris–HCl, pH 7.4 and 0.1% BSA prior to separating them on a 1–2 M sucrose gradient (*Table* 2).

If you are preparing fungal mitochondria on a large scale it is usually best to use some kind of grinding mill instead of the hand shaking method described in step (iii); a typical grinding mill is shown in *Figure 1*. However when using this type of mill, care must be taken not to grind the cells too much as otherwise the mitochondria can become irreversibly contaminated with material from other cell organelles. In the

Figure 1. Grinding mill for breaking open fungal cells. The cells mixed with glass beads are placed in the cooled, water-jacketed chamber and the rotation of the hard rubber impeller inside the chamber breaks open the cells.

authors' experience (20) it is advantageous to use a somewhat larger size of beads (typically 1 mm for yeast), than is used for the hand shaking method.

One feature of yeast is that some strains have mitochondria that are respirationally defective as a result of changes in the nuclear or mitochondrial DNA (see Section 8 of Chapter 7). Such mitochondria can be purified in the usual way because they band at a density similar to normal mitochondria although the appearance of the band, especially in terms of colour, may be different because of a lack of cytochromes. In the case of yeast grown anaerobically there is some evidence of slight changes in density of the pro-mitochondria, but even so similar purification protocols can be used (23).

2.4 Isolation of mitochondria from plant tissues

It is significantly more difficult to obtain pure mitochondria from plant tissues than from either animal or fungal cells. The reasons for this are that not only can the presence of the plant cell wall material interfere with the isolation procedures, but also the presence of plastids in plant cells presents a real threat to the isolation of pure mitochondria for

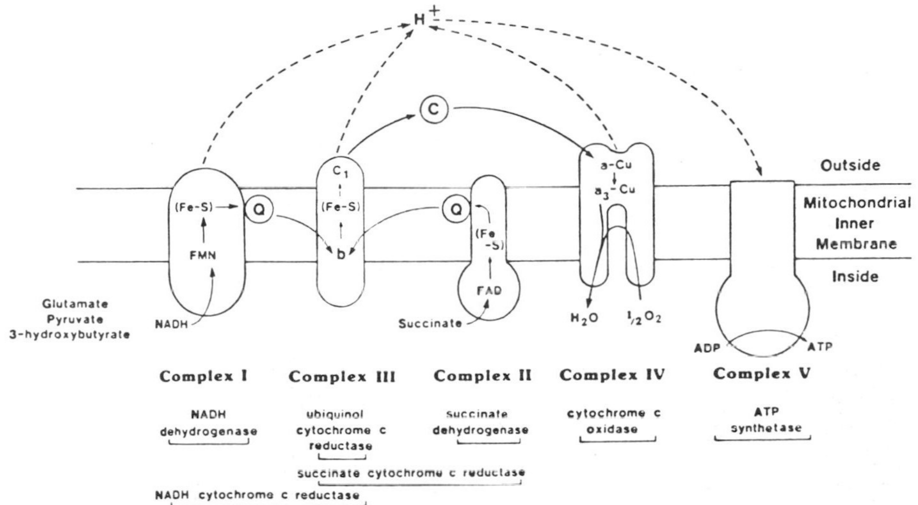

Figure 2. The proteins of oxidative phosphorylation. The electron transport proteins (complexes I–IV) and the ATP synthetase (complex V) are shown in their probable orientation in the mitochondrial inner membrane. The lipid-soluble electron shuttle co-enzyme Q is depicted as Q and the water-soluble electron transfer protein cytochrome *c* as C. Co-enzyme Q is shown here, for simplicity, as transferring electrons from complexes I and II to complex III in a linear fashion. In reality the situation is more complex and the reader is referred to (ref. 26) for a more complete explanation of electron transfer in this part of the chain. The sites of proton translocation at 'coupling sites' I (complex I), II (complex III) and III (complex IV) are shown. The activities measured from the chain can include the entire sequence and, in coupled mitochondria, the ATP synthetase. This is the case when the O_2 electrode is used for assay. For example if NADH is used then 'NADH oxidase' activity is measured. Artificial electron acceptors can be added to pick up electrons earlier in the chain and their reduction monitored spectrophotometrically. Exogenous cytochrome *c* is often used for this purpose. Some of the typical activity measurements made and the parts of the chain they span are also shown.

studies of the respiratory and enzymatic activities of mitochondria, as well as analysis of their molecular biology. Another problem particular to plants is that some tissues (e.g. pigmented tissues or tissues that readily form melanin) are rich in phenolic compounds that can bind to mitochondria and inactivate them. The addition of an anti-oxidant such as 20 mM ascorbic acid to the homogenization medium helps to prevent oxidation of the phenolic compounds which can also be scavenged by the addition of 0.6% polyvinylpyrrolidone (PVP).

Although all plant cells have a rigid cell wall, the variety of plant tissues is quite diverse ranging from tubers to green or etiolated leaf tissue. As in the case of fungi, cells can be broken open either directly or after the cells have been protoplasted. The problem with protoplasting is that it requires prolonged incubation (2–20 h) and the overall yield of protoplasts can be quite low (15%). The alternative method is to use a high-speed blender, although care must be taken to avoid excessive damage to the mitochondria; usually 2–3 sec is sufficient for most types of plant tissues. If possible, it is advisable to use etiolated tissues as a source of mitochondria. Here a typical experimental protocol for the isolation of plant mitochondria is given.

(i) Chop the plant tissue using a knife or scissors, roots must be cleaned of all soil particles and leaves should be de-ribbed. About 100–200 g is a convenient amount of tissue to start with.

(ii) Homogenize 100–200 g of tissue in 500 ml of cold 0.3 M mannitol, 4 mM cysteine, 1 mM EDTA, 30 mM Mops–KOH, pH 7.8 containing 0.2% BSA. For green tissue also include 0.6% PVP. The tissue should be homogenized for 2–3 sec using a Waring Blendor at low speed or a Polytron with a PT 35K probe.
(iii) Filter the homogenate through six layers of cheesecloth and centrifuge it at 3000 *g* for 5 min at 5°C to pellet cell walls, nuclei and incompletely homogenized tissue.
(iv) Carefully decant the supernatant and centrifuge it at 12 000 *g* for 20 min at 5°C to pellet the mitochondria. Discard the supernatant.
(v) Resuspend the mitochondria for further purification.

The presence of plastids in this pellet is an added complication and it is usual to purify mitochondria either by repeating steps (ii) and (iii) or by isopycnic centrifugation. Unfortunately it is difficult to obtain good separation of plastids and mitochondria on isopycnic sucrose gradients. An alternative approach is to use Percoll or Nycodenz gradients. For Percoll gradients carry out the following steps.

(i) Prepare a linear gradient of 8–60% (v/v) Percoll containing 0.25 M sucrose, 0.2% BSA and 10 mM Mops–KOH, pH 7.2 using a gradient maker.
(ii) Load the crude mitochondria suspended in 0.25 M sucrose, 0.2% BSA and 10 mM Mops–KOH, pH 7.2 onto the gradient and centrifuge the gradients at 37 000 *g* for 20 min at 5°C.
(iii) Remove the mitochondrial band using a Pasteur pipette, dilute the suspension by the addition of 2 vols of 0.25 M sucrose, 0.2% BSA, 10 mM Mops–KOH, pH 7.2 and pellet the mitochondria by centrifugation at 25 000 *g* for 10 min at 5°C.
(iv) Wash the mitochondria several times in isotonic medium to remove as much Percoll as possible.

Preliminary experiments by one of the authors using metrizamide and Nycodenz have also indicated that these two types of gradient medium also give better resolution of plant mitochondria than does sucrose although more work is required before one can be sure whether these media are better than Percoll. Detailed descriptions of the several variations of this general isolation procedure for a wide selection of plant tissues are given elsewhere (24).

3. PURITY AND FUNCTIONAL CHARACTERIZATION OF INTACT MITOCHONDRIA

The mitochondrial electron transport chain conserves the energy of electron transfer in the form of a proton gradient which is then used to drive the synthesis of ATP via the ATP synthetase. This function is critically dependent on the inner membrane being impermeable to protons. For a simple scheme of oxidative phosphorylation showing this relationship between electron transfer and ATP production see *Figure 2*. A more detailed discussion of energy conservation is given elsewhere (26). In mitochondria with an intact inner membrane electron transfer can only proceed if the proton gradient is continuously dissipated by the flow back into the mitochondrion via the ATP synthetase. In practical terms this means that in intact mitochondria, electron transfer, which can be measured by O_2 consumption, can only occur if ADP is also present. This phenomenon is described as 'coupling'. The extent of coupling in a mitochondrial preparation can be easily measured as the respiratory control ratio (RCR) and is the most rapid and effective method for determining mitochondrial functional integrity. The consumption of oxygen in the presence of substrate and ADP (state 3 respiration) is measured and compared with the rate of respiration after the ADP has been consumed (state 4 respiration). Mitochondria which show no difference in state 3 and state 4 respiration are 'uncoupled', a state which can also be induced by some reagents (e.g. dinitrophenol) appropriately termed uncouplers. In addition to measures of mitochondrial function, structural criteria for purity can be assessed by isopycnic centrifugation (Section 2.1.1) or electron microscopy (Chapter 2).

3.1 The oxygen electrode

Mitochondrial respiration is most conveniently and quickly measured using a 'Clarke' type oxygen electrode. This consists of a silver/silver chloride reference anode surrounding (generally) a platinum cathode. These electrodes are immersed in saturated KCl solution and separated from the reaction vessel by a thin Teflon membrane that is permeable to oxygen but which prevents electrode poisoning. The electrodes are polarized at a voltage of 0.6 V. At the platinum cathode electrons reduce oxygen molecules to water.

$$4H^+ + 4e^- + O_2 \rightarrow 2H_2O$$

The chloride anions migrate to the anode and release electrons.

$$4Ag + 4Cl^- \rightarrow 4AgCl + 4e^-$$

The overall result is that a transfer of electrons from the cathode to the anode occurs causing a current to flow between the two electrodes which can be measured in an external circuit. The current is proportional to the partial pressure of oxygen in the sample and, as the response is linear, only two calibration points are necessary. The current which flows for air-saturated water at 20°C is about a few microamps. The current generated is very temperature dependent and it is therefore important to operate the electrode at constant temperature.

It can be appreciated from the above equations that as oxygen is consumed by the electrode, the solution close to the electrode becomes progressively anaerobic. It is therefore necessary to stir the solution vigorously to ensure that the bulk solution under study is continuously brought into contact with the electrode. The most convenient way to do this is with a magnetic stirrer. Small magnetic stirrer bars can easily be made by enclosing pins or pieces of paper clip in sealed capillary tubing.

A suitable circuit for polarizing the electrodes and allowing measurements to be made is shown in *Figure 3*. For output to a chart recorder, R is chosen such that

$$R = 2 \times \text{span of recorder (in mV)} \times 10^3\ \Omega$$

The electrodes may conveniently be enclosed in a small chamber of approximately 3 ml volume. This chamber is isolated from contact with the atmosphere by a close-fitting cap. Solutions are introduced into the chamber and their oxygen concentrations measured directly once the apparatus has been calibrated. (Suitable electrodes polarizing and measuring circuits may be purchased from Rank Brothers, High Street, Botisham, Cambridge, UK.)

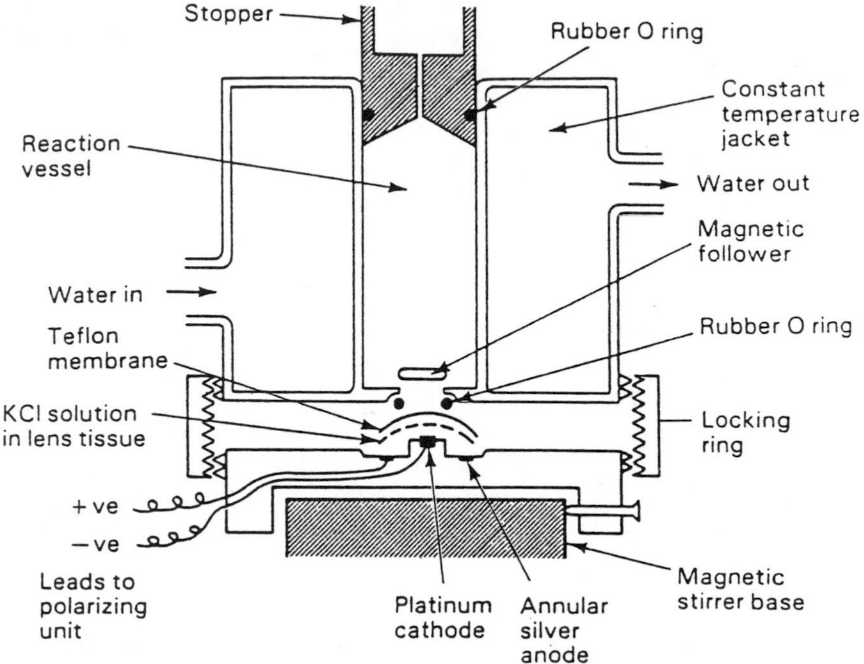

Figure 3. Oxygen electrode. A suspension of mitochondria (2.0−3.0 ml) in buffer is introduced into the water-jacketted reaction chamber which is then closed by a tight-fitting stopper. This stopper is gently pushed down to remove the air space completely and care is taken to exclude all air bubbles from solution, the conical depression in the base of the stopper helping in this task. Additions of substrates, uncouplers, etc. are made through the narrow part in the stopper using a syringe capable of delivering microlitre quantities. The solution is stirred continuously with a magnetic stirrer. The current flowing in the external, polarizing circuit is proportional to the dissolved oxygen concentration in the solution in the reaction chamber.

3.1.1 *Calibration of the oxygen electrode*

(i) Pipette 3 ml of distilled water into the electrode chamber, add a few crystals of sodium dithionite ($Na_2S_2O_4$). The oxygen concentration rapidly falls to zero. The output of the electrode to the chart recorder also falls and the position of the pen on the chart recorder is set to zero.

(ii) Wash out the electrode chamber several times with distilled water to remove sodium dithionite. Pipette 3 ml of distilled water, previously equilibrated with air at 20°C into the electrode chamber, itself thermostatted at 20°C.

(iii) Set the chart recorder to 90% of its deflection by suitable choice of sensitivity. This level now corresponds to 260 μM dissolved oxygen. This provides two known oxygen concentrations, that is 0 and 260 μM from which other concentrations can be determined.

3.2.1 *Measurements of the respiratory control ratio (RCR)*

The conditions described below are suitable for the oxygen electrode described previously and assume a final volume of 2.5 ml.

(i) Add homogenization buffer (see Section 2.2.1).

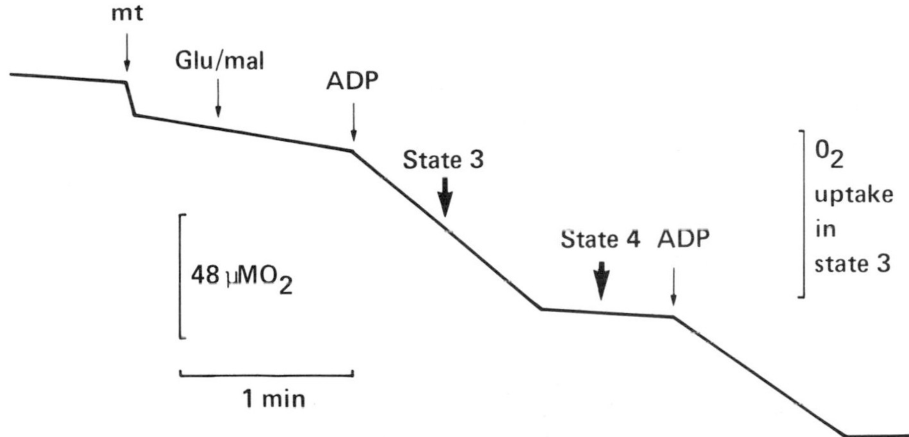

Figure 4. Respiratory control in rat-liver mitochondria. Mitochondria (mt) are added to 2.4 ml of isolation medium in the oxygen electrode followed by glutamate/malate to a final concentration of 2.5 mM. Record the O_2 uptake for a few minutes and then add ADP to a final concentration of 180 μM. State 3 and state 4 respiration are shown as is the O_2 consumed in state 3 respiration for the calculation of the P/O ratio.

Table 4. Functional characteristics of beef-heart and rat-liver mitochondria.

Substrate	*RCR*		*P/O*	
	BHM	*RLM*	*BHM*	*RLM*
Glu/mal	8–10	10–16	2.5–2.8	2.5–3.0
Succinate	3–5	4–6	1.5–1.8	1.6–2.0
	BHM		*O_2 consumption/mg protein sonicated BHM*	
NADH (600 μM)	50–100			500–600 μM O_2/min/mg

RCR = respiratory control ratio, P/O = ATP synthesized divided by oxygen (O_2) consumed in state 3 respiration. BHM = beef heart mitochondria. RLM = rat-liver mitochondria.

(ii) Add mitochondria to give a final concentration of 0.2–0.5 mg/ml and measure the background rate for 1–2 min.
(iii) Add substrate. For glutamate/malate, to a final concentration of 2.5 mM from a stock solution of 80 mM adjusted to pH 7.4 with KOH, and for succinate a final concentration of 5 mM from a stock solution of 160 mM. Measure the rate of oxygen uptake for 1–2 min.
(iv) Add ADP to a final concentration of 180 μM from a stock solution of 11.6 mM. Measure the oxygen uptake for 10 min.
(v) Calculate the RCR by dividing the rate of oxygen uptake in state 3 by the rate of oxygen uptake in state 4 (see *Figure 4*). for the P/O ratio calculate the oxygen consumed from the start of state 3 respiration to its end (see *Figure 4*). The P/O ratio is then the ratio of the nanomoles of ADP added divided by the nanograms

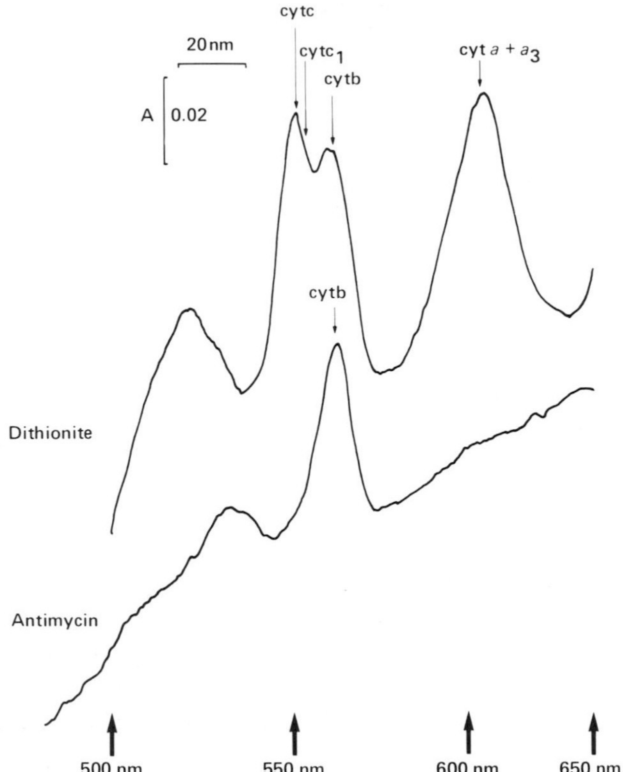

Figure 5. Visible absorption spectra of mitochondrial cytochromes. Beef-heart mitochondria were suspended to a concentration of 1 mg/ml and the reduced minus oxidized spectra of the dithionite-reduced and antimycin-blocked succinate-reduced samples taken. The cytochrome content was calculated as described in *Table 4*.

of atoms of oxygen utilized during state 3 respiration. Typical values are given in *Table 4*.

Since NADH is unable to cross the intact mitochondrial inner membrane an alternative functional estimate of integrity is to measure the rate of NADH oxidation in intact and sonicated mitochondria. In a good preparation then the endogenous rate of NADH oxidation should be low (*Table 4*).

(i) Add mitochondria and buffer to the sample chamber of the oxygen electrode followed by NADH to a final concentration of 600 μM and measure the oxygen uptake.

(ii) Sonicate the mitochondria (15 mg/ml) for six 10-sec bursts on ice. Then use for assay as described in step (i). An increase in rate of 6-fold in the oxygen consumption compared with that before sonication indicates an intact mitochondrial preparation (see *Table 4*).

3.3 Spectral characterization

Some of the proteins of the inner mitochondrial membrane contain haem prosthetic groups that give characteristic visible absorption spectra and are called the cytochromes.

Table 5. Calculation for cytochrome content in mitochondria (28).

Wavelength pairs (nm)	*Cytochrome*
550–535	cyt *c*
554–540	cyt c_1
563–577	cyt *b*
605–630	cyt *a*

Substitute these absorbances into the following equations:

$$\alpha_1 = \frac{\Delta A_{550-535}}{25.1}$$

$$\alpha_2 = \frac{\Delta A_{554-540} - 7.78\ \alpha_1}{15.6}$$

$$\alpha_3 = \frac{\Delta A_{563-577} + 1.39\ \alpha_1 - 1.48\ \alpha_2}{27.5}$$

$$\alpha_4 = \frac{\Delta A_{605-630} + 0.26\ \alpha_1 + 0.48\ \alpha_2 - 0.153\ \alpha_3}{13.13}$$

$C_4 = \alpha_4$ = concentration of cytochrome *a*, divide by two to give cytochrome *c* oxidase, cyt aa_3 nmol/mg

$C_3 = \alpha_3 + 0.0144\ C_4$ = cyt *b*. nmol/mg

$C_2 = \alpha_2 - 0.45\ C_3 - 0.0523\ C_4$ = cyt c_1 nmol/mg

$C_1 = \alpha_1 - 0.41\ C_2 + 0.25\ C_3 - 0.0275\ C_4$ = cyt *c* nmol/mg

Cytochromes *a* and a_3 are redox components of the terminal electron acceptor cytochrome *c* oxidase. These are often termed cytochrome *a* in many reports but are in fact two distinct centres. Cytochrome *b* and cytochrome c_1 are both components of the inner membrane protein ubiquinol cytochrome *c* reductase. There are two cytochrome *b*s for every cytochrome c_1. Cytochrome *c* is the water-soluble mobile redox protein which shuttles electrons from ubiquinol cytochrome *c* reductase to cytochrome *c* oxidase. A detailed discussion of the inner membrane structural proteins of oxidative phosphorylation is given elsewhere (27). All of the cytochromes have characteristic absorption spectra in the visible region in their reduced state. This can be conveniently used to estimate the relative cytochrome content in mitochondria (27,28).

The procedure we will describe is for recording the reduced–oxidized difference visible spectrum of the cytochromes using a double-beam spectrophotometer but can be adapted for use in single-beam instruments by memorizing the oxidized spectrum as the baseline.

(i) *Spectrum of reducible cytochrome b.* Dilute the mitochondria to 1 mg/ml and place in the reference and sample cuvettes. After recording the baseline from 500 to 650 nm, antimycin is added to a final concentration of 2 μM, from a 500 μM stock solution in ethanol, followed by 2 mM sodium succinate, from a 50 mM stock solution, to the sample cuvette. The reducible cytochrome *b* peak appears at 563 nm after 1 min (*Figure 5*).

(ii) *Spectrum of all reduced cytochromes.* Determine the baseline as described in (i) and then add a few grains of sodium dithionite to the sample cuvette and record the spectrum (*Figure 5*). Note the red shift in cytochrome *b* which is induced by antimycin. The relative amount of the cytochromes can be calculated by measuring the absorbances at the given wavelength pairs and substituting in the equations shown in *Table 5* (28). Typical values for beef heart mitochondria are: 0.48 nmol/mg cytochrome *c* oxidase, 0.48 nmol/m cytochrome *b*, 0.65 nmol/m cytochrome *c* and 0.19 nmol/mg cytochrome c_1.

3.4 Analysis of purity on the basis of DNA

For the many studies of the molecular biology of mitochondria, especially of the sequence and expression of mitochondrial DNA, the purity of the mitochondrial preparations has been judged by the analysis of the mitochondrial DNA. In some organisms, particularly lower eukaryotes, the buoyant density of the mitochondrial DNA is different from the other DNA of the cell. Alternatively, when the small genome of mitochondria is digested with appropriate restriction nucleases usually a defined set of restriction fragments is formed which give sharp bands on polyacrylamide gels (see Section 5.1 of Chapter 7); any nuclear contamination of the mitochondria is revealed as a background smear of bands while, in plant cells, the presence of plastid DNA would result in the presence of extra bands.

It should be noted that the presence of homologous DNA sequences in the nucleus and mitochondria and, in the case of plants, the chloroplast and mitochondria, prevents one from using hybridization techniques as a criteria for the purity of mitochondria.

4. REFERENCES

1. Hiura,S., Mori,M., Amaya,Y. and Tatibana,M. (1982) *Eur. J. Biochem.*, **122**, 641.
2. Tzagoloff,A. (1982) *Mitochondria.* Plenum Press, New York.
3. Nicholls,D.G. (1979) *Biochim. Biophys. Acta,* **416**, 1.
4. Singer,T.P., Von Korf,R.W. and Murphy,D.L., eds (1979) *Monamine Oxidase: Structure and Function.* Academic Press, New York.
5. Bates,E.J. and Saggerson,E.D. (1979) *Biochem. J.*, **182**, 751.
6. Hay,R., Bohni,P. and Schatz,G. (1984) *Biochim. Biophys. Acta,* **779**, 65.
7. Bernandi,P. and Azzone,G.F. (1982) *Biochim. Biophys. Acta,* **679**, 19.
8. Klingenberg,M. (1970) *FEBS Lett.*, **6**, 145.
9. Olson,M.S., Scholz,R., Buffington,C.K., Dennis,S.C., Padma,A., Patel,T.K., Naymack,P.P. and DeBuysere,M.S. (1981) In *The Regulation of Carbohydrate Formation and Utilization in Mammals.* University Park Press, Baltimore, p. 153.
10. Moyes,C.D., Moon,T.W. and Ballantyne,J.S. (1986) *J. Exp. Zool.*, **237**, 119.
11. Kovacevic,Z. and McGivan,J.D. (1983) *Physiol. Rev.*, **63**, 547.
12. Gamble,J.G. and Lehninger,A.L. (1973) *J. Biol. Chem.*, **243**, 610.
13. Bremer,J. (1983) *Physiol. Rev.*, **63**, 1420.
14. Richter,C. and Frei,B. (1985) In *Oxidative Stress.* Sies,H., (ed.), Academic Press, New York, p.221.
15. Siess,E.A. (1983) *Hoppe-Seyler's Z. Physiol. Chem.*, **364**, 279.
16. Mela,L. and Seitz,S. (1979) In *Methods in Enzymology.* Fleischer,S. and Packer,L. (eds), Academic Press, New York, Vol. 55, p. 39.
17. Whitfield,C.D., Bostedor,R., Goodrum,D., Haak,M. and Chu,E.H.Y. (1981) *J. Biol. Chem.*, **256**, 6651.
18. Rickwood,D. (1984) In *Centrifugation—A Practical Approach.* (2nd edition) Rickwood,D. (ed.) IRL Press, Oxford, UK, p.1.
19. Parry,E.M. and Parry,J.M. (1984) In *Mutagenicity Testing—A Practical Approach.* Venitt,S., Parry,J.M. (eds), IRL Press, Oxford, UK, p.119.
20. Rickwood,D. and Hayes,A. (1984) *Prep. Biochem.*, **14**, 163.
21. Lambowitz,A.M., Smith,E.W. and Slayrian,E.W. (1972) *J. Biol. Chem.*, **247**, 4859.
22. Lang,B., Burger,B., Doxiadis,I., Thomas,D.Y., Bandlow,W. and Kaudewitz,F. (1977) *Anal. Biochem.*, **77**, 110.
23. Schatz,G. and Kovac,L. (1974) In *Methods in Enzymology.* Fleischer,S. and Packer,L. (eds), Academic Press, New York, Vol. 31, p. 627.
24. Moore,A.L. and Proudlove,M.O. (1983) In *Isolation of Membranes and Organelles from Plant Cells.* Hall,L.J. and Moore,A.L. (eds), Academic Press, London, p.153.
25. Loomis,W.D. (1974) In *Methods in Enzymology.* Fleischer,S. and Packer,L. (eds), Academic Press, New York, Vol. 31, p. 528.
26. Nicholls,D.G. (1982) *Bioenergetics,* Academic Press, New York.
27. Capaldi,R.A. (1982) *Biochim. Biophys. Acta,* **694**, 291.
28. Vanneste,W.H. (1966) *Biochim. Biophys. Acta,* **113**, 175.

CHAPTER 2

Electron microscopy of mitochondria

R.D.A.LANG

1. INTRODUCTION

Electron microscopy includes a range of techniques (e.g. thin-sectioning, negative staining, freeze-etching) each of which provides different information about the structure of the specimen. Obtaining reproducible results, and understanding possible artifacts that may arise, requires considerable skill and experience. The aim of this chapter is to discuss the principles of the techniques and to show what each can reveal about mitochondrial structure. All of the methods that will be discussed here are used to prepare specimens for transmission electron microscopy. Details of the operation of microscopes will not be given. These vary between different manufacturers' instruments and the principles of the design and use of electron microscopes are well covered in other books (e.g. ref. 1).

1.1 **Fundamental principles**

Two basic factors in transmission electron microscopy impose constraints on the design of specimen preparation methods. Firstly, the microscope column, in which the specimen is held, must be kept under vacuum. Secondly, biological material consists mostly of low atomic number atoms; their nuclei do not scatter electrons to any great extent and thus give rise to low contrast images.

Some techniques of specimen preparation involve drying, or dehydrating and fixing, the specimen and increasing its contrast by staining it with the salts of heavy metals. Both thin-sectioning and negative staining adopt this approach, although in other respects they are very different procedures. An alternative approach is to form a metal replica of the specimen and then to view this, rather than the specimen itself, in the microscope. This approach solves both the dehydration and the contrast problems simultaneously, and is used in freeze-etching.

1.2 **Choice of method**

Assuming that the necessary apparatus for the various techniques is available, the choice of the most suitable method depends on the information that is required. If one wishes to assess the purity of a mitochondrial fraction and the gross morphology of the mitochondria, thin-sectioning is suitable. Negative staining can be applied to whole mitochondria (2), but is most commonly used for high resolution studies of the structure of isolated proteins or of membranes containing regular arrays of proteins or protein complexes.

Freeze-etching reveals extensive views of the surfaces and interiors of membranes and, if rapid freezing methods (see Section 4.3) are employed, avoids the need for any chemical modification, such as fixation, of the mitochondrial membranes.

2. THIN SECTIONING

2.1 **Principles**

The specimen is first chemically fixed to protect it from structural damage caused by the subsequent procedures. The specimen is sometimes stained at this stage. Next, the specimen is dehydrated; all of the water is removed and replaced by an inert liquid (e.g. ethanol or acetone) which is miscible with both water and the embedding medium. Then the specimen is infiltrated with an embedding medium which is subsequently polymerized to form a solid matrix supporting the specimen. This matrix is cut into very thin sections which are mounted on copper grids. The sections are then stained with solutions of heavy metal salts and the grids are viewed in the electron microscope.

2.2 **Experimental procedures**

Since there are many published 'recipes' for this procedure, I shall give one here which has given good results with isolated rat liver mitochondria (based on ref. 3).

(i) Add glutaraldehyde to a final concentration of 3% to a suspension of mitochondria. Some authors add fixative after centrifuging the mitochondrial suspension. However, when they are in a pellet, the mitochondria are likely to become anaerobic before they are fixed. Anaerobic mitochondria show rapid contraction of their inner compartments (3,4).

(ii) After 60 min, centrifuge the mitochondrial suspension at 10 000 *g* for 10 min to pellet the mitochondria. Remove the supernatant and then wash the pellet three times with isotonic sucrose (0.25 M) in 10 mM sodium phosphate buffer, pH 7.4.

(iii) Add a solution containing 2% osmium tetroxide in 10 mM sodium phosphate buffer and leave overnight. [Note that osmium tetroxide is volatile and very toxic as well as being an excellent fixative. The vapour should not be inhaled or allowed to come into contact with the eyes, so steps (iii) and (iv) should be done in a fume cupboard].

(iv) Remove the osmium tetroxide solution. Replace it with 75% ethanol. This reduces the remaining osmium tetroxide to osmium dioxide, which forms a precipitate in the alcohol. After 10 min remove the alcohol and replace it with another few millilitres of 75% ethanol.

(v) After 30 min, remove the alcohol and replace it with 95% ethanol. Leave for 30 min.

(vi) Repeat step (v) three times using 100% ethanol.

(vii) For the last step in dehydration, replace the alcohol with 100% ethanol which has been specially dried by standing over hygroscopic beads (available from Agar-Aids, Stansted Essex). Again, leave the fixed pellet in the dried ethanol for 30 min.

(viii) Equilibrate the pellet for 30 min in a 1:1 mixture of 1,2-epoxypropane (EPP; also known as propylene oxide) and the embedding medium, Epon 812 (also called 'Epikote Resin 812'). A mixture of the resin and two hardening agents, dodecenyl succinic anhydride (DDSA) and methyl nadic anhydride (MNA) is used. A diamine catalyst (generally N-benzyl-N N-dimethylamine) is added just before use. Wear gloves when handling the resin since contact with skin can cause allergic reactions.

(ix) Pour off the 1:1 mixture and replace it with full-strength resin. Soak for several hours, then change the resin again. Repeat this step several times (the last one overnight) to ensure full infiltration of the embedding medium.

(x) Transfer the pellet to a gelatine drug capsule (available from Agar-Aids Stansted Essex) (used as a mould to form the final block) and fill the capsule with fresh resin mixture. Place the top on the capsule.

(xi) Place the capsule in a rack and incubate it for 48 h in an oven at 60°C, to polymerize the resin completely. Then take the capsule, remove its lid and place it in hot water for about 1 h to dissolve the gelatine. Dry the block.

(xii) Next, trim the rounded end of the block (where the mitochondrial pellet is situated) using a razor blade and mount the block in an ultramicrotome. Then cut sections according to the manufacturers' instructions. The sections are picked up on grids (which may be coated; Section 3.2.2) and post-stained with 1% uranyl acetate (w/v in water), rinsed with distilled water and dried.

When the grids have dried, they may be viewed in the electron microscope.

3. NEGATIVE STAINING

3.1 Principle of the method

Negative staining is the simplest and fastest method of preparing specimens for transmission electron microscopy. A sample suspension is mixed with an aqueous solution of an electron-dense salt (the negative stain) and the mixture is dried on a grid. When viewed in the electron microscope, samples (e.g. proteins, membranes or organelles) appear electron-transparent against a dark background of stain (*Figure 1*).

3.2 Experimental procedures

3.2.1 *Preparation of negative stain solutions*

(i) *Ammonium molybdate, 5%*

(a) Crush crystals of ammonium molybdate using a pestle and mortar.

(b) Dissolve 5 g of the crystals in approximately 50 ml of distilled water.

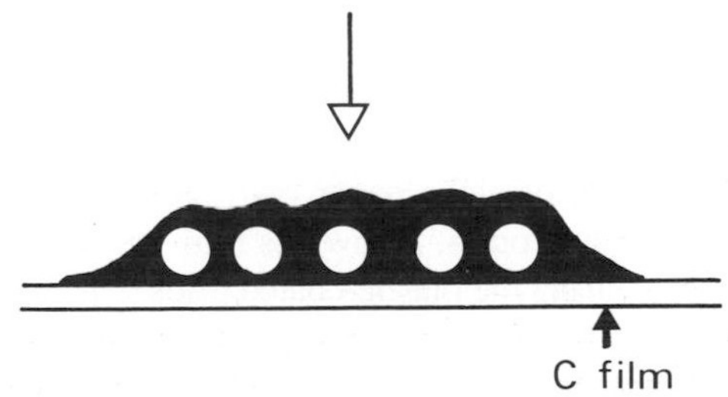

Figure 1. Diagram showing the principle of negative staining. Circles represent spherical organelles which have not been penetrated by the opaque stain. The arrow indicates the direction of the electron beam.

(c) Bring the pH of the solution to pH 6.9 by adding 1 M KOH.
(d) Add water to bring the volume to 100 ml then filter the solution and store it in a refrigerator.

Ammonium molybdate is a particularly useful negative stain for use with osmotically-sensitive organelles, such as intact mitochondria, since solutions at the concentrations used for negative staining are iso-osmotic with media commonly used for the isolation of organelles. For example, 2% ammonium molybdate has the same osmolality as 0.25 M sucrose (2, 5).

(ii) *Sodium phosphotungstate, 5%*

(a) Dissolve 5 g of phosphotungstic acid in 60 ml of distilled water.
(b) Adjust the pH of the solution to about pH 7.3 by adding 1 M NaOH.
(c) Add distilled water to bring the volume to 100 ml then filter the solution and store it in a refrigerator.
(d) Dilute the stock solution to 2% before use.

(iii) *Uranyl acetate*

1% uranyl acetate is commonly used for negative staining of isolated proteins such as mitochondrial F_1-ATPase (6). It has also been used to stain crystalline arrays of cytochrome oxidase that are formed when mitochondrial inner membranes are extracted with detergents (e.g. 7, 8). The pH of 1% uranyl acetate is about pH 4.2.

3.2.2 *Coating of grids*

Since negative staining involves drying liquid samples directly onto electron microscope grids, the spaces between grid bars must be covered with a transparent support film. Thin carbon, or carbon-coated plastic, films are used.

(i) The stock solution from which plastic support films are made is 0.25% celloidin (or Formvar) in *n*-butyl acetate.
(ii) Using a Pasteur pipette, spread 2–4 drops of celloidin solution on a clean glass slide and drain off the excess solution. Allow the slide to dry. The top surface of the slide will thus be coated with a thin plastic film.
(iii) Score the edges of the film with a sharp scalpel or razor blade.
(iv) Float the film onto the surface of some clean distilled water in a trough. To do this, gently lower the slide into the water keeping the slide's top surface almost parallel to the surface of the water. You should be able to see that the film detaches from the slide and floats.
(v) Using a pair of jeweller's forceps, place copper electron microscope grids shiny side up all over the film.
(vi) Carefully lower a piece of clean paper onto the film and use it to remove the film (and grids) from the trough. Allow this film–grids–paper sandwich to dry.
(vii) Transfer the paper bearing the plastic-coated grids to a vacuum coating apparatus and evaporate a thin layer of carbon onto the plastic film.
(viii) The coated grids may now be used. Separate them carefully from the paper using forceps and taking care not to tear the films on each grid.

(ix) Alternatively, for very high resolution work, the plastic films may be removed leaving carbon support films. Place the grids (with the carbon films uppermost) onto drops of amyl acetate for about 60 min to dissolve the plastic film. Then allow the grids to dry in air.

3.2.3 *Negative staining*

The aim of this procedure is to cover a large area of the grid with a uniformly thin layer of negative stain containing the specimens. The specimens should be frequent enough to be found easily without being clumped together. Specimen clumping is more likely to be a problem in the case of membranous samples, such as mitochondria, than for isolated proteins. One has to experiment to find the best concentration of sample to use. However, as a rough guide the protein concentration of a mitochondrial suspension should be less than about 4 mg/ml, while for isolated proteins a suitable starting concentration should be much lower (e.g. $\sim$0.1 mg/ml).

There are two general ways of adsorbing samples onto the grid and then washing and staining them. In the first method, the grid is held, carbon film upwards, in a pair of forceps and the various solutions are added using a Pasteur pipette. Alternatively, drops of the various solutions are placed onto a hydrophobic surface, such as a piece of dental wax, and the grid (carbon film downwards) is placed onto the surface of each drop in turn. The following steps apply to both these methods.

(i) Allow the sample to adsorb onto the grid for between 1 and 5 min.
(ii) Wash off excess sample suspension with distilled water or negative stain solution. This step is particularly important if the suspending medium contains high concentrations of sucrose or salts since if these solutes are not removed, they cause bubbles to form in the stain when the electron beam passes through the specimen.
(iii) Stain the sample with several drops of negative stain solution.
(iv) Remove excess stain from the grid by holding it in a pair of forceps and applying filter paper to the edge of the grid. Any liquid that gets between the tips of the forceps should be removed by placing a small piece of filter paper between the tips. If this is not done, the grid will be drawn between the blades of the forceps by liquid surface tension when the forceps are opened.
(v) Allow the grid to dry at room temperature. The specimen may then be viewed in a transmission electron microscope. *Figure 2* shows a mitochondrion negatively stained with ammonium molybdate.

4. FREEZE-ETCHING

4.1 **Principles of the technique**

The freeze-etch technique involves freezing the specimen, fracturing it and producing a high resolution metal replica of the frozen, fractured surface. The replica, not the original specimen, is viewed in a transmission electron microscope so staining of the specimen is not required. After fracturing, the specimen may be etched, by allowing some of the ice to sublime away, so that parts of the non-aqueous components of the specimen (membranes and proteins) are left standing proud of the surface. Etching is

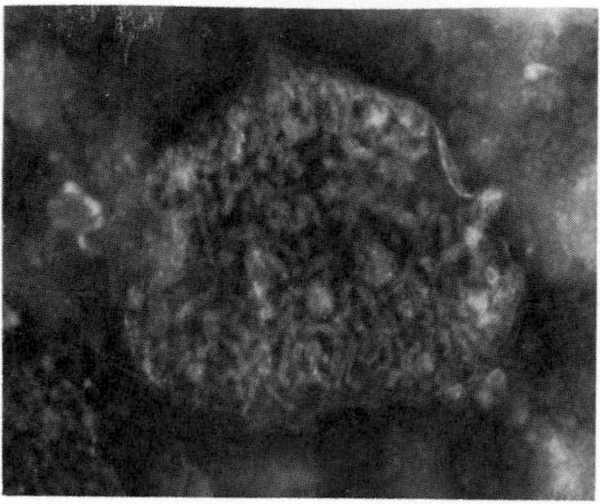

Figure 2. Isolated rat liver mitochondrion, negatively stained with ammonium molybdate. ×30 000.

optional; if it is omitted, the specimen is said to be freeze fractured. Etching and replication, and usually fracturing, are performed under high vacuum in order to avoid contamination of the specimen surface and to produce a high quality replica.

Several freeze-etch machines are currently manufactured such as those made by Balzers High Vacuum, Lichtenstein. These machines are modern vacuum evaporation units which incorporate a remotely operated knife, or microtome, and a specimen table whose temperature can be very accurately controlled and maintained automatically using pressure-fed liquid nitrogen. The knife is used to fracture the specimens.

The major advantages of freeze-etching over other specimen preparation techniques for electron microscopy are that the specimen is maintained in its hydrated state, with minimal chemical treatment, and that extensive views of membranes are revealed. The technique reveals important structural details of membranes which cannot be demonstrated clearly by other ultrastructural methods. In the late 1960s, strong evidence was obtained showing that some membranes are fractured through their hydrophobic regions, between the two halves of the lipid bilayer (9). Since that time, it has become generally, though not universally (10), accepted that freeze-fracturing exposes internal faces of most biological membranes, including those of mitochondria. Furthermore, there is a great deal of evidence that the particles that appear on membrane fracture faces (intramembranous particles or IMPs) represent proteins or lipoprotein complexes within the lipid bilayers (11). Analysis of the frequencies and size distribution of IMPs provides quantitative data that can be used to compare different membranes or the same membranes under different conditions. This will be discussed further in Section 4.4.

4.2 **Experimental procedure**

Freeze-etching falls conveniently into six separate steps which will be discussed here in order to show what effects each stage may have on the final results.

4.2.1 *Pre-treatment and freezing*

Generally, the best structural preservation of the specimen is achieved by the highest freezing rates. Ideally, the water in the specimen should be frozen to a 'glass' with no ice crystals (vitrification). The necessary freezing rate for vitrification is

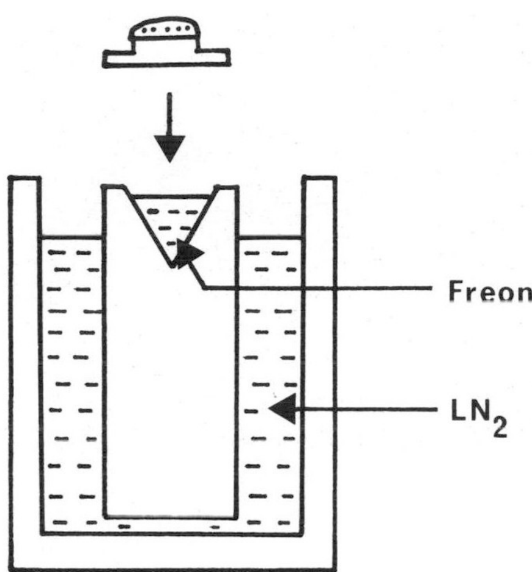

Figure 3. Diagram of standard freezing. The specimen (e.g. a mitochondrial suspension) is mounted on a gold support which is then plunged into liquid Freon, cooled by liquid nitrogen (LN_2).

10 000−20 000°C/sec (12, 13) and vitrified water may not exist at temperatures warmer than −140°C since recrystallization occurs above this temperature (14). Truly vitrified water is never formed in frozen biological specimens but very small ice crystals that do not disrupt the structure of membranes can be tolerated.

The critical freezing rate can be reduced to 100−1000°C/sec by impregnation of the specimen with a cryoprotectant such as glycerol. The cryoprotectant lowers the freezing point of water and raises the ice recrystallization temperature. Thus, the critical temperature interval, in which ice crystal formation occurs, is reduced and can be passed successfully at a lower freezing rate. Unfortunately, glycerol can cause changes in membrane structure and swelling of organelles such as mitochondria (15). Because of this, specimens are usually fixed chemically, for example with glutaraldehyde, before being treated with glycerol and then frozen. This is the major disadvantage of conventional freezing for freeze-etching since the ideal of purely physical fixation is not attained. Considerable effort has been devoted to designing rapid freezing techniques to overcome these problems. Rapid freezing methods which are applicable to mitochondria will be discussed separately (Section 4.3).

In conventional (standard) freezing, the specimen (~ 1 mm^3) is mounted on a gold support and plunged into a coolant at its melting point (*Figure 3*). Suitable coolants include chlorodifluoromethane (Freon 22) or propane, cooled by liquid nitrogen. Liquid nitrogen cannot be used as the direct coolant since its boiling point is very low (−196°C) and specimens become enveloped in nitrogen gas which slows down heat transfer (the Leidenfrost effect). Freon has a melting point of −158°C and a boiling point of −30°C, at normal atmospheric pressure. After freezing, the specimen is stored under liquid nitrogen until it is transferred to the pre-cooled specimen table of the freeze-etch machine.

During freezing, the water and solute separate into two phases; pure ice crystals surrounded by a eutectic mixture. This separation depends on the freezing rate and standard freezing is too slow to prevent it, although high concentrations of glycerol ($>25\%$)

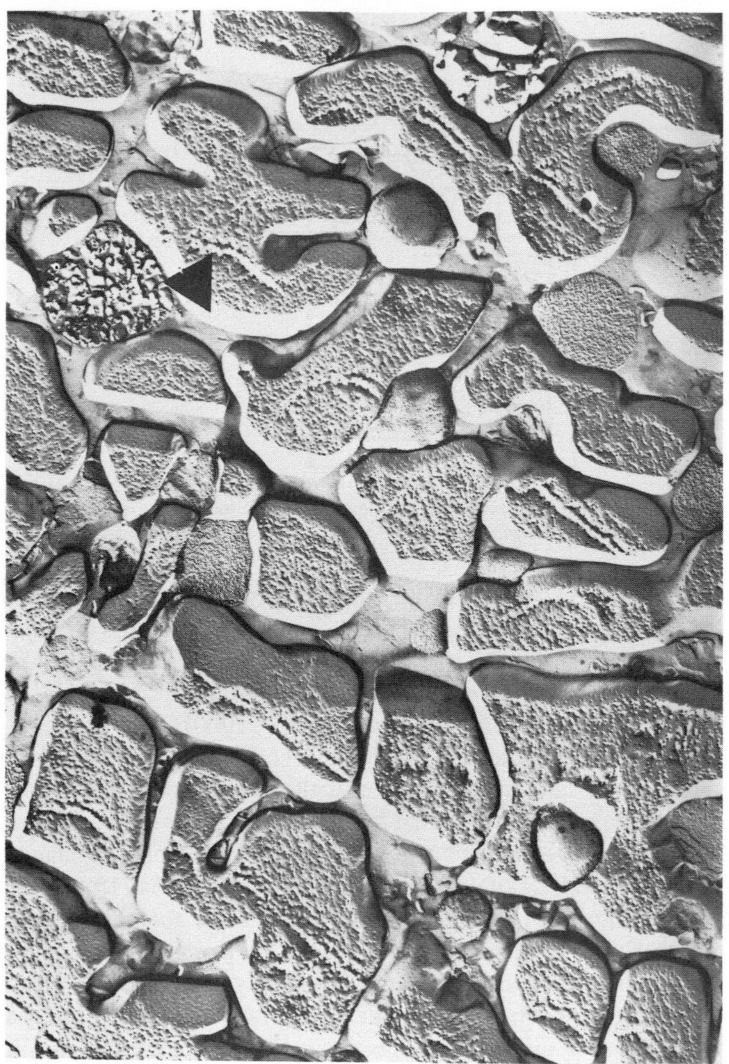

Figure 4. Rat liver mitochondria suspended in 10% glycerol, frozen in Freon then freeze-fractured and etched for 2 min before replication. Etching has removed ice from the large ice crystals, leaving the solute (glycerol) as a ridge-like network containing mitochondria. The mitochondrion indicated by the arrow contains small etched ice crystals showing that it has been damaged since it has been penetrated by the glycerol solution. ×10 000.

minimize it. Suspended particles can become concentrated in the frozen solute phase and growing ice crystals may deform organelle structure (*Figure 4*).

4.2.2 *Fracturing*

The specimen is often fractured with a knife mechanism under high vacuum, inside the machine. The knife is kept as cold as, or colder than, the specimen to prevent contamination of the fracture surface. Alternative methods of fracturing employ special hinged specimen holders that can be opened by moving the knife arm against a lever. In these devices, the specimen is sandwiched between two specimen supports. When the hinged device is opened, the specimen supports separate, thereby fracturing the specimen. Such devices retain both halves of the fractured specimen, one half being turned over, so that both fracture faces may be replicated simultaneously and complementary replicas of a single fracture plane may be examined. In addition, many of these devices can accommodate several specimens, thereby increasing the productivity of the technique and also allowing comparisons to be made between different specimens under identical etching and replication conditions.

Fracturing is known to cause plastic deformation of some biological structures (e.g. ref. 16) but the extent to which membrane components, such as IMPs, are plastically deformed is unknown.

4.2.3 *Etching*

Etching is the controlled sublimation of ice from the freshly fractured specimen surface. It reveals fine details of the specimen and the outer surfaces of membranes (9, 17). The etching rate depends on the specimen temperature (which determines the vapour pressure of the water), provided that the pressure in the vacuum chamber is low enough. At −100°C, ice will be removed to a depth of 100 nm if etching is continued for 1 min. Above this temperature (i.e. warmer), etching becomes uncontrollably fast, while below −110°C it will be very much reduced and condensation of water onto the specimen may occur, depending on the amount of water vapour in the vacuum chamber (18).

Etching therefore requires precise control of the specimen temperature and it is preferable to use as high an experimental vacuum as possible. As etching occurs, the partial vapour pressure of water around the specimen rises. A metal surface, such as the knife arm, which is cooled by liquid nitrogen to −196°C, is positioned close to the specimen during etching in order to trap molecules leaving the specimen and to prevent contamination of the surface.

Dissolved solutes greatly reduce the etching rate, so deep etching can only be achieved when specimens are suspended in very dilute solutions. Etching is much shallower within a specimen such as an isolated mitochondrion than in the surrounding medium and, in glycerinated specimens, it is almost negligible. The etching stage is therefore often omitted.

4.2.4 *Replication*

The replica should reflect faithfully the fine details of the specimen; it must scatter electrons sufficiently to produce adequate contrast in the final image and it must not be affected by reagents used to clean it. Platinum/carbon (Pt/C) films meet these requirements. By evaporating Pt/C at an angle (typically 45°) to the specimen surface, a replica is obtained in which the surface topography is shown in shadow relief. The replica is strengthened by evaporating a film of pure carbon onto it, normal to the fracture plane. In rotary shadowing, the specimen is rotated while Pt/C is evaporated at a much lower angle (e.g. 10°). Rotary shadowing may provide more structural detail than unidirectional shadowing.

Evaporation of the replica may be achieved by electrical resistance heating of two pointed carbon rods which, in the case of the Pt/C shadowing, have a small coil of platinum wire around their points. Electron beam guns give more reproducible results, however, and are rapidly becoming standard equipment in freeze-etch machines.

4.2.5 *Cleaning the replica*

After the specimen has been removed from the freeze-etch machine, the replica must be detached from the specimen and cleaned to remove all traces of organic material. Replicas of cell or organelle suspensions are generally easier to clean than replicas of intact tissues, which sometimes adhere tenaciously to the replica. Replicas are very fragile and frequently break into small pieces at this stage. Sodium hypochlorite (bleach) and strong acids are the cleaning agents most commonly used. The replica is usually detached from the specimen by lowering the specimen into water (or glycerol solution) until the replica floats off. The replica may then be transferred between cleaning solutions using a platinum wire loop. Finally, it is washed in distilled water.

4.2.6 *Viewing*

After cleaning and washing, the replica is picked up on a grid (which may be coated if the pieces of replica are very small) allowed to dry and is then examined in a transmission electron microscope. Replicas are stable in the electron beam, although very high beam intensities may cause reaggregation of the shadowing material, resulting in increased granularity and some loss of resolution.

4.3 **Rapid freezing methods**

There are several techniques for freezing specimens very rapidly so that the use of chemical fixatives and cryoprotectants is unnecessary. Only those that are applicable to suspensions of isolated mitochondria, membranes or liposomes will be discussed here.

4.3.1 *Spray-freezing*

(i) *Principles.* In this technique (19, 20), very high freezing rates are achieved by greatly reducing the size of the specimens. A liquid sample (e.g. a mitochondrial suspension) is sprayed into liquid propane, cooled by liquid nitrogen. The propane is then removed, under vacuum, leaving a fine powder of frozen specimen droplets. The droplets are then glued together in an inert medium, mounted on a specimen support and freeze-etched following standard procedures.

The coolant used must have a low melting point, a high thermal conductivity and a high specific heat. Its boiling point must be high so that the Leidenfrost effect is avoided. Since it is to be evaporated under vacuum, at low temperature, its vapour pressure must be high. Finally, the viscosity of the coolant must be low enough to allow the specimen droplets to penetrate the surface. Propane meets these requirements (20).

After spraying, the frozen droplets have to be processed so that they can be freeze etched, but they must not reach a temperature at which recrystallization of ice could cause damage. Gross morphological changes do not occur at temperatures below −70°C (20). Since etching is performed at −100°C, an inert binding medium (glue) is required which is solid at −100°C but liquid below −70°C. The compound used as the glue is *n*-butylbenzene, which has a sublimation rate at −100°C comparable to that of ice (20).

(ii) *Procedure for spray-freezing.* Although it is feasible to construct spray-freezing apparatus in a departmental workshop (21), anyone who wishes to use this technique will probably have access to one of the units produced commercially by Balzers High Vacuum.

The spray-freezing unit consists of a large box with a hinged transparent lid. Inside the box is a large metal block, the temperature of which is controlled and maintained by pressure-fed liquid nitrogen. This block acts as a heat sink and it contains cavities designed to hold: a bottle of butylbenzene; a specimen freezing container; specimen supports; and various tools such as forceps and platinum wire loops. This arrangement ensures that all these items can be maintained at −85°C, the optimum temperature for working with the butylbenzene 'glue'.

In addition, the unit is provided with a means to remove the propane under vacuum from the specimen-freezing container. Below the box is a rotary vacuum pump which is connected, via a flexible tube and a control valve, to a plastic lid that fits over the specimen freezing container, making a gas-tight seal.

Full details of the procedure are given in the instruction manual for the spray-freezing unit, so I shall give only a brief summary of the main steps here. The first stage, spray-freezing itself, is done outside the unit. When handling cryogenic liquids, one should wear protective gloves and goggles. In addition, because this procedure employs propane, there is a risk of fire or explosions. Spray-freezing should therefore be carried out in a well-ventilated laboratory away from naked flames and using the appropriate safety equipment.

(i) Switch on the unit and set the temperature control to −85°C. Place the brass specimen-freezing container (SFC) into a Dewar flask containing liquid nitrogen.

(ii) When the SFC is cold (i.e. when the liquid nitrogen stops boiling) condense propane in the conical cavity of the SFC. To do this, you will need a propane gas cylinder connected via a pressure-reducing valve and a length of flexible tubing to a thin metal tube. Use this to direct a gentle stream of propane gas into the cavity in the SFC. After a short time, the cavity will be filled with liquid propane.

(iii) Cover the SFC with a pre-cooled cover which has a hole in it with a diameter slightly smaller than that of the opening of the cavity. (This prevents the ice which forms during spraying from falling into the propane.)

(iv) Spray a small volume (0.3−0.5 ml) of mitochondrial suspension from a height of about 10 cm into the liquid propane. To do this, use the airbrush supplied with the unit connected to a compressed air line. (The air pressure should be set to 40−60 kPa, i.e. 6−9 p.s.i.)

(v) Remove the cover from the SFC and replace it with a pre-cooled lid. Remove the SFC from the Dewar flask and transfer it to the cold block in the unit. Close the lid to prevent condensation of water vapour on the cold surfaces.

The remaining steps are carried out inside the unit, with the lid closed. The operators' arms are inserted into rubber sleeves which are built into the sides of the unit.

(vi) Remove the lid from the SFC and replace it with the plastic cover connected to the vacuum line. Switch on the rotary pump and then slowly open the control

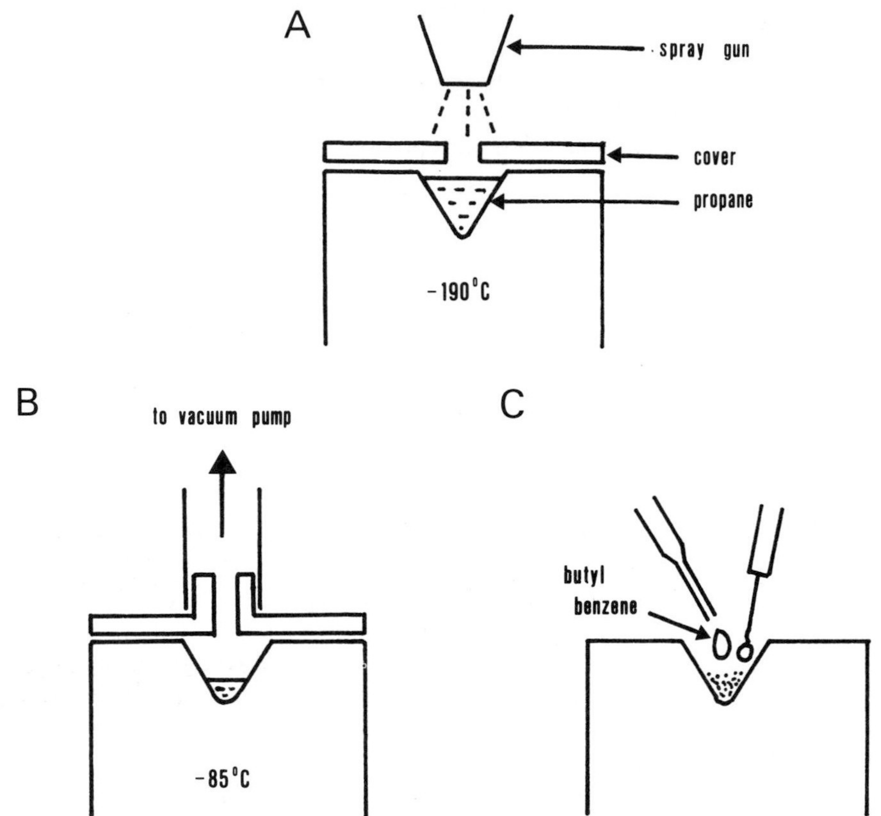

Figure 5. Spray-freezing. (**A**) A mitochondrial suspension is sprayed into liquid propane. (**B**) After warming to $-85°C$, the propane is evaporated off. (**C**) Butylbenzene is added and mixed with the frozen droplets of mitochondrial suspension. This mixture is then placed on specimen supports which are then placed into liquid nitrogen, which freezes the butylbenzene.

valve. Make sure that the cover and the SFC form a tight seal. You will be able to see the level of propane in the SFC falling as it evaporates.

(vii) When all of the propane has been removed, close the vacuum control valve, switch off the pump and remove the cover from the SFC. You should be able to see the frozen specimen droplets which appear as a fine whitish powder in the cavity in the SFC.

(viii) Using the Pasteur pipette provided, place 2–3 drops of cold butylbenzene in the cavity. Mix the specimen droplets with the butylbenzene using a cooled platinum wire loop. Then use the loop to transfer portions of the mixture onto cold specimen supports.

(ix) Using cold forceps, drop the specimen supports into a Dewar flask containing liquid nitrogen. This freezes the butylbenzene.

The specimens are stored under liquid nitrogen until freeze-etching is performed. The main stages of spray-freezing are shown in *Figure 5*, and freeze-fracturing of spray-

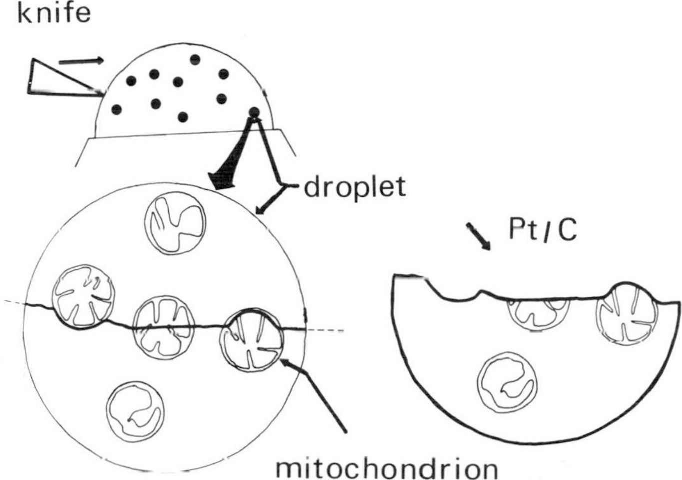

Figure 6. Freeze-fracture of spray-frozen mitochondria. The specimen (in the freeze-etch machine) is fractured using a remotely controlled knife. The fracture passes through some of the frozen droplets of mitochondrial suspension, embedded in the butylbenzene, thereby cleaving some of the mitochondria. A replica, composed of a mixture of platinum and carbon (Pt/C) is evaporated onto the fractured surface.

frozen mitochondria is shown schematically in *Figure 6*. Note that when replicas of spray-frozen specimens are cleaned, they are first floated onto acetone which dissolves the butylbenzene.

4.3.2 *Propane jet freezing and sandwich freezing*

In propane jet freezing (22), the specimen (e.g. a mitochondrial suspension) is placed between two gold specimen supports as used in standard freeze-etching. This sandwich is placed in a holder between two nozzles through which liquid propane is squirted onto both sides of the sandwich. The application of propane jets greatly increases the rate of heat exchange at the specimen surfaces. Balzers High Vacuum produces a commercial version of the apparatus for this technique. A simpler and less expensive method, in which propane is squirted onto one side of the specimen sandwich, has also proved successful (17).

Very rapid freezing (i.e. sufficiently fast for cryoprotection not to be necessary) of specimen sandwiches can be achieved without using jets of coolant (23). The approach here is to keep the mass of the specimen supports as low as possible by making them from thin copper foil. A very small volume (1 μl) of sample is placed between two such supports and is then plunged quickly into Freon cooled by liquid nitrogen.

Samples frozen by the above methods are freeze-fractured (in a freeze-etch machine) by separating the two halves of the sandwich.

4.4 **Interpretation and analysis of freeze-etch replicas**

4.4.1 *Identification of membrane faces*

Since mitochondria possess two membranes, freeze-fracture exposes four possible fracture faces, two for each membrane, and etching (and sometimes fracturing) may reveal

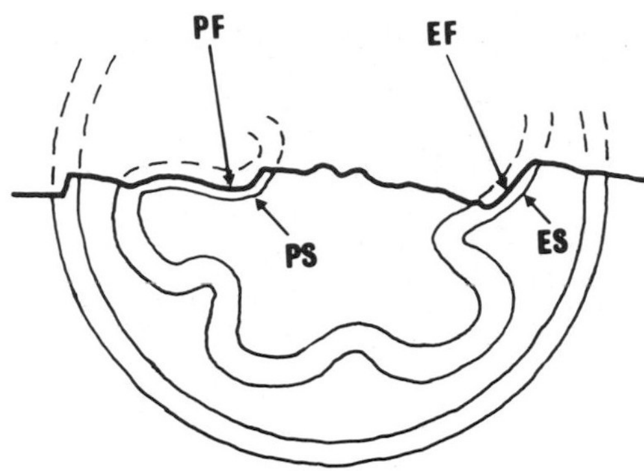

Figure 7. Labelling of the fracture faces and surfaces of the mitochondrial inner membrane, according to an agreed nomenclature (24).

Figure 8. Spray-freeze-fractured mitochondrion. The fracture has passed over the outer, or P, surface of the outer membrane (OPS) which is relatively smooth. The fracture has then deviated into the interior of the inner membrane, exposing the P face (IPF) which is covered with IMPs. ×90 000.

the true outer surfaces of the membranes (again two for each membrane). There is an agreed nomenclature for the labelling of membrane fracture faces and surfaces (24), as shown for the mitochondrial inner membrane in *Figure 7*.

For the inner membrane, the half of the membrane next to the mitochondrial matrix is designated the 'protoplasmic' half or P. Thus, PF is the inner fracture face and PS is the surface of the matrix side of the inner membrane. The outer half of the mem-

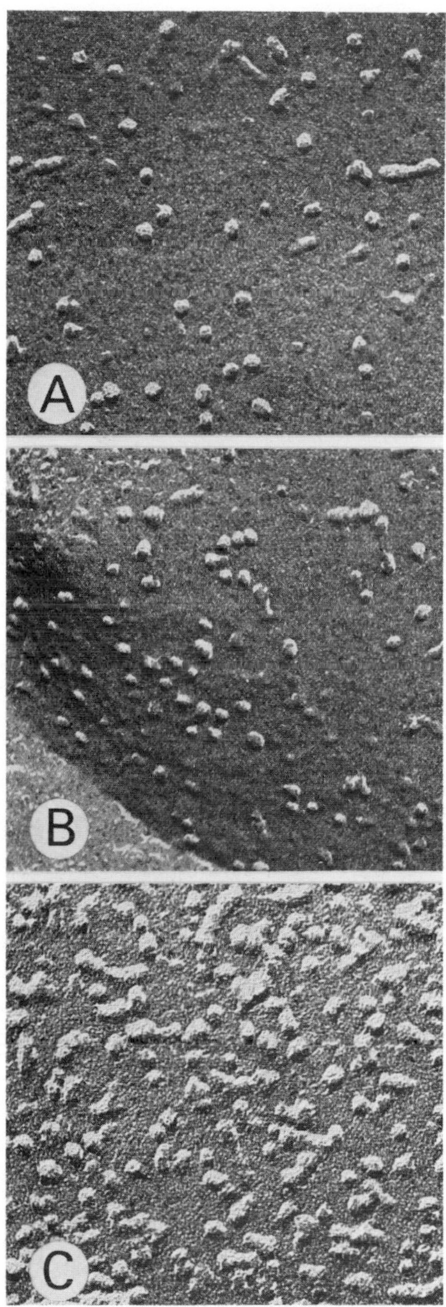

Figure 9. Phospholipid membranes containing ATPase isolated from yeast mitochondria (25). The initial ratio of protein:lipid used to form the membranes was varied: (**A**) 0.75 mg/10 mg; (**B**) 1 mg/10 mg; (**C**) 2 mg/10 mg, leading to an increase in the frequencies of IMPs in the freeze-fractured membranes.

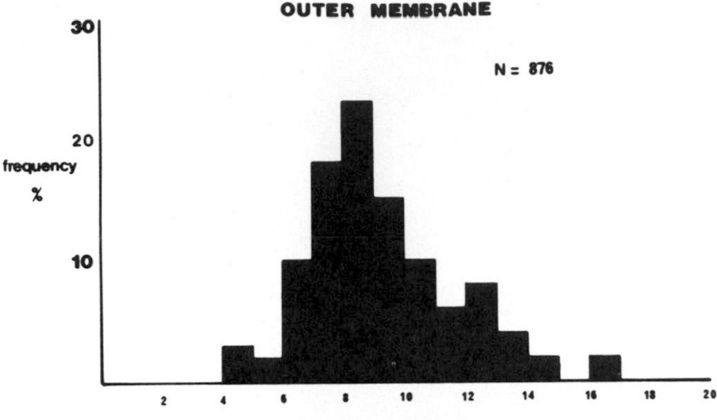

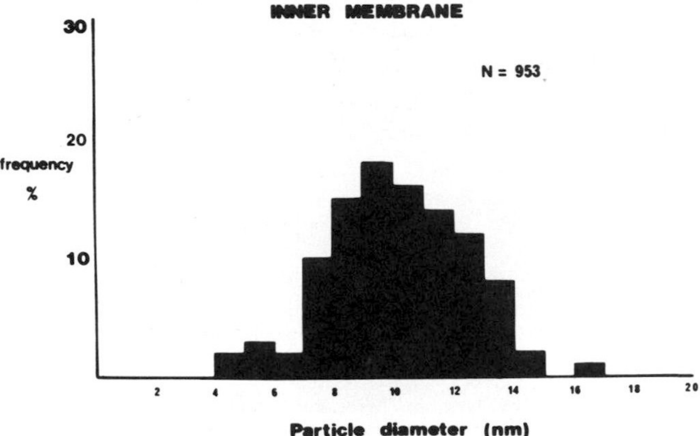

Figure 10. Size−frequency histograms for IMPs on the P-faces of the outer and inner membranes of rat liver mitochondria. The inner membranes contain a higher proportion of larger particles.

brane, next to the intermembrane space, is the external or E half. Thus, EF and ES are the outer fracture face and outer surface, respectively.

In the case of the outer membrane, PF and PS refer to the half of the membrane next to the cytoplasm, i.e. the outer half of the membrane. ES and EF refer to the inner half, next to the intermembrane space. *Figure 8* shows a freeze-fractured mitochondrion with membrane faces labelled.

4.4.2 *Analysis of IMPs*

The particles (IMPs) that appear on membrane fracture faces have been used to provide quantitative information about (and comparisons between) membranes. Considerable attention has been paid to this aspect of the analysis of micrographs of freeze-fracture replicas. The field will not be reviewed here but the major types of information that can be obtained will be summarized.

(i) *IMP frequencies*. The numbers of IMPs per unit area of membrane fracture face have been determined for many types of membranes. Generally, higher IMP frequencies are found in membranes with high ratios of protein to lipid (11, 25). Pure phospholipid membranes are usually smooth (*Figure 9*) and IMPs are found when isolated proteins are introduced into such membranes (e.g. 25).

(ii) *IMP sizes*. Measurements of the diameters of IMPs are used to produce size–frequency histograms such as those in *Figure 10*. This, like counting IMPs, is a fairly tedious procedure even with the aid of computer-based image analysers. However, it does reveal differences between membranes that may reflect true structural differences. For example, the higher proportion of large IMPs in the mitochondrial inner membrane (*Figure 10*) probably reflects the presence in the inner membrane of large electron transport complexes.

(iii) *Lateral distributions of IMPs*. Various experimental perturbations, such as cooling of membranes or electrophoresis (26), cause IMPs to aggregate. The degree to which IMPs are aggregated or randomly distributed can be determined by statistical analysis of the positions of IMPs on membrane fracture faces. Computer-based techniques are used to store and analyse these data (e.g. refs 27, 28).

5. REFERENCES

1. Meek,G.A. (1976) *Practical Electron Microscopy for Biologists*. John Wiley.
2. Munn,E.A. (1968) *J. Ultrastruct. Res.*, **25**, 362.
3. Webster,K.A. and Bronk,J.R. (1978) *J. Bioenerg. Biomembr.*, **10**, 23.
4. Lang,R.D.A. and Bronk,J.R. (1978) *J. Cell Biol.*, **77**, 134.
5. Muscatello,U. and Horne,R.W. (1968) *J. Ultrastruct. Res.*, **25**, 73.
6. Tsuprun,V.L., Mesyanzhinova,I.V., Kozlov,I.A. and Orlova,E.V. (1984) *FEBS Lett.*, **167**, 285.
7. Frey,T.G., Chan,S.H.P. and Schatz,G. (1978) *J. Biol. Chem.*, **253**, 4389.
8. Henderson,R., Capaldi,R.A. and Leigh,J.S. (1977) *J. Mol. Biol.*, **112**, 631.
9. Branton,D. (1971) *Phil. Trans. R. Soc. Lond. B*, **261**, 133.
10. Sjostrand,F.S. and Candipan,R.C. (1985) *J. Ultrastruct. Res.*, **91**, 38.
11. Verkleij,A.J. and Ververgaert,P.H.J.Th. (1978) *Biochim. Biophys. Acta*, **515**, 303.
12. Dowell,L.G. and Rinfret,A.P. (1969) *Nature*, **188**, 1144.
13. Moor,H. (1971) *Phil. Trans. R. Soc. Lond. B*, **261**, 121.
14. Yannas,I. (1968) *Science*, **160**, 298.
15. Niedermeyer,W. and Moor,H. (1976) *Proc. 6th Eur. Congr. Electron Microsc.*, **2**, 108.
16. Dunlop,W.F. and Robards,A.W. (1972) *J. Ultrastruct. Res.*, **40**, 391.
17. Knoll,G., Oebel,G. and Plattner,H. (1982) *Protoplasma*, **111**, 161.
18. Dunlop,W.F., Parish,G.R. and Robards,A.W. (1972) *Proc. 5th Eur. Congr. Electron Microsc.*, **1**, 248.
19. Bachmann,L. and Schmitt,W.W. (1972) *Proc. Natl. Acad. Sci. USA*, **68**, 2149.
20. Bachmann,L. and Schmitt-Fumian,W.W. (1973) in *Freeze-etching – Techniques and Applications*. Benedetti,E.L. and Favard,P. (eds), Societe Francaise de Microscopie Electronique, Paris. p. 73.
21. Lang,R.D.A., Crosby,P. and Robards,A.W. (1976) *J. Microsc. (Oxford)*, **108**, 101.
22. Moor,H., Kistler,J. and Muller,M. (1976) *Experientia*, **32**, 805.
23. Gulik-Krzywicki,T. and Costello,M.J. (1978) *J. Microsc. (Oxford)*, **112**, 102.
24. Branton,D., Bullivant,S., Gilula,N.B., Karnovsky,M.J., Moor,H., Muhlethaler,K., Northcote,D.H., Packer,L., Satir,B., Satir,P., Speth,V., Staehelin,L.A., Steere,R.L. and Weinstein,R.S. (1975) *Science*, **190**, 54.
25. Lang,R.D.A. and Brown,E.. (1981) *Eur. J. Cell Biol.*, **26**, 129.
26. Sowers,A.E. and Hackenbrock,C.R. (1985) *Biochim. Biophys. Acta*, **821**, 85.
27. de Laat,S.W., Tertoolen,L.G.J. and Bluemink,J.G. (1981) *Eur. J. Cell Biol.*, **23**, 273.
28. Appleyard,S.T., Witkowski,J.A., Ripley,B.D., Shotton,D.M. and Dubowitz,V. (1985) *J. Cell Sci.*, **74**, 105.

CHAPTER 3

Transport across membranes

A.DAWSON, M.KLINGENBERG and R.KRÄMER

1. CATION TRANSPORT: INTRODUCTION

The most fundamental cation-transporting systems, present in all mitochondria, are the H^+ pumps associated with the electron-transport chain and the mitochondrial ATPase system. Together, these constitute the components of the ATP-synthesizing machinery. Although the mechanism of H^+ transport is not well understood (for example, electron transport-driven H^+ movements may be either by Mitchellian loops or by H^+ pumps or both), the measurement of the stoichiometries of H^+ per oxygen atom (O), or H^+ per ATP, and of the magnitude of the electrochemical proton gradient, are frequently required to understand the mechanisms of action of compounds on mitochondrial metabolism and the mechanisms of other mitochondrial-transporting systems.

Ca^{2+} transport systems are also commonly, but not universally, found in mitochondria. Most mammalian mitochondria have at least two different mechanisms of transporting Ca^{2+} between matrix and cytosol. The physiological function of mitochondrial Ca^{2+} transport is now widely believed, in many tissues, to be involved in the control of intramitochondrial Ca^{2+} (1), rather than, as previously thought, an intracellular Ca^{2+} store. Some mitochondria, from plants and from some species of insects, have rather weak Ca^{2+}-transporting abilities.

Many mitochondria have other cation-translocating systems, for example passive Na^+/H^+ exchange, and energy-dependent K^+/H^+ exchange. The Na^+/H^+ exchange system is very active in mammalian mitochondria and can be detected by passive swelling techniques. It is difficult to measure by any other means, and its nature is not known. K^+/H^+ exchange is very ill-defined. It may just reflect a slow passive entry of K^+ with extrusion of H^+ as the membrane potential drops in response to K^+ entry. Its main impact on the experimenter is usually to infuriate him or her by putting large-scale pH drifts into pH records from mitochondrial suspensions. The Na^+/H^+ and K^+/H^+ antiport systems will not be dealt with further in this chapter, except in the manner in which they interact with other measurements.

2. PROTON TRANSPORT

2.1 **Fundamental concepts**

Proton (H^+) pumping, either by electron flow or by ATP hydrolysis, is electrogenic, that is, the movement of H^+ from the matrix of the mitochondrion to the outside space leaves an excess negative charge in the matrix space. It can be calculated (2) that, in the absence of any compensating ion movements, the transport of 1 nmol of H^+ per mg of protein across the membrane would result in a membrane potential of 200 mV (negative inside) and a pH gradient of 0.05 units (alkaline inside). This is an illustration of the fact that the proton-motive force (or the proton electrochemical potential), $\Delta\tilde{\mu}_{H^+}$, is comprised of two components, a membrane potential ($\Delta\psi$) and a pH gradient (ΔpH). The $\Delta\tilde{\mu}_{H^+}$, the Gibbs free energy change for 1 mole of H^+ moving down its electrochemical gradient across the membrane, is given by

$$\Delta\tilde{\mu}_{H^+} = \Delta\psi - \frac{2.3RT}{F}\Delta pH \qquad \text{Equation 1}$$

$$\text{or, at 30°C: } \Delta\tilde{\mu}_{H^+} = \Delta\psi - 60\,\Delta pH, \qquad \text{Equation 2}$$

In Equation 2, the values of $\Delta\tilde{\mu}_{H^+}$ and $\Delta\psi$ are expressed in millivolts.

The concepts behind Equation 2 have important consequences for the measurement of H^+ pumping and H^+ gradients across the mitochondrial membrane. The 'back pressure' of H^+ acting on either electron transport or ATP hydrolysis to impede further H^+ pumping is a function of $\Delta\tilde{\mu}_{H^+}$ which, in turn is a function of both $\Delta\psi$ and ΔpH. In measurements of H^+/O or H^+/ATP stoichiometries it is important that the magnitude of $\Delta\tilde{\mu}_{H^+}$ does not get so large that it impedes further H^+ pumping. For this reason, as will be described below, such measurements are usually carried out under so-called 'level-flow' conditions, where $\Delta\tilde{\mu}_{H^+}$ is clamped essentially at 0 mV by the presence of a K^+ concentration which is similar to the internal K^+ concentration inside mitochondria ($\sim$ 100 mM), and the ionophore valinomycin, which allows rapid electrogenic K^+ movement across the membrane. For measurements of $\Delta\tilde{\mu}_{H^+}$, however, the consequence of Equation 2 is that both $\Delta\psi$ and ΔpH must be measured, preferably simultaneously, but at least under identical experimental conditions. The reason for this is that $\Delta\psi$ and ΔpH can vary in a reciprocal fashion, depending on whether or not a counter-ion is present for H^+. As an example, consider the situation where the medium surrounding the mitochondria contains a small amount of Ca^{2+}. The Ca^{2+}ions (as will be seen later) enter mitochondria passively, down the membrane potential. As they do so, the membrane potential decreases, due to the influx of positive charge. This allows more H^+ to be pumped out, to restore the value of $\Delta\tilde{\mu}_{H^+}$. This may continue until essentially all of the Ca^{2+} in the medium has been accumulated. The end result is that the $\Delta\tilde{\mu}_{H^+}$ is comprised of a relatively smaller $\Delta\psi$ component, and a relatively larger ΔpH component. The converse applies if the medium contains a weak acid, such as acetate, which can enter the mitochondria passively as the undissociated species, AH, but is not permeable in the charged form A^-. Because the inside of the mitochondrion is alkaline compared with the outside, AH enters to maintain an internal concentration of AH equal to the external AH. Inside, it dissociates to A^- and H^+, thereby decreasing ΔpH, but not affecting $\Delta\psi$ because no charge has passed across the membrane. The decrease in ΔpH decreases $\Delta\tilde{\mu}_{H^+}$, therefore more H^+ is pumped out to maintain $\Delta\tilde{\mu}_{H^+}$ constant. A new steady-state is set up where $\Delta\psi$ is increased, ΔpH is decreased and $\Delta\tilde{\mu}_{H^+}$ remains constant. At the steady-state, acetate is present at a higher concentration in the mitochondrial matrix than outside, although the concentration of AH is the same on both sides of the membrane. The concentration of A^- on each side of the membrane is determined by the pH in each compartment, as given by the Henderson–Hasselbach equation:

$$pH = pK_a + \log \frac{[A^-]}{[AH]}$$

Although the above examples are primarily designed to show that $\Delta\psi$ and ΔpH can change relative to one another while $\Delta\tilde{\mu}_{H^+}$ remains constant, they also serve to show that, under appropriate conditions, ion distributions can be used to measure $\Delta\psi$ and ΔpH. A cation which can move across the mitochondrial membrane only as the charged species will distribute itself according to the membrane potential difference, $\Delta\psi$, as described by the Nernst equation:

$$\Delta\psi = \frac{2.3RT}{nF} \,.\, \log \frac{[C^{n+}]_o}{[C^{n+}]_i}$$

The subscripts 'o' and 'i' refer to the concentrations of the species C^{n+} outside and inside the mitochondrion. Similarly, a weak acid, AH, where the anion is impermeant, will distribute itself according to the pH difference as determined by the Henderson–Hasselbach equation, allowing measurement of ΔpH. For sub-mitochondrial particles which are orientated inside-out (i.e. the proton pumps are directed into the interior of the vesicle rather than into the outside medium), permeant anions accumulate inside driven $\Delta\psi$ and weak bases accumulate inside driven by ΔpH, since the direction of both $\Delta\psi$ and ΔpH is reversed compared with intact mitochondria.

2.2 Measurement of H^+-pumping and stoichiometries

2.2.1 *Methodology*

The only readily available and satisfactory method of measuring the quantity of H^+ transported by mitochondria during the utilization of a given amount of oxygen is the 'pulse' technique. An anaerobic suspension of mitochondria is rapidly presented with a small amount of oxygen. Electron flow starts and H^+ is pumped out but before $\Delta\tilde{\mu}_{H^+}$ can become of sufficient size to exert a back-pressure on electron flow, the added oxygen is exhausted and electron flow stops again. The H^+ appearing in the external medium is measured with a rapidly-responding pH electrode and a pH meter coupled to a pen recorder. For the experimental system to be reliable, several criteria must be satisfied.

(i) H^+-pumping must give rise to a readily measurable pH change (~ 0.05 units), so a very lightly buffered medium should be used.

(ii) The concentration of mitochondrial protein must be high, to ensure that the oxygen pulse is consumed very quickly, to minimize problems due to back-diffusion of H^+ into the mitochondria during the pulse.

(iii) The value of $\Delta\tilde{\mu}_{H^+}$ during the pulse must be minimized, so a high concentration of a permeant cation species must be present to prevent high values of $\Delta\psi$ developing. (For most experiments this is most readily accomplished by using a medium containing at least 50 mM K^+, and valinomycin to render the inner mitochondrial membrane permeable to K^+. Alternatively, for some mitochondria, Ca^{2+} at a concentration of about 0.1 mM, can be used).

(iv) Oxygen, rather than any other parameter, for example substrate or $\Delta\tilde{\mu}_{H^+}$, must be limiting, or, looked at from a more empirical point of view, the amount of H^+ pumped out must be proportional to the amount of oxygen added.

(v) The amount of H^+ carried back into the mitochondria by rapidly moving weak acid species (H_2CO_3, acetate) or proton-coupled anion movements (e.g. phosphate) must be minimized. The practical consequences of this, in relation to phosphate movements will be illustrated experimentally in Section 2.2.3.

(vi) Finally, and perhaps most importantly, strict anaerobiosis must be preserved except during the oxygen pulse. The consequences of oxygen leaks are very serious for two reasons; firstly, partial energization of the mitochondria will render ratios of H^+ per oxygen meaningless; secondly, in the presence of oxygen, high K^+ and valinomycin there will be extensive accumulation of K^+, leading to mitochondrial swelling, structural change and uncoupling.

The measurement of H^+/ATP ratios, using ATP pulses can, in theory, be done using a precisely similar technique. However, measurements of this sort on intact mitochondria are complicated by the operation of the ADP^{3-}/ATP^{4-} translocator and the Pi^-/OH^- antiporter, which together result in the effective movement of 1 H^+ out of the mitochondrial matrix for every ATP hydrolysed, irrespective of any H^+ translocation by the ATPase. In addition, ATP hydrolysis gives rise to 'scalar' (i.e. non-vectorial) protons:

$$ATP^{4-} + H_2O \longrightarrow ADP^{3-} + Pi^{2-} + H^+$$

This results in a net pH change in the medium due to the ATP hydrolysis itself, again independent of any proton pumping by the ATPase.

Effects due to operation of the translocators can be overcome by the use of inverted sub-mitochondrial particles, where ATP has direct access to the ATPase. The net pH change due to the ATP hydrolysis can either be corrected for, for example by comparison of results obtained in the presence and absence of a proton-translocating uncoupler, or, more elegantly, by working at a lower pH, below the second p*K* of phosphate, so that there is no net pH change during ATP hydrolysis. A pH of 6.1 has been found to be suitable for this (3).

2.2.2 *Apparatus*

The standard form of apparatus for the measurement of H^+-pumping reactions is shown in *Figure 1*. The prime requirement is for a thermostatted vessel with a lid bored out to take a small combined pH and reference electrode and a small hole to allow insertion of a microsyringe needle. It is also convenient to include an oxygen electrode in the setup, so that anaerobiosis can be checked. This can either be fitted through the lid (as shown), or the whole apparatus can be based on an oxygen electrode vessel where the oxygen electrode is at the bottom. Since it is impossible to seal the lid completely, the lid assembly has a flow of O_2-free N_2 passing over it to prevent O_2 diffusing into the apparatus from the atmosphere.

Any pH meter is suitable, as long as it has a recorder output giving at least 10 mV per pH unit. When coupled to a 1 mV potentiometric recorder this will give 0.1 pH unit full-scale deflection. It is also useful to have a back-off voltage in the circuit, so

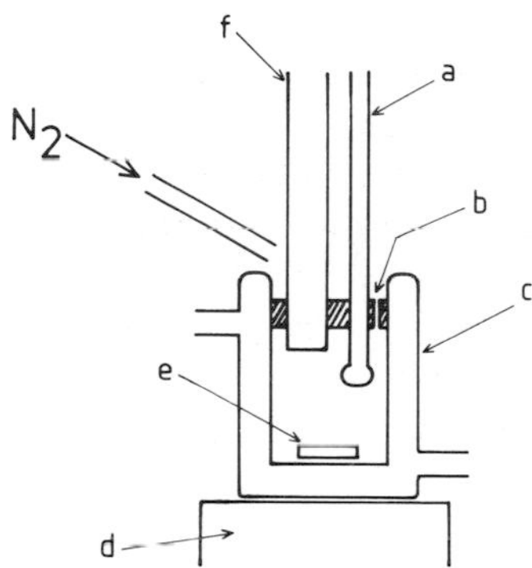

Figure 1. Apparatus for measuring H^+/O ratios. (**a**) Small combined pH and reference electrode; (**b**) lid with small hole as inlet port; (**c**) reaction vessel (capacity ~5 ml) with thermostatted jacket; (**d**) magnetic stirrer; (**e**) stirring flea; (**f**) oxygen electrode. A flow of nitrogen is directed across the top of the reaction vessel.

that the recorder can be brought on scale without using the buffer adjustment control of the pH meter.

2.2.3 *Method*

(i) Place 5.0 ml of a solution containing 100 mM KCl,1 mM Hepes−KOH, pH 7.0 and the oxidizable substrate in the reaction vessel and flush with N_2 to remove most of the oxygen in the solution. In the case illustrated here, 2.5 mM succinate (adjusted to pH 7.0 with KOH) is the substrate. Rotenone, (1 μg/ml final concentration) is present to block oxidation of endogenous NAD-linked substrates.

(ii) When most of the O_2 has been removed, add the mitochondrial suspension (~10 mg mitochondrial protein).

(iii) Apply the lid, carefully removing trapped air bubbles while simultaneously avoiding breaking the pH electrode.

(iv) Start the N_2 flow across the apparatus.

(v) When the oxygen electrode indicates complete anaerobiosis, and not before, add 2 μl of 1 mg/ml valinomycin.

(vi) There is a short period of pH drift (2−3 min). As soon as this has stopped, start the chart recorder running at about 10 cm/min and check that a flat trace is being produced.

(vii) With a microsyringe through the hole in the lid, rapidly add 25 μl of air-saturated (at 30°C) unbuffered 100 mM KCl. If the system is working properly, there should be a rapid decrease of pH about 0.01 units, followed by a slower return to the original pH (see *Figure 2*).

(viii) Repeat the procedure, adding 50 μl of air-saturated KCl. This should produce a rapid acidification twice the size of the previous one.

(ix) To calibrate the pH change in terms of nmol H^+, add 1 μl to 0.1 M HCl (i.e. 100 nmol). This should produce an essentially instantaneous pH change, and enables results to be calculated in terms of the amount of H^+ without having to worry about buffering capacity.

To illustrate the effect of blocking the phosphate transporter, *Figure 2* also shows

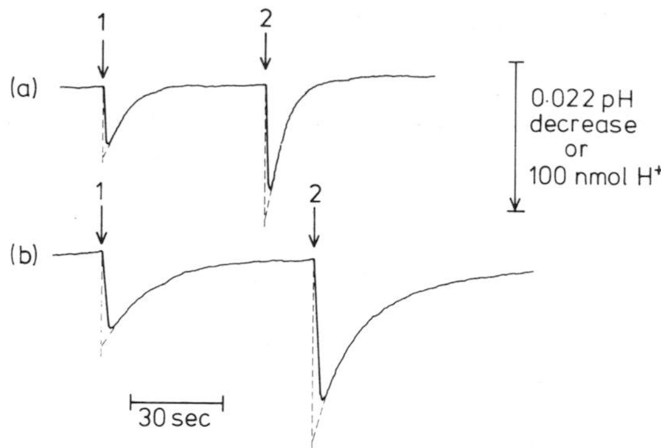

Figure 2. Traces of O_2 pulses for rat-liver mitochondria. For trace (**a**) the reaction medium contained in a final volume of 5.2 ml: 100 mM KCl, 1 mM Hepes–KOH, pH 7.0, 2.5 mM potassium succinate, 5 μg of rotenone, 10 mg of mitochondrial protein and 2 μg of valinomycin (added after anaerobiosis). At the arrows marked 1, 25 μl of aerated 100 mM KCl was added, and at the arrows marked 2, 50 μl of aerated 100 mM KCl was added. In trace (**b**) the protocol was precisely the same as for trace (**a**) except that 150 nmol mersalyl were present in the incubating medium. For trace (**a**), addition of 25 μl of KCl is equivalent to adding 11.5 ng atoms of oxygen, and the extrapolated H^+ pulse height is equivalent to 47 nmol H^+. H^+/O is 47/11.5 = 4.1. For the 50 μl KCl addition, the equivalent values are 23 ng atoms of oxygen, and 93 nmol H^+, giving H^+/O = 4.0. For trace (**b**) the 25 μl addition gave an extrapolated H^+ pulse height of 65 nmol H^+, giving H^+/O = 5.7 and the 50 μl addition gave 125 nmol H^+, H^+/O = 5.4.

a trace from an experiment where the initial incubation mixture included 15 μl of 10 mM mersalyl. This increases the extent of the acidification and slows down H^+ re-entry, since, with the phosphate transporter blocked, phosphate movements cannot take place to compensate for the pH gradient.

In theory, it is possible to do many oxygen pulses on the same incubation mixture. In practice, after 5–10 min, the rate of H^+ re-entry after the pulse starts increasing substantially as the state of the mitochondria deteriorates. To get a true value for H^+ ejected in response to the oxygen pulse it is necessary to extrapolate back the initial rate of decay of the H^+ gradient to the time of addition of the oxygenated-KCl (as shown in *Figure 2*). The higher the decay rate, the greater the error involved in doing this. The solubility of oxygen in 100 mM KCl at 30°C is 0.46 μg atoms/ml. Knowing the extrapolated value of H^+ ejected (in nmol), and the amount of oxygen added (in ng atoms), the H^+/O ratio is easily calculated, as shown in the legend to *Figure 2*. The H^+/O ratio should be independent of the amount of oxygen added, and it is well worth doing quite a wide range of KCl additions to check that this is so. The pH after the oxygen pulse should return to the same value as it was before the pulse. For some substrates and for ATP pulses there can be net production or utilization of H^+, and the pH will not return to the initial value. For example when reduced cytochrome *c* is the substrate protons are taken up in accordance with the following:

$$2 \text{ cyt } c\ (Fe^{2+}) + 1/2\ O_2 + 2H^+ \longrightarrow 2 \text{ cyt } c\ (Fe^{3+}) + H_2O$$

Net pH changes of this sort must be allowed for in calculating the H^+/O ratio, which should measure only H^+ transported across the membrane, not scalar H^+ changes in the medium as a whole.

2.3 Measurement of proton-motive force

2.3.1 *Methodology*

Most of this section will be devoted to a consideration of the various ion distribution techniques available, all of which are based on the theory described in Section 2.1. However, it is worth mentioning that other methods do exist, based on spectral shifts of dyes which stack in the mitochondrial membrane under the influence of a membrane potential. Safranine is frequently used for this purpose, the spectral shift induced by a membrane potential (negative inside) being measured in a dual wavelength spectrophotometer at 533−511 nm. The spectral shift as a function of membrane potential is calibrated using the Nernst potential produced by various external K^+ concentrations in the presence of valinomycin. The technique works very well for mammalian mitochondria. However, the authors found it unsatisfactory for locust flight muscle mitochondria, where the safranine spectral shift did not appear to reflect membrane potential but rather the accumulation of dye in the matrix. The use of this method on mitochondria from 'unusual' sources therefore requires validation by other techniques. In passing, it is worth noting that many photosynthetic energy-transducing membranes contain endogenous pigments (e.g. carotenoids) which change their spectral properties in response to membrane potential.

In Section 2.1, it was pointed out that cations which move through the mitochondrial inner membrane as the charged species accumulate in the mitochondrial matrix in response to membrane potential in a manner which can often be described by the Nernst equation. Similarly, weak acids which can pass through the membrane as the undissociated acid, but not as the anion, accumulate according to ΔpH. Determination of $\Delta\psi$ and ΔpH using the ion distribution technique requires firstly a suitable combination of permeant ions and secondly some way of measuring their distribution between the mitochondrial matrix and the medium.

Ions which have been commonly used for determination of $\Delta\psi$ in mitochondria are Rb^+ in the presence of valinomycin, and the lipophilic cations tetraphenylphosphonium (TPP) and triphenylmethylphosphonium (TPMP). All of these move across the membrane in a positively charged form and therefore accumulate on the electrically negative side of the membrane. The thiocyanate anion has been used on inverted submitochondrial particles and other vesicular systems where the membrane potential is positive inside.

For the determination of ΔpH, the weak acid species acetate and 5,5′-dimethyloxazolidine-2,4-dione (DMO) have been found to accumulate on the alkaline side of the membrane (i.e. inside mitochondria) according to ΔpH, while the weak base methylamine is accumulated on the acid side of the membrane (i.e. inside inverted submitochondrial particles or mitochondria with a reversed ΔpH).

Measurement of the distribution of the probe between the medium and mitochondria can be done in several different ways. Continuous flow dialysis (4) and ion-specific electrodes (used for TPP ref. 5) allow a continuous measure of the concentration of free probe in the medium and therefore, by subtraction, a measure of the intra-vesicular probe concentration. Both of these methods have been used successfully, but do require rather specialized apparatus (for electrode work) or suffer from rather slow time response (flow dialysis). Most measurements of $\Delta\psi$ and ΔpH have been carried out using discontinuous techniques, involving separation of the mitochondria from the medium. Two methods commonly used for doing this are filtration and centrifugation through oil, and examples of both are described here. The main criterion for success is that the separation method must be sufficiently rapid to avoid redistribution of the probe during the separation process. Pelleting of mitochondria in a centrifuge can lead to anaerobiosis in the pellet and consequent changes in $\Delta\psi$ and ΔpH. Additionally, the probes must not be metabolized or passively bound to the mitochondria. To allow calculation of the results, it is necessary to know how much of the incubation medium remains associated with the mitochondria after separation, and also the intra-mitochondrial volume. Both of the techniques described below suffer from artifacts when the proton-motive force is low, but this is not usually an insuperable problem.

The filtration method which will be described is the very elegant technique described by Nicholls (6). The medium contains three probes. $^{86}Rb^+$ (+ valinomycin) is accumulated in response to the membrane potential, [^{3}H]acetate is accumulated in response to a ΔpH (alkaline inside) and excluded by a ΔpH (acid inside) and [^{14}C]methylamine is accumulated in response to a ΔpH (acid inside) and excluded by a ΔpH (alkaline inside). Under any given incubation condition, either methylamine or acetate will be accumulated and the other will be largely excluded. This latter can then be used to calculate the amount of medium retained on the filter. The use of the three isotopes $^{86}RB^+$, ^{14}C and ^3H which have widely separated energy spectra for radiation allows the determination of all three simultaneously in a scintillation counter capable of counting in three windows. The main limitation of the method lies in the presence of valinomycin, which has to be added to allow Rb^+ to cross the mitochondrial membrane. Valinomycin also allows K^+ to permeate, so to prevent massive swelling of the mitochondria during the experiment a medium with a low K^+ concentration must be used (either sucrose or LiCl can be used as the osmotic support). However, since mitochondria contain about 120 mM K^+, this in turn means that at low values of proton-motive force, the $\Delta\psi$ term is determined by the Nernst potential of K^+ and does not drop to essentially zero as might be expected from the chemiosmotic hypothesis. However, under these conditions, as will be shown below, ΔpH reverses so that a true value for $\Delta\tilde{\mu}_{H^+}$ is observed.

The centrifugation method to be described uses TPP as the permeant cation, and DMO as the permeant weak acid. Both of these compounds are available labelled only with ^{14}C, and they cannot therefore be determined simultaneously. Incubations have to be carried out in parallel, one series with labelled TPP and unlabelled DMO, and one series with labelled DMO and unlabelled TPP. In both cases [^{3}H]sucrose or inulin can be included to allow determination of extra-mitochondrial space present after separation. Apart from the obvious disadvantage that $\Delta\psi$ and ΔpH are not determined simultaneously, the main advantage of TPP over $^{86}Rb^+$–valinomycin as a probe is that it is present at very low concentrations and therefore does not significantly perturb $\Delta\psi$ by imposing a Nernst potential. It does, however, bind significantly to mitochondria, even in the absence of a membrane potential and this can result in artifactually high values for $\Delta\psi$. These can be corrected using K^+ (+ valinomycin) diffusion potentials to calibrate the system.

2.3.2 *Determination of intra-mitochondrial volume*

This is a necessary prerequisite for determination of $\Delta\psi$ and ΔpH, since the intra-mitochondrial concentration of the probe has to be determined. The apparatus requirements are a dual-channel scintillation counter and small microcentrifuge, with very high acceleration rate, high '*g*' ($\sim$12 000 *g*) and able to take 1.5 ml polypropylene tubes. The method described here uses [^{14}C]sucrose to determine the extra-mitochondrial space and ^3H_2O to determine total water space. Separation of the mitochondria and medium is done by centrifugation through an oil layer. The buoyant density of the oil mixture used is adjusted to be intermediate between 10% (w/v) perchloric acid (PCA) and 100 mM KCl. Since sucrose-based media are denser than 100 mM KCl, it would be necessary to alter the oil composition if the intra-mitochondrial volume in such a medium was to be determined (see also Section 4.3.3).

(i) Set up a series of 1.5 ml polypropylene microcentrifuge tubes each containing as the bottom layer 150 μl of 10% (w/v) PCA and as the upper layer 500 μl of a mixture of 1-bromododecane + dinonylphthalate (1.7:1 v/v).

(ii) The mitochondrial incubation mixture should resemble as closely as possible that used for membrane potential measurements. As an example, in a volume of 2.0 ml:
   100 mM KCl, 5 mM Hepes–KOH, pH 7.4
   2.5 mM Tris–phosphate, pH 7.4
   10 μM TPP
   10 μM DMO
   10 μg of rotenone
   10 μM EGTA
   2.0 μCi (74 kBq) of ^3H_2O
   0.2 μCi (7.4 kBq) of [^{14}C]sucrose
   2.5 mM Tris–succinate, pH 7.4
   Finally add 10 mg of mitochondrial protein.

(iii) After 2 and 4 min, remove duplicate 300 μl aliquots of the incubation mixture and layer them onto the oil layer in the microcentrifuge tubes. This is quite a simple operation but care should be taken in not adding the 300 μl too fast, otherwise it penetrates the oil layer and in mixing with the perchlorate layer leads to completely erroneous results.

(iv) Centrifuge the tubes for 1 min at 12 000 *g*, after which remove 100 μl samples of the supernatants (above the oil layer) and 100 μl of the perchlorate layers and add to scintillation vials containing water-soluble scintillation fluid.

(v) Measure the ^3H and ^{14}C radioactivity in each vial using a dual-label programme on a scintillation counter. Clearly the best way to do this is to use an automatic quench correction and isotope-sorting programme which will give the answer in terms of ^3H d.p.m. and ^{14}C d.p.m. However, if this is not available it will be necessary to do a manual correction for ^{14}C in the ^3H channel using suitable [^3H] and [^{14}C]standards.

The calculation of the results is very straightforward. The rationale is to work out the 'spaces' for ^3H_2O and [^{14}C]sucrose in the mitochondrial pellet from the amounts

Table 1. Calculation of intra-mitochondrial volume.

^3H d.p.m./100 μl supernatant	220 000
^{14}C d.p.m./100 μl supernatant	22 000
^3H d.p.m./100 μl perchlorate layer	6100
^{14}C d.p.m./100 μl perchlorate layer	340
^3H d.p.m./ml supernatant	2.2×10^6
^{14}C d.p.m./ml supernatant	2.2×10^5
Total ^3H d.p.m. in perchlorate layer	9150
Total ^{14}C d.p.m. in perchlorate layer	510
^3H space in perchlorate (μl)	4.16
^{14}C space in perchlorate (μl)	2.32
Sucrose-impermeable space (μl)	1.84
Matrix space (μl/mg mitochondrial protein)	1.23

of ^3H and ^{14}C in the perchlorate layer. Assuming that ^3H_2O can completely penetrate the mitochondria and [^{14}C]sucrose is essentially excluded from the matrix space, the difference between these two represents sucrose–inaccessible space, or matrix space.

$$^3H_2O \text{ space (ml)} = \frac{\text{Total } ^3H \text{ counts in perchlorate layer}}{^3H \text{ counts in 1 ml of supernatant}}$$

$$[^{14}C]\text{sucrose space (ml)} = \frac{\text{Total } ^{14}C \text{ counts in perchlorate layer}}{^{14}C \text{ counts in 1 ml of supernatant}}$$

A single sample result, with the details of the calculation, is shown in *Table 1*.

In practice, it is necessary to make a series of determinations to get some idea of the error in the results. It is quite important to establish that there is no systematic time-dependent change in matrix volume (hence determinations after 2 and 4 min of incubation time), which could indicate significant swelling or shrinking of the mitochondria. Sucrose is generally regarded as not penetrating the matrix space for rat-liver mitochondria, but this may not be the case for all mitochondrial types, so it may be necessary to find other extra-mitochondrial space markers (e.g. inulin). Note also that the rat-liver mitochondria used in the experiment shown in *Table 1* were prepared in a sucrose-based medium, so that the addition of cold carrier sucrose to the experimental medium was done by adding the mitochondrial suspension.

An optimistic view of mitochondrial space determination is that since, in the equations for $\Delta\psi$ and ΔpH, the mitochondrial volume appears as a log term, the values of $\Delta\psi$ and ΔpH are not very sensitive to errors in matrix volume determination. A 2-fold error in volume causes only a 30% error in $\Delta\psi$ or ΔpH. In ionic media such as isotonic KCl and LiCl, many types of mitochondria have a matrix volume of about 1 μl/mg protein, while smaller values are generally obtained in isotonic sucrose media (e.g. 0.4 μl/mg for rat-liver mitochondria).

2.3.3 *Determination of $\Delta\psi$ and ΔpH by filtration*

The method described here follows closely that described by Nicholls (6) and you should refer to the original paper for all the background information on the validity of this technique. It uses $^{86}Rb^+$-valinomycin to determine membrane potential, [^{14}C]methylamine and [^3H]acetate to determine ΔpH. It requires a three-channel scintillation counter, 0.45 μm mixed cellulose ester filters (Millipore) and a filtration device. The filtration device which we have used consists of a hard polypropylene tube, i.d. 8 mm, o.d. 10 mm, connected via a trap to a water vacuum pump. (The use of a conventional porous support under the filter would encourage spreading of the medium across the filter and retention of medium on the under face.)

Application of gentle suction as the filter is placed on the open end of the tube results in a small depression into which the sample to be filtered can be placed. Using this unsupported filter, the retained volume of medium in the filter after completion of filtration is usually less than 10 μl. However, using an unsupported filter does require some skill, because if too much suction is applied the filter ruptures and flies into the air, leaving a trail of radioactive debris behind it.

(i) Set up the following mixtures (the total volumes are indicative only, depending on how many determinations are to be carried out):
    (a) 3 ml of 250 mM sucrose, 10 mM Tris–HCl, pH 7.6, 0.05 ml of 2 mM RbCl, containing 5 μCi/ml (185 kBq/ml) of $^{86}Rb^+$;
    (b) 3 ml of 250 mM sucrose, 10 mM Tris–HCl, pH 7.6, 0.1 ml of 0.6 mM methylamine (containing 25 μCi/ml, 0.9 mBq/ml, of [^{14}C]methylamine);
    (c) 3 ml of 250 mM sucrose, 10 mM Tris–HCl, pH 7.6, 0.01 ml of 10 mM sodium acetate (containing 3 mCi/ml, 100 mBq/ml, of [^{3}H]acetate).

(ii) To measure the way in which the isotopes spread across the three counting channels, apply duplicate 15 μl samples of each of the above to separate membrane filters and put them straight into scintillation vials containing scintillant.

(iii) Mix the remainder of (a), (b) and (c) together to make 9 ml of cocktail containing all three isotopes. To this, add 5 μl of valinomycin (1 mg/ml), 50 μl of 100 mM KCl and, if succinate is to be used as the substrate, 4 μl of rotenone (1 mg/ml).

(iv) For each experimental run, take 2 ml of the cocktail and add 40 μl of the mitochondrial suspension (at ~100 mg mitochondrial protein/ml). After incubation for 2 min at the required reaction temperature (30°C in the case of the results described here), remove a 150 μl sample and apply to a filter on the suction line.

(v) When filtration is complete (<10 sec), remove the filter and place in a scintillation vial.

(vi) Add the required oxidizable substrate to the incubation mixture (e.g. 10 μl of 0.5 M Tris-succinate, pH 7.6) and at suitable time intervals (e.g. 45 sec), remove 150 μl samples and filter them, the filters being immediately placed in scintillation vials. The radioactivity of duplicate 150 μl samples of the total incubation mixture should also be measured.

The results are calculated on the basis of the following equations:

$$\Delta\psi = 59 \log_{10} \frac{(\mathrm{Rb})_i}{(\mathrm{Rb}^+)_o}$$

and

$$\Delta\mathrm{pH} = \log_{10} \frac{\{\mathrm{acetate}^-\}_o}{\{\mathrm{acetate}^-\}_i}$$

$$= \log_{10} \frac{\{\mathrm{methylamine}^+\}_i}{\{\mathrm{methylamine}^+\}_o}$$

Table 2. Calculation of isotope distribution for $^{86}Rb^+$, ^{14}C and ^{3}H.

	c.p.m. channel (1)-^{3}H	*c.p.m. channel (2)-^{14}C*	*c.p.m. channel (3)-$^{86}Rb^+$*
15 μl of $^{86}Rb^+$ solution	66	929	1606
15 μl of ^{14}C solution	5043	7942	36
15 μl of ^{3}H solution	61 269	139	23
Average of duplicate 150 μl sample of incubation mixture	244 072	37 177	5398
Incubation mixture ^{14}C in channel (2) corrected for $^{86}Rb^+$		$37\ 177 - \frac{5398 \times 929}{1606} = 34\ 054$	
Incubation mixture ^{3}H corrected for ^{14}C in channel (1)	$244\ 072 - \frac{34\ 054 \times 5043}{7942}$ $= 222\ 448$		
Filter minus succinate	13 142	2353	500
Filter ^{14}C in channel (2) corrected for $^{86}Rb^+$		$2353 - \frac{500 \times 929}{1606} = 2063$	
Filter ^{3}H in channel (1) corrected for ^{14}C	$13\ 142 - \frac{2063 \times 5043}{7942}$ $= 11\ 832$		
Filter plus succinate	13 364	2293	1187
^{14}C in channel (2) corrected for $^{86}Rb^+$		1606	
^{3}H in channel (1) corrected for ^{14}C	12 344		

The p*K*s of acetate and methylamine are 4.72 and 10.62, respectively. At pH values near neutrality, the contribution of the unionized species of each is so small that it can be neglected and

$$\{\text{acetate}^-\} = \{\text{acetate}_{\text{TOTAL}}\}$$
$$\{\text{methylamine}^+\} = \{\text{methylamine}_{\text{TOTAL}}\}$$

The first stage in calculating the results is to work out the radioactivity of $^{86}Rb^+$, ^{14}C and ^{3}H in the incubation mixture and retained on the filters. The method for doing this is shown in *Table 2*. The second stage is to work out the effective space for each isotope on the filter.

For each isotope

$$\frac{s}{n} = \frac{V}{N-n}\,;\; s = \frac{V}{(N/n-1)}$$

where s = apparent space for isotope on filter (μl); n = retained counts for isotope; V = volume of filtrate (μl), N = total amount of radioactivity applied to filter (c.p.m.) To a first approximation V can be taken as the total volume of incubation mixture applied to the filter. If the volume of filtrate retained in the filter is less than 10 μl, and 150 μl are applied, the error is less then 6%. Note, however, that the volume of retained filtrate is essentially independent of the total volume added, so that the use of samples smaller than 150 μl results in a greater error.

The extra-mitochondrial space is taken as the smallest of the methylamine and acetate spaces. If the mitochondria are alkaline inside, methylamine is effectively excluded and becomes the extra-mitochondrial space marker, while if the mitochondria are acid inside acetate is excluded and becomes the extra-mitochondrial space marker. There is some deviation from this ideal behaviour at very small values of ΔpH, and under these circumstances the equation given by Nicolls (6) should be used.

For each isotope, having determined the total space (s) on the filter, the extra-mitochondrial volume (v) on the filter and knowing the intra-mitochondrial volume (m) the accumulation ratio is given by:

$$\frac{\{\text{isotope}\}_i}{\{\text{isotope}\}_o} = \frac{s-v}{m}$$

$$\Delta\psi = 59\,\log_{10}\frac{\{\text{Rb}^+\}_i}{\{\text{Rb}^+\}_o}$$

$$\Delta\text{pH} = \log_{10}\frac{\{\text{acetate}\}_o}{\{\text{acetate}\}_i} = \log_{10}\frac{\{\text{methylamine}\}_i}{\{\text{methylamine}\}_o}$$

$\Delta\tilde{\mu}_{H^+} = \Delta\psi - 59\,\Delta\text{pH}$.

These latter stages of the calculation are shown in *Table 3*.

Three final points are worth noting about this method. Firstly the isotopes in the cocktail are set up so that the radioactivity of ^{3}H is much greater than ^{14}C which in turn is much greater than $^{86}Rb^+$ radioactivity. This is to minimize errors in the carry over calculation, which can be large, and cumulative, unless this sort of protocol is adopted. Secondly, to reiterate what was pointed out earlier, the value of Δψ obtained in the absence of substrate is quite large due to the K^+ diffusion potential arising from high K^+ inside and low K^+ outside in the presence of valinomycin. Thirdly, if the K^+

Table 3. Calculation of $\Delta\psi$ and ΔpH from data in *Table 2*.

		−Succinate	*+Succinate*
Counts on filter	$^{86}Rb^+$	500	1187
	^{14}C	2063	1606
	^{3}H	11 832	12 344
Space for	$^{86}Rb^+$ (μl)	$\frac{150}{\left[\frac{5398}{500}\right]-1} = 15.3$	$\frac{150}{\left[\frac{5398}{1187}\right]-1} = 42.3$
Space for	^{14}C(μl)	$\frac{150}{\left[\frac{34\,054}{2063}\right]-1} = 9.67$	$\frac{150}{\left[\frac{34\,054}{1606}\right]-1} = 7.42$
Space for	^{3}H(μl)	$\frac{150}{\left[\frac{222\,448}{11\,832}\right]-1} = 8.43$	$\frac{150}{\left[\frac{222\,448}{12\,344}\right]-1} = 8.81$
Intra-mitochondrial volume on filter (0.4 μl/mg protein)		0.12 μl	0.12 μl
Accumulation ratio $^{86}Rb^+$		$\frac{15.3 - 8.43}{0.12} = 57.2$	$\frac{42.3 - 7.42}{0.12} = 290.7$
Accumulation ratio ^{14}C		$\frac{9.67 - 8.43}{0.12} = 10.3$	–
Accumulation ratio ^{3}H		–	$\frac{8.81 - 7.42}{0.12} = 9.9$
$\Delta\psi$ (mV)		59 log 57.2 = 104	59 log 290.7 = 145
ΔpH		log 10.3 = 1.01	$\log\left[\frac{1}{9.9}\right] = -0.996$
59 ΔpH		59.6	−58.7
$\Delta\tilde{\mu}_{H^+}$ (mV)		104 − 59.6 = 44	145 + 58.7 = 204

concentration in the medium is increased, it has the effect of decreasing $\Delta\psi$, increasing ΔpH but also increasing the matrix volume, due to energy-dependent K^+ accumulation in the presence of valinomycin. If K^+ concentration is too high, this can give rise to uncoupling.

2.3.4 *Determination of* $\Delta\psi$ *and ΔpH by centrifugation through oil*

One of the difficulties encountered in the use of the filtration method is that, as can be seen from *Table 3*, the determination of ΔpH depends on measurements of a small difference between two large numbers, because the retained extra-mitochondrial volume on the filter is quite large. The great advantage of centrifugation through oil is that the adherent extra-mitochondrial medium is very much smaller.

The probes used in this technique are TPP and DMO, both of which are only available as ^{14}C-labelled compounds, so determinations of $\Delta\psi$ and ΔpH are carried out in parallel, using [^{3}H]inulin as an extra-mitochondrial space marker. The general methodology of centrifuging through oil was described in Section 2.3.2 and the techniques used here are precisely similar, with one exception. In our hands, TPP tends to leach out of rat-liver mitochondria on their way through the oil layer, and recovery in the perchlorate layer is poor. For TPP accumulation it is therefore better to measure TPP disappearance from the supernatant than appearance in the perchlorate layer. The latter is thus redundant in the [^{14}C]TPP series of tubes, as is [^{3}H]inulin.

(i) The incubation medium for $\Delta\psi$ determination contains:
  150 mM LiCl (or KCl), 5 mM Hepes–LiOH (or KOH), pH 7.4
  1 μg/ml rotenone
  2.5 mM Tris–phosphate, pH 7.4
  10 μM DMO
  100 μM inulin
  10 μM TPP
  0.18 μCi/ml (7 kBq/ml) of [^{14}C]TPP (31.4 Ci/mol, 1.21 TBq/mol)
  5 mg/ml mitochondrial protein

(ii) Start respiration by adding 2.5 mM (final) Tris–succinate (N.B. omit rotenone if other substrates are used).

(iii) At suitable time intervals, for example at 2 min intervals, remove triplicate 300 μl samples and add to 1.5 ml microcentrifuge tubes, containing 0.5 ml of bromododecane/dinonylphthalate (1.7:1 v/v).

(iv) Centrifuge for 1 min at 12 000 *g* and remove 100 μl samples of the supernatants to scintillation vials. Also take triplicate 100 μl samples of the total incubation mixture for scintillation counting.

For rat-liver mitochondria at 25°C, TPP accumulation reaches equilibrium within 60 sec and remains stable for about 10 min.

(i) For measurement of ΔpH, the incubation mixture is as follows:
  150 mM LiCl (or KCl), 5 mM Hepes–LiOH (or KOH), pH 7.4
  2.5 mM Tris–phosphate, pH 7.4
  10 μM TPP
  10 μM DMO
  0.18 μCi/ml (7 kBq/ml) of [^{14}C]DMO (55 Ci/mol, 2 TBq/mol)
  2 μg/ml rotenone
  0.6 μCi/ml (22 kBq/ml) of [^{3}H]inulin
  100 μM inulin
  5 mg/ml mitochondrial protein.

(ii) Initiate the reaction by adding 2.5 ml (final) of Tris–succinate.

(iii) At suitable time points, for example 2 min intervals, remove 300 μl samples from the reaction medium, layer onto microcentrifuge tubes containing 150 μl of 10% PCA and 500 μl of bromododecane/dinonylphthalate (1.7:1 v/v) and centrifuge for 1 min at 12 000 *g*.

(iv) After centrifugation remove 100 μl samples of the supernatant and of the perchlorate layer for scintillation counting.

(v) Count the samples on a dual channel ^{3}H/^{14}C programme with quench correction.

Table 4. Calculation of $\Delta\psi$ and ΔpH using TPP and DMO.

(a) TPP d.p.m. in 100 μl of incubation medium		39 600
TPP d.p.m. in 100 μl of supernatant		14 500
TPP in 0.5 mg of mitochondria		25 100
Intra-mitochondrial volume for 0.5 mg mitochondria		0.6 μl
Accumulation ratio	$\frac{25\ 100}{0.6} \times \frac{100}{14\ 500}$	
	$= 288$	
$\Delta\psi$	$= 59 \log_{10} 288$	
	$= 145$ mV	
(b) 100 μl DMO supernatant ^{3}H	132 108 d.p.m.	
^{14}C	38 262 d.p.m.	
100 μl perchlorate layer ^{3}H	1717 d.p.m.	
^{14}C	3192 d.p.m.	
Inulin space in perchlorate layer	$\frac{1717 \times 100}{132\ 108} = 1.3\ \mu l$	
DMO space in perchlorate layer	$\frac{3192 \times 100}{38\ 262} = 8.34\ \mu l$	
Intra-mitochondrial DMO space	$8.34 - 1.3 = 7.04\ \mu l$	
Intra-mitochondrial volume in 100 μl perchlorate layer	$1.5 \times 1.2 \times \frac{100}{150} = 1.2\ \mu l$	
$[DMO]_i/[DMO]_o$	$= 5.87$	
pH_i	$= 8.20$	
ΔpH	$= 0.80$	
59 ΔpH	$= -47.2$ mV	
$\Delta\tilde{\mu}_{H^+}$	$= 145+47.2 = 192$ mV	

Under these conditions, at 25°C, a steady-state DMO distribution is obtained within 2 min, and this remains constant for at least 10 min.

As for $^{86}Rb^+$ distribution,

$$\Delta\psi = 59 \log \frac{[TPP]_i}{[TPP]_o}$$

However, for determination of intra-mitochondrial pH, the situation is rather more complicated with DMO than with acetate. The p*K*a of DMO is 6.32, so that at pH 7.4 approximately 10% of DMO is in the protonated state. This means that the distribution ratio for DMO has to be corrected for the amount of undissociated DMO on both sides of the membrane. The following equation is used (7):

$$pH_i = \log_{10} \left\{ \frac{[DMO]_i}{[DMO]_o} (10^{pK_a} + 10^{pH_o}) - 10^{pK_a} \right\}$$

Sample results and the calculations are shown in *Table 4*. Note that in KCl and LiCl media, the matrix volume is 1.2 μl for 1 mg of protein.

As in the case of $^{86}Rb^+$ distribution, TPP distribution gives erroneous results at low values of $\Delta\psi$, although for a different reason. You will recall that, in the case of $^{86}Rb^+$, the presence of valinomycin and the necessarily low external K^+ concentration give

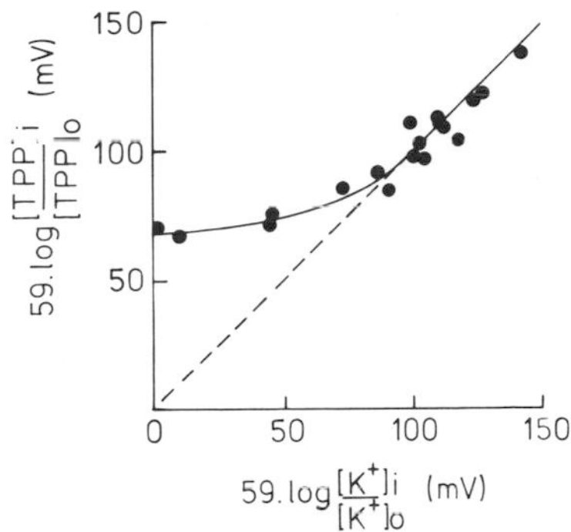

Figure 3. Relationship between TPP distribution and K^+ distribution for rat-liver mitochondria. TPP distribution was measured using [^{14}C]TPP and centrifugation through oil. The medium contained 150 mM LiCl, 5 mM Hepes–LiOH, pH 7.4, 1 μg/ml oligomycin, 1 μg/ml rotenone, 1 μg/ml antimycin (to ensure that no energy-driven ion movements occurred), 5 mg/ml mitochondria, 10 μM TPP, 0.18 μCi/ml (6.7 KBq/ml) of [^{14}C]TPP, 100 μM inulin and 0.6 μCi/ml (22 KBq/ml) of [^{3}H]inulin. In addition, K^+ in the medium was varied by adding between 0.2 and 100 mM KCl. At concentrations of KCl of 5 mM and above, the LiCl concentration was correspondingly reduced to maintain osmotic balance. K^+ permeability of the mitochondria was ensured by addition of 0.5 μg/ml valinomycin. The dotted line on the graph is the theoretical line assuming that both K^+ and TPP partitioned between the inside and outside of the mitochondria according to the Nernst equation. The data are reproduced from I.W.Warhurst, PhD thesis (1983), University of East Anglia, Norwich, UK.

rise to a K^+ diffusion potential. In the case of TPP, an artificially high matrix TPP concentration arises due to passive binding of TPP to the mitochondria. This can be measured, and to some extent compensated for, by using K^+ and valinomycin, in the absence of any electron transport (e.g. + rotenone and antimycin A) to set up membrane potentials of known value (8). The internal K^+ concentration is assumed to be constant at 120 mM. TPP distribution is measured in exactly the same way as is described above, except for the inclusion of rotenone, antimycin and oligomycin in the medium and a varying K^+ concentration plus valinomycin. The relationship between TPP distribution and K^+ diffusion potential is shown in *Figure 3*. Due to passive binding, TPP distribution seriously overestimates $\Delta\psi$ at values less than 90 mV, and, unless a correction is made, this will give rise to large apparent values of $\Delta\tilde{\mu}_{H^+}$ in the presence of uncoupler. In contrast to the $^{86}Rb^+$-valinomycin technique, the overestimated $\Delta\psi$ values will not be compensated by a reversal of ΔpH, since they do not arise from a true membrane potential.

3. CALCIUM TRANSPORT

3.1 **Fundamental concepts**

In mammalian mitochondria, three principal mechanisms of Ca^{2+} transport have been described. The route for Ca^{2+} uptake is an electrogenic uniport, where Ca^{2+} enters the mitochondria in response to the (negative inside) membrane potential. This Ca^{2+} uniporter is inhibited by very low concentrations of ruthenium red. The uniporter is very fast and the rate of uptake can be limited by electron transport. If the uniporter were the only mechanism transporting Ca^{2+}, the Ca^{2+} gradient would reach equilibrium with the membrane potential. However, there are two routes catalysing

Ca^{2+} efflux, using different driving forces, so that electrochemical equilibrium is not reached and Ca^{2+} cycles in and out of the mitochondria, giving a steady-state accumulation level when the rate of uptake equals the rate of efflux. The two routes of efflux are a Ca^{2+}/Na^{+} exchange system (9) and a more ill-defined Na^{+}-independent electroneutral Ca^{2+} exchanger, which ultimately results in $Ca^{2+}/2H^{+}$ exchange (10). The contribution that these two routes make to Ca^{2+} efflux differs in different types of mitochondria. In rat-liver mitochondria, the Ca^{2+}/Na^{+} exchanger is quite slow compared with the electroneutral exchanger, while in mitochondria from heart and brain the electroneutral Ca^{2+} efflux is essentially absent and Ca^{2+} efflux is almost entirely Na^{+} dependent. Neither the electroneutral Ca^{2+} efflux system nor the Ca^{2+}/Na^{+} exchange are inhibited by ruthenium red.

When Ca^{2+} enters mitochondria via the electrogenic uniporter, $\Delta\psi$ decreases and ΔpH increases. If phosphate is present (and it usually is to some extent, due to endogenous phosphate in the mitochondria), it will enter in response to the increased ΔpH, restoring ΔpH and $\Delta\psi$ to their previous value. Calcium phosphate precipitates inside the mitochondrial matrix, leading to very extensive Ca^{2+} accumulation. Usually, when 200–400 nmol of Ca^{2+} per mg of protein have been accumulated in the presence of phosphate, structural change occurs, resulting in uncoupling and therefore a decreased $\Delta\tilde{\mu}_{H^+}$. Acetate can substitute for phosphate in supporting massive Ca^{2+} accumulation, but in this case, since calcium acetate is soluble, swelling and uncoupling result at much lower Ca^{2+} accumulation levels ($\sim$100 nmol Ca^{2+}/mg protein). When uncoupling occurs, the $\Delta\tilde{\mu}_{H^+}$ decreases and Ca^{2+} efflux takes place by reversal of the electrogenic uniporter. It is essential, therefore, when measuring efflux by Ca^{2+}/Na^{+} or electroneutral exchange, that loading is limited to an extent where $\Delta\tilde{\mu}_{H^+}$ is maintained at normal levels.

Most ordinary laboratory solutions, stored in glass containers, contain at least 10 μM Ca^{2+}. In the cell, mitochondria are normally exposed to free Ca^{2+} concentrations of $10^{-7}-10^{-6}$M. Furthermore, *in vitro* most mammalian mitochondria will not accumulate much Ca^{2+} until the external free Ca^{2+} increases to greater than 1 μM. These circumstances conspire to make it very difficult to measure mitochondrial Ca^{2+} transport systems under anything resembling physiological conditions. It seems very likely that the function of the mitochondrial transporting systems is the control of intramitochondrial Ca^{2+}-sensitive enzymes (10), and that, at least in liver, the normal *in vivo* mitochondrial Ca^{2+} content is very low (2–4 nmol/mg protein). The usual strategy is to measure the transport systems under circumstances where measurement is relatively easy, for example using Ca^{2+} loadings of the order of 20–40 nmol per mg and then try to apply the mechanisms found to the *in vivo* situation.

Although *in vivo*, mitochondria may contain very small quantities of Ca^{2+}, by the time they have been isolated and exposed to the traces of Ca^{2+} present in solutions, the endogenous Ca^{2+} content has usually increased to about 10 nmol/mg protein. This, and the Ca^{2+} contamination of solutions, must be taken into account when setting up reaction conditions for measuring mitochondrial Ca^{2+} transport.

3.2 **Methodology**

Several techniques are available to measure mitochondrial Ca^{2+} movements. All have limitations of one sort or another, and it is important to select the most suitable method

for the parameter to be measured. The principle techniques used are:
(i) metallochromic indicators,
(ii) Ca^{2+}-sensitive electrodes and
(iii) radioisotopic measurements using $^{45}Ca^{2+}$.

Metallochromic indicators are compounds which change their spectral properties when they bind metal ions. The ones most frequently used for mitochondrial systems are Arsenazo III (11) and murexide (12). These compounds cannot enter or bind to mitochondria and they therefore monitor changes in extra-mitochondrial Ca^{2+}. They respond to changes in free Ca^{2+}, rather than total Ca^{2+}. The Ca^{2+} range in which the indicators work depends on the stability constant for the indicator$-Ca^{2+}$ complex under the precise reaction conditions used. In general, Arsenazo III works best in the range $10^{-5}-10^{-6}$ M free Ca^{2+}, while murexide is much less sensitive. The great advantage of metallochromic indicators over all other methods is their very fast kinetic response, limited theoretically by the speed of association or dissocation of Ca^{2+} with the indicator. Although they respond to changes in free Ca^{2+}, it is actually quite difficult to use them to measure absolute free Ca^{2+} concentrations. Particularly in the case of Arsenazo III, the dissociation constant is strongly affected by ionic strength, pH and Mg^{2+}, so that the calculations involved are complex (13). Clearly, also, the ionic strength, pH and Mg^{2+} must be tightly controlled during the experiment. Since mitochondrial suspensions are turbid, and the turbidity may change due to mitochondrial swelling or shrinking, it is highly desirable to use a dual-wavelength spectrophotometer to monitor the spectral change. The method is, of course, incapable of measuring intra-mitochondrial Ca^{2+} except by difference, that is the Ca^{2+} lost to or gained from the outside medium. It therefore gives no information about the total intra-mitochondrial Ca^{2+} loading unless the amount of intra-mitochondrial Ca^{2+} present at the beginning of the experiment is known.

The second commonly used technique is the Ca^{2+}-specific electrode. The advantage of this method is the ability to measure absolute values of free Ca^{2+} concentrations in the extra-mitochondrial medium, without the need for complex calculations or precise values for Ca^{2+} complex stability constants. The method suffers from the same problem as the indicator method—that it does not give an accurate measure of the intra-mitochondrial Ca^{2+} load. Additionally, the electrode response is typically not very fast, so that it is not suitable for measuring rapid changes in Ca^{2+} concentration, such as may occur with Ca^{2+} uptake at high external free Ca^{2+} concentrations in the absence of a Ca^{2+} buffer. Some commercial Ca^{2+} electrodes show a very non-linear response to Ca^{2+} concentration in the biologically interesting range ($5 \times 10^{-6}-10^{-7}$ M) and also suffer considerable interference from Mg^{2+}. These problems can be overcome by the use of much better Ca^{2+}-sensitive membranes which are now available commercially. Their source and use is discussed in the following relevant section.

Isotopic methods use $^{45}Ca^{2+}$, (a β-emitter of similar energy to ^{14}C, half-life 152 days) coupled with a rapid separation of mitochondria from the medium, for example filtration or centrifugation through oil. Time resolution is a problem here, and can be improved by using an 'inhibitor stop', where ruthenium red, one of the lanthanide ions or Mg^{2+} is used to block Ca^{2+} movement, whereupon the separation of mitochondria and medium can then proceed in a more leisurely fashion. There are, however, some further considerations. Because of Ca^{2+} contamination in mitochondria and solution, it is necessary to ensure isotopic equilibration between intra- and extra-mitochondrial Ca^{2+} if net Ca^{2+} movements are to be measured. Otherwise there will be a contribution from $^{45}Ca-Ca$ exchange. In order to make the results quantitative, total Ca^{2+} must be measured in the incubation mixture by atomic absorption spectrophotometry. The ^{45}Ca technique is, under these circumstances, the best method for measuring total Ca^{2+} movements with a reasonably true measure of intra-mitochondrial Ca^{2+}. If the effects of external free Ca^{2+} are to be measured, the free Ca^{2+} has to be calculated from a knowledge of the Ca^{2+}-chelating properties of the various species in the incubation medium. This can entail rather formidable calculations, best done using a computer. If a Ca^{2+} electrode is available, however, it can be measured directly.

The experimental protocols to be described are no more than illustrative examples of the sort of reaction conditions required for mammalian mitochondria. They are designed to indicate the strengths and weaknesses of the three different techniques, but it is clearly possible to carry out some experiments using any one of the three.

3.3 Metallochromic indicators

The spectrum of Arsenazo III and its Ca^{2+} and Mg^{2+} complexes is shown in *Figure 4*. On a more mundane level, uncomplexed Arsenazo III is pink, while the Ca^{2+} complex is blue, so that it is rather easy to see if the initial incubation medium is heavily contaminated with Ca^{2+}. The dissociation constant of Arsenazo III is of the order of 30 μM (11) (varying with pH and ionic strength) so that at free Ca^{2+} concentrations of that order and above the colour change becomes markedly non-linear with Ca^{2+} concentration. If work at free Ca^{2+} concentrations of greater than about 10^{-5} M is planned it is better to use murexide which has a much larger dissociation constant.

Many commercial preparations of Arsenazo III are very impure and heavily contaminated with Ca^{2+}. It is possible to remove the latter by passage through a Chelex

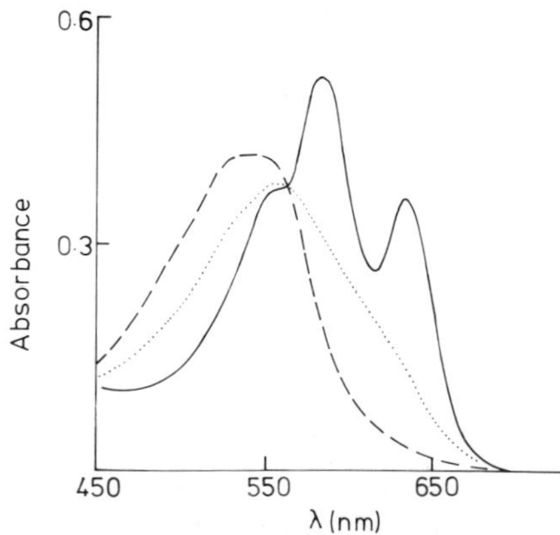

Figure 4. Spectra of Arsenazo III and its Ca^{2+} and Mg^{2+} complexes. The spectra were determined in 100 mM Hepes$-$KOH, pH 7.0 with 10 μM Arsenazo III. Dashed line, plus 1 mM EGTA; solid line, plus 1 mM $CaCl_2$; dotted line, plus 1 mM $MgCl_2$.

100 column, but on the whole it is quicker, easier and more cost-effective to buy the highly purified sodium salt supplied by Sigma.

For measurement of the Arsenazo III absorbance the authors use an Aminco DW2 dual-wavelength spectrophotometer, the lid of which has been adapted to take a small overhead battery-driven stirrer motor (*Figure 5*). The stirring vane goes into a circular, 1 cm diameter, cuvette (in fact, a 1 cm diameter glass specimen vial), and stirs in the top of the reaction medium so that it does not interfere with the light beam. Additions to the reaction medium are made with a microsyringe through a small, angled hole in the lid. The overall mixing and response time is of the order of 0.25 sec. Use of a square reaction vessel considerably lengthens the response time, due to the relatively unstirred corners. Without the stirring device, the necessity for manual mixing and removal and replacement of the lid causes the loss of quite a lot of useful information. In fact, it defeats the object of using a metallochromic indicator, which is to follow rapid transients. Calibration of Ca^{2+} movements is best done by the addition of known amounts of Ca^{2+} to the reaction mixture at some convenient stage in the experiment, and measurements of the resulting step changes. This allows the conversion of absorbance changes into changes in total Ca^{2+} concentration in the extra-mitochondrial medium. Use of an internal calibration of this type overcomes problems that may occur due to variation in Arsenazo III response as experimental parameters such as pH, ionic strength and Mg^{2+} concentration are changed.

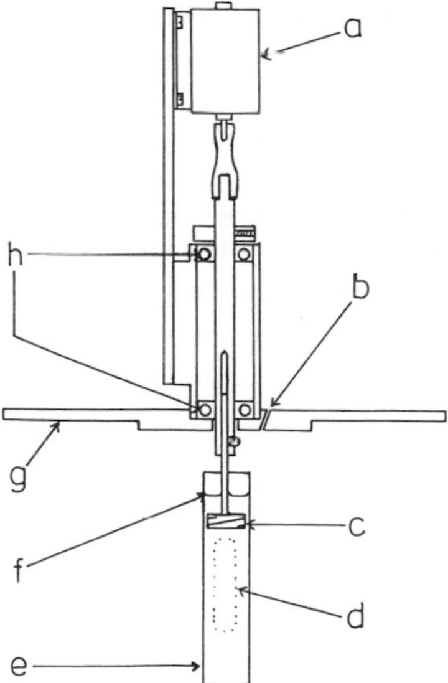

Figure 5. Lid incorporating a stirring device for a dual-wavelength spectophotometer. This particular version is designed to fit an Aminco DW2 spectrophotometer. (**a**) 4.5 V d.c. motor; (**b**) addition port; (**c**) vane; (**d**) limit of illuminated area; (**e**) 1 cm diameter circular cell; (**f**) meniscus level; (**g**) optical compartment lid; (**h**) ball races. The d.c. motor is coupled by a small flexible link to the driving shaft, and the stirring vane shaft is coupled to the other end of the driving shaft by a grub screw allowing the height of the stirring vane to be altered. The motor speed is controlled by a small rheostat in series with three 1.5 V batteries. The addition hole (**b**) is angled so that a microsyringe needle put through it enters the top of the reaction cell. The stirring vane (**c**) has helical grooves cut in the edge to aid mixing.

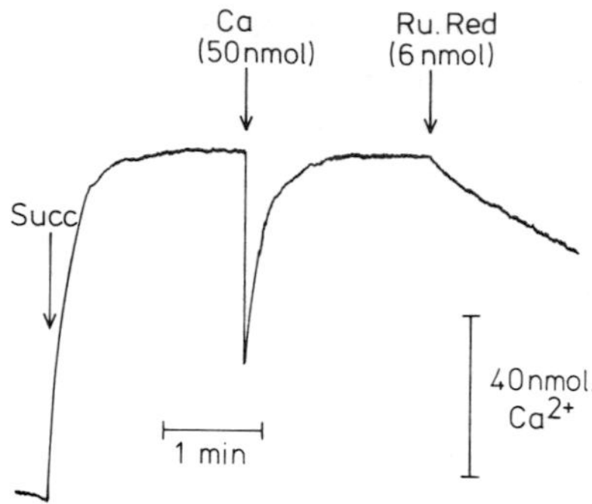

Figure 6. Dual-wavelength spectrophotometer tracing of Arsenazo III absorbance changes during mitochondrial Ca^{2+} movements. The mitochondria used were isolated from rat liver. Reaction conditions were precisely as described in the text, and the cell compartment was thermostatted at 30°C.

A typical incubation medium for a Ca^{2+} uptake and release experiment might contain:

250 mM sucrose, 20 mM Tris−HCl, pH 7.6
0.2 μg/ml rotenone
1.2 mg/ml mitochondria
8 μM Arsenazo III

The authors usually use the wavelength pair 685−665 nm, and the recorder output is set to 0.05 absorbance units full scale. A typical trace for liver mitochondria is shown in *Figure 6*. There is an initial increase in medium Ca^{2+} as endogenous Ca^{2+} leaks out of the mitochondria. Addition of succinate (2.5 mM final concentration) causes rapid Ca^{2+} uptake, until a steady-state is reached, where the rate of Ca^{2+} uptake balances the rate of Ca^{2+} efflux by the electroneutral exchange route. Further Ca^{2+} can be loaded without a change in the steady-state external Ca^{2+}. If ruthenium red is added to block the electrogenic uptake route, the efflux rate becomes visible. For heart and brain mitochondria, efflux does not occur significantly until Na^+ (~ 10 mM) is added. The record was calibrated in this case by a series of step additions of Ca^{2+}. If you are working in the right Ca^{2+} concentration range for the indicator, this calibration should produce a 'staircase' of equal steps. If the steps steadily decrease in size the free Ca^{2+} concentration is too high relative to the K_D value for the indicator.

Apart from the factors mentioned above, pH, Mg^{2+} and ionic strength, another important influence on the sensitivity of the method is the presence of Ca^{2+} chelating species. These may be added to stabilize Ca^{2+} concentrations for a particular purpose [e.g. ethyleneglycobis (β-aminoethyl ether) tetraacetic acid (EGTA), N-hydroxyethyl ethylene diamine triacetic acid (HEDTA), nitrilotriacetate (NTA)] but may also be added incidentally, for example ATP. ATP is as good at binding Ca^{2+} as it is Mg^{2+}, so that addition of ATP to drive Ca^{2+} uptake can result in firstly a large artefact as Ca^{2+} binds to ATP, and secondly a change in sensitivity of the indicator response to total Ca^{2+} concentration. In circumstances where ATP is being hydrolysed at a high rate (e.g. in the presence of an uncoupler), the Ca^{2+} binding capacity due to ATP, and hence free Ca^{2+}, will be changing continuously independent of any movement of Ca^{2+} into or out of the mitochondria, so great caution is required. This particular problem can be surmounted by including an ATP-regenerating system (e.g. creatine phosphate and creatine phosphokinase) in the incubation mixture.

3.4 Ca^{2+}-sensitive electrodes

3.4.1 *Apparatus*

Several Ca^{2+}-sensitive electrode systems are available commercially, and some of these, for example, the Radiometer version (type 2112Ca) have been used very successfully in the past for measuring mitochondrial Ca^{2+} movements. However, even the best of these have serious limitations, such as non-linearity and poor time response at free Ca^{2+} concentrations of less than 10^{-6} M. A much improved Ca^{2+}-sensitive electrode membrane has, however, been developed by Simon *et al.* (14) and is available commercially from W.Möller (Gubelstrasse 37, Postfach 8795, 8050 Zurich). It is supplied as either small sheets or as 7 mm diameter circular discs. The membrane consists of polyvinylchloride (PVC) with the neutral Ca^{2+} ionophore ETH 1001 and tetraphenyl boron dissolved in it. The system we use is a Radiometer Ca^{2+}-sensitive electrode which has had the Radiometer Ca^{2+}-sensitive membrane replaced by a Simon membrane. We remove the Radiometer membrane from the end of the barrel by shaving it off with a razor blade, then glue on the replacement with the aid of high molecular weight PVC powder (Fluka) dissolved in tetrahydofuran to make a viscous glue. The glue is painted round the plastic end of the electrode barrel, the new membrane is pressed on and sealed with more glue round the edge. Care must be taken to avoid getting any glue on the central part of the membrane through which the Ca^{2+} concentration is sensed. We use Radiometer Ca^{2+}-electrode filling solution as the internal electrolyte and a standard Ag/AgCl reference electrode. An electrode system of this sort gives an essentially linear response down to at least 10^{-7} M Ca^{2+}, has a reasonable response time and shows minimal interference by physiological Mg^{2+} concentrations. The electrode output is, in common with other systems, affected by ionic strength, so that calibration must be carried out at the working ionic strength. The electrodes dip in to a stirred, thermostatted glass (or preferably plastic) pot of about 5 ml capacity.

Because Ca^{2+} is divalent, Ca^{2+} electrodes only give half the slope of a normal H^+ electrode (i.e. 29 mV/decade instead of 59 mV/decade). The Radiometer PHM63 pH meter which we use is supplied with a doubling device, so that the meter can be calibrated to read directly in pCa[$-\log$(free Ca^{2+})]. This is very convenient, but failing this any pH meter can be used in millivolt output mode. All that is then required is a calibration table to convert the millivolt output into free Ca^{2+} concentration. The pH meter is coupled to a potentiometric recorder in exactly the same way as that described in Section 2.2.

3.4.2 *Calibration and the use of Ca^{2+} buffers*

Because of the contamination of solutions with Ca^{2+}, and the response properties of Ca^{2+} electrodes, it is very difficult to calibrate Ca^{2+} electrodes of any type at Ca^{2+} concentrations of less than 10^{-4} M without the use of Ca^{2+} buffers (15). Ca^{2+} buffers operate in a totally analogous fashion to H^+ buffers. A known concentration of a Ca^{2+}-chelating agent is added to the solution, together with a known amount of Ca^{2+}, and then, knowing the Ca^{2+}-chelator dissociation constant it is possible to calculate the free Ca^{2+} concentration. The amount of Ca^{2+} added can be large compared with possible endogenous Ca^{2+} contamination, so that free Ca^{2+} becomes independent of the latter.

The particular Ca^{2+} chelator selected depends on a variety of factors, but the major one is that it should buffer Ca^{2+} in the right range under the particular experimental conditions. This means that the Ca^{2+}-chelator dissociation constant has to be similar to the free Ca^{2+} required (in the same way that for a pH buffer, the buffer is best near its p*K* value). A range of Ca^{2+}-chelators has been used for mitochondrial studies, the most frequently used, in ascending order of affinity for Ca^{2+}, being: NTA, HEDTA and EGTA. These all have the property of binding H^+ as well and the effective dissociation constant for Ca^{2+} is very pH-sensitive in the physiological pH range. Mg^{2+} also perturbs the Ca^{2+} buffering range, although the effect is rather slight for HEDTA and EGTA. Association constants for H^+, Ca^{2+} and Mg^{2+} can be found elsewhere (16).

It is possible to calculate free Ca^{2+} manually for a simple system containing Ca^{2+} and one Ca^{2+}-chelating agent. It gets somewhat more difficult if Mg^{2+} is present, and essentially impossible if more than one Ca^{2+}-chelating species is present in addition (e.g. ATP plus HEDTA).

In the simplest case, for HEDTA, if K_1, K_2 and K_3 refer to the successive association constants for H^+, K_4 to the association constant for Ca^{2+} binding to $HEDTA^{3-}$ and K_5 to the association constant for Ca^{2+} binding to $HEDTA.H^{2-}$:

$$[Ca^{2+}]_{free} = \frac{[Ca^{2+}]_{bound}}{[HEDTA]_{free}\left[\frac{K_4}{X} + \frac{K_5}{Y}\right]}$$

$[HEDTA]_{free}$ = total concentration of HEDTA species not complexed with Ca^{2+};

$$X = 1 + K_1[H^+] + K_1K_2[H^+]^2 + K_1K_2K_3[H^+]^3$$

$$Y = 1 + \frac{1}{K_1[H^+]} + K_2[H^+] + K_2K_3[H^+]^2$$

To a first approximation:

$$\text{If } [Ca^{2+}]_{free} << [Ca^{2+}]_{total}$$

$$[Ca^{2+}]_{bound} = [Ca^{2+}]_{total}$$

$$\text{and } [HEDTA]_{free} = [HEDTA]_{total} - [Ca^{2+}]_{bound} = [HEDTA]_{total} - [Ca^{2+}]_{total}$$

Note that this means that:

$$[Ca^{2+}]_{free} = \frac{[Ca^{2+}]_{total}}{([HEDTA]_{total} - [Ca^{2+}]_{total})\left[\frac{K_4}{X} + \frac{K_5}{Y}\right]}$$

and $[Ca^{2+}]_{free}$ depends only on the ratio of $[Ca^{2+}]_{total}$ to $[HEDTA]_{total}$ and is indepen-

Table 5. The use of HEDTA as a Ca^{2+} buffer.

Total Ca^{2+} (mM)	*Free Ca^{2+} (μM)*			
	pH 7.0	*pH 7.0 + 1 mM Mg^{2+}*	*pH 7.5*	*pH 7.5 + 1 mM Mg^{2+}*
0.1	0.428	0.733	0.135	0.350
0.2	0.961	1.68	0.303	0.835
0.3	1.64	2.93	0.519	1.52
0.4	2.55	4.64	0.806	2.51
0.5	3.81	7.06	1.21	3.99
0.6	5.66	10.7	1.80	6.31
0.7	8.65	16.5	2.79	10.2
0.8	14.2	26.6	4.71	17.5
0.9	26.6	46.5	9.80	33.4

All data is calculated for 1 mm HEDTA, using the following values for log association constants (16).

HEDTA + H^+	9.72	HEDTA.H + Ca^{2+}	1.38
HEDTA.H + H^+	5.25	HEDTA + Mg^{2+}	5.78
HEDTA.H.H + H^+	2.04	HEDTA.H + Mg^{2+}	1.43
HEDTA + Ca^{2+}	8.14		

dent of their absolute concentrations. This is entirely analogous to a pH buffer, where the buffer concentration does not greatly affect pH. Again, by analogy, one can use dilute Ca^{2+} buffers to stabilize free Ca^{2+} at a known value and then measure free Ca^{2+} changes as Ca^{2+} is taken up or released by the mitochondria, or concentrated Ca^{2+} buffers to fix the free Ca^{2+} irrespective of any Ca^{2+} movements into or out of mitochondria.

The more complicated situations where Mg^{2+} and two or more Ca^{2+}-chelating species are present are best dealt with by a computer, using iterative procedures. Two programs are frequently used, one based on the algorithm of Storer and Cornish-Bowden (17), the other using the method of Perrin and Sayce (18). These programs run on small microcomputers such as an Apple IIe. To give some idea of the system involved, *Table 5* shows values of free Ca^{2+}, using HEDTA as the buffer, at a series of total Ca^{2+}, H^+ and Mg^{2+} concentrations.

To calibrate the Ca^{2+} electrode, it is a good idea to set it at a pCa^{2+} of 3, using 1 mM $CaCl_2$ in a medium of the same ionic strength and pH as is going to be used for measuring Ca^{2+} movements, and then to use a series of Ca^{2+}-HEDTA buffers across the working range to check the electrode slope in the free Ca^{2+} concentration range of interest. Referring to *Table 5*, at pH 7.0 and 1 mM HEDTA, 0.3 mM Ca^{2+} gives a pCa^{2+} of 5.79, 0.5 mM gives a pCa^{2+} of 5.42 and 0.7 mM gives a pCa^{2+} of 5.06.

3.4.3 *Measurement of Ca^{2+} uptake and release with a Ca^{2+} electrode*

Firstly, a word of warning. Many lipophilic agents, for example uncouplers, ionophores, that are routinely used in mitochondrial studies can have deleterious (but fortunately not usually irreversible) effects on Ca^{2+} electrode membranes. Before using them it

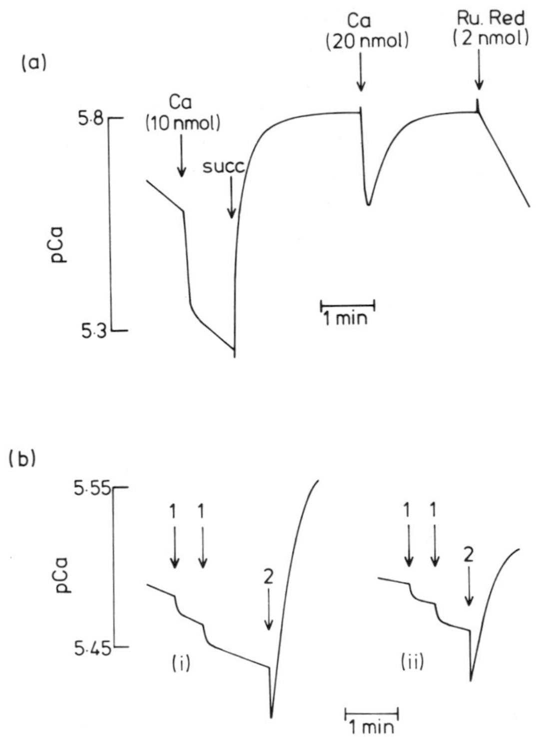

Figure 7. Mitochondrial Ca^{2+} movements measured using a Ca^{2+} electrode. (**a**) In the absence of a Ca^{2+} buffering system. The incubation medium and conditions were as described in the text, using rat-liver mitochondria. (**b**) In the presence of a Ca^{2+} buffer. Free Ca^{2+} concentration changes were slowed down by the presence of 0.5 mM HEDTA, allowing initial rate measurements to be made. For both (i) and (ii), the arrows marked 1 show the addition of 20 nmol $CaCl_2$, and the arrows marked 2 show the addition of 20 μl of 0.5 M potassium succinate, pH 7.0. For (i) the rat-liver mitochondrial protein concentration was 1.90 mg/ml, and for (ii) it was 0.95 mg/ml. The temperature was 30°C and other conditions were as described in the text.

is therefore essential to check that they do not perturb the electrode response seriously. The extent of interference of this sort varies with the type of membrane used.

Two types of experiment can be done very successfully using the electrode technique. The first is experiments similar to that shown in Section 3.3, with measurement of steady-state external pCa^{2+}, and rates of Na^{+}-dependent and independent Ca^{2+} release. Rates of uptake are, under these conditions, usually too fast to measure with the electrode. However, the second type of measurement is to use Ca^{2+} buffers to slow down the pCa^{2+} changes to within the electrode response time and then measure initial rates of uptake as a function of pCa^{2+}.

An experiment of the first type is shown in *Figure 7a*. In this case, the experimental medium contained 3.8 ml of 100 mM KCl, 20 mM Hepes−KOH, pH 7.0; 4 μl of rotenone (1 mg/ml) and 0.1 ml of rat-liver mitochondria (76 mg/ml).

Full-scale deflection on the recorder was set to about 1 pCa^{2+} unit. As in the case of the metallochromic indicator experiment, after the addition of mitochondria there is a downward drift as endogenous Ca^{2+} comes out of the mitochondria. The trace was calibrated for total Ca^{2+} by the addition of 10 μl of 1 mM Ca^{2+} (i.e. 10 nmol total). 20 μl of 0.5 M potassium succinate, pH 7.0 were added, resulting in the rapid uptake of Ca^{2+} into the mitochondria. The final steady-state free Ca^{2+} obtained in the experiment was 1.26 μM, corresponding to a pCa^{2+} of 5.90. Addition of a further 20 nmol of Ca^{2+} results in a downward deflection, followed by a rapid return to the same steady-state free Ca^{2+} level. Note that the downward deflection caused by 20 nmol Ca^{2+} is rather less than that caused by the calibration pulse of 10 nmol Ca^{2+}. This is because in the former case a large proportion of the added Ca^{2+} is accumulated by the mitochondria within the electrode response time. *Figure 7a* also shows the effect of adding ruthenium red (2 nmol). This produces a linear rate of Ca^{2+} efflux due to the ruthenium red-insensitive efflux pathway. By reference to the Ca^{2+} calibration pulse, the rate of this efflux is calculated as 1.39 nmol Ca^{2+}/min/mg mitochondrial protein. Note that ruthenium red is rather 'sticky' and is difficult to remove from the walls of the reaction vessel.

The second type of protocol, where changes in pCa^{2+} are slowed down by the inclusion of a Ca^{2+} buffer, is shown in *Figure 7b*. The reaction medium in this case contained 3.8 ml of 100 mM KCl, 20 mM Hepes−KOH, pH 7.0; 4 μl of rotenone (1 mg/ml); 20 μl of 100 mM HEDTA−KOH, pH 7.0; 40 μl of 100 mM potassium phosphate, pH 7.0 and 70 μl of 10 mM $CaCl_2$.

This mixture gave a pCa^{2+} of 5.48. The pCa^{2+} value is, of course, readily altered by increasing or decreasing the amount of $CaCl_2$ added.

For the first trace in *Figure 7b*, 7.6 mg of rat-liver mitochondrial protein was added, followed by two calibration pulses each of 20 nmol Ca^{2+}. Uptake was started by adding 20 μl of 0.5 M potassium succinate, pH 7.0. The initial rate of Ca^{2+} uptake is fairly linear for about the first 15 sec of the trace, and has a value, in this case, of 1.05 nmol Ca^{2+}/sec/mg protein. The figure also shows a trace at half the mitochondrial concentration, which shows that the initial rate of uptake is also halved (i.e. remains the same per mg protein). It is essential to test the system in this way, because if the electrode response is limiting, the rate of uptake will not be proportional to protein concentration.

The presence of phosphate in the incubation mixture extends the linear portion of the initial rate, although it is perfectly possible, using this type of system, to measure uptake rates in the absence of any added accompanying anion.

3.5 Accumulation of Ca^{2+} studied using $^{45}Ca^{2+}$

As described in Section 3.2, techniques using the radioactive Ca^{2+} isotope can be used very successfully for measuring kinetics of Ca^{2+} uptake and release, particularly if combined with an inhibitor-stop method to improve time resolution. A method of this sort is described elsewhere (19). Because it is a discontinuous assay and is somewhat more labour-intensive than metallochromic indicators or Ca^{2+}electrodes, it is perhaps not the method of choice if the other two techniques are available.

The experiment described below to illustrate the method is, however, rather different. $^{45}Ca^{2+}$ is used to measure intra-mitochondrial Ca^{2+} as a function of extra-mitochondrial Ca^{2+} at values of the latter in the physiological range. At the low loadings obtained under these conditions, chemical analysis of mitochondrial pellets by atomic absorption spectrometry becomes difficult and $^{45}Ca^{2+}$ is used as a tracer for mitochondrial Ca^{2+} content.

In our hands, rat-liver mitochondria contain about 0.5 − 1.0 nmol of Ca^{2+}/mg of protein which is not removable by prolonged exposure (>60 min) of the mitochondria to EGTA and which is essentially non-exchangeable with $^{45}Ca^{2+}$ added outside. Quite what this Ca^{2+} is we do not know, but while accepting that it is there we do not include it in any of the calculations below.

At low external Ca^{2+} concentrations (e.g. in the presence of HEDTA or EGTA outside), isotopic equilibrium with exchangeable mitochondrial Ca^{2+} is achieved in about 10 min at 30°C in a KCl-based medium in the presence of rotenone. The principle difficulty in setting up the experiment lies in making an estimate of endogenous Ca^{2+} levels, which are not determined until the end of the experiment. It is therefore a good idea to use a range of HEDTA and/or added Ca^{2+} concentrations to find the right conditions.

(i) Set up the following incubating mixtures:
   (a) 5.4 ml of 100 mM KCl, 5 mM Hepes−KOH, pH 7.5
   0.01 ml of 1 mg/ml rotenone
   0.06 ml of 10 mM HEDTA
   0.5 μl of $^{45}Ca^{2+}$ (1 mCi/ml, 37 MBq/ml).
   (b) 5.4 ml of 100 mM KCl, 5 mM Hepes-KOH, pH 7.5
   0.01 ml of 1 mg/ml rotenone
   0.03 ml of 10 mM HEDTA
   0.5 μl of $^{45}Ca^{2+}$ (1 mCi/ml, 37 MBq/ml).

(ii) Equilibrate each mixture to 30°C, then to each add 0.6 ml of a rat-liver mitochondrial suspension containing 50 mg protein per ml.

(iii) Incubate the mixtures for 10 min to achieve isotopic equilibration.

(iv) After 10 min, remove 0.5 ml samples of each and layer onto 0.5 ml of bromododecane/dinonylphthalate (1.7:1 v/v) in microcentrifuge tubes.

(v) Centrifuge for 1 min at 12 000 *g*.

(vi) Add 0.1 ml of 0.5 M potassium succinate to each incubation mixture, and remove successive 0.5 ml samples from each at 1, 3, 5 and 10 min after adding succinate, centrifuging through oil immediately after removal. There should be at least 2 ml of each of the incubation mixtures left, which will be used for measuring total Ca^{2+} and total $^{45}Ca^{2+}$.

(vii) From each of the microcentrifuge tubes, remove duplicate 0.1 ml samples of the supernatant, for scintillation counting. Suck off the remainder of the supernatant, and the oil layer, to leave the pellet. Carefully wipe the walls of the tube to remove adhering supernatant, without disturbing the pellet. Resuspend each pellet in 0.75 ml of 1% sodium dodecyl sulphate (SDS), and remove duplicate 0.2 ml samples for scintillation counting. Also count duplicate 0.2 ml samples of the original incubation mixtures.

(viii) For measurement of total Ca^{2+}, take duplicate 1.0 samples of the original incubation mixtures and add to each 0.5 ml of 5% (w/v) trichloroacetic acid (TCA). (Note that the TCA solution must be made up and stored in plastic vessels, since it leaches Ca^{2+} out of glassware giving very high blank values.) Centrifuge off the precipitated protein and assay Ca^{2+} in the supernatant by atomic absorption spectrophotometry.

Table 6. Measurement of Ca^{2+} accumulation using $^{45}Ca^{2+}$.

Incubation (a) Total Ca^{2+} = 27.1 μM
Total $^{45}Ca^{2+}$ = 14 826 c.p.m./0.2 ml
Ca specific activity = 2735 c.p.m./nmol

Time	*c.p.m. (0.2 ml pellet)*	*c.p.m. (0.1 ml supernatant)*	*Pellet Ca^{2+} (nmol/mg)*	*Supernatant Ca^{2+} (μM)*
0	866	6523	0.47	23.8
1	1552	5735	0.85	21.0
3	2731	5260	1.49	19.2
5	3044	5162	1.66	18.9
10	2856	5130	1.56	18.8

Incubation (b) Total Ca^{2+} = 24.6 μM
Total $^{45}Ca^{2+}$ = 14 950 c.p.m./0.2 ml
Ca specific activity = 3038 c.p.m./nmol

Time	*c.p.m. (0.2 μl pellet)*	*c.p.m. (0.1 μl supernatant)*	*Pellet Ca^{2+} (nmol/mg)*	*Supernatant Ca^{2+} (μM)*
0	578	6466	0.28	21.3
1	4302	3664	2.11	12.1
3	4122	3658	2.02	12.0
5	4297	3548	2.11	11.7
10	4502	3612	2.21	11.9

Pellet Ca^{2+} (nmol/mg) is calculated from:

$$\frac{\text{c.p.m. for } 0.2\ \mu\text{l of pellet}}{(Ca^{2+}\ \text{specific activity}) \times (\text{protein in } 0.2\ \text{ml pellet})}$$

Supernatant Ca^{2+} (μM) is calculated from:

$$\frac{(\text{c.p.m. for } 0.1\ \text{ml supernatant}) \times 10}{(Ca^{2+}\ \text{specific activity})}$$

In the experiment for which the results are shown in *Table 6* total Ca^{2+} was found to be 27 μM for incubation mixture (a) and 24.6 μM for incubation mixture (b). Note, however, that this is a variable figure which must be determined for each individual incubation mixture, since Ca^{2+} contamination of apparatus, pipettes, etc. is very difficult to control.

In incubation (a), on addition of succinate the mitochondrial Ca^{2+} increases from 0.47 nmol/mg to a steady-state loading level of 1.57 nmol/mg (average of 3, 5 and 10 min values), while the total extra-mitochondrial Ca^{2+} decreases from 23.8 to 19 μM. At 0.1 mM HEDTA, this corresponds to a change from 3.7×10^{-7} M free Ca^{2+} to 1.8×10^{-7} free extra-mitochondrial Ca^{2+}.

For incubation (b), intra-mitochondrial Ca^{2+} reaches a steady accumulation level of about 2.1 nmol/mg after 3 min, with a corresponding extra-mitochondrial Ca^{2+} of 12.0 μM. At 0.05 mM HEDTA this corresponds to an extra-mitochondrial free Ca^{2+} of 3.7×10^{-7} M.

It is clearly possible, by adding extra Ca^{2+}, or by using a variety of HEDTA concentrations, or both, to construct a loading curve for any type of mitochondria at a range of steady-state extra-mitochondrial Ca^{2+} concentrations. From these experiments, at free Ca^{2+} concentrations of the order of 10^{-7} M, rat-liver mitochondrial Ca^{2+} loadings are very small. Under the above conditions, the carryover of supernatant into the pellet is negligible. However, if required this could be corrected for by incorporating [^{3}H]sucrose or [^{3}H]inulin in the incubation mixture as described in Section 2.3.

4. METHODS FOR MEASURING METABOLITE TRANSPORT IN MITOCHONDRIA

4.1 **Introduction**

The transport of metabolites in mitochondria is measured mostly by following the distribution between the intra- and extra-mitochondrial spaces. The metabolites can be assayed in either one or both spaces enzymatically and/or radioactively. Recording methods have only been applied by using the turbidity changes due to volume increase in the case of a massive net uptake of metabolites (20). Under these conditions unphysiologically high concentrations of metabolites are used and no quantitative data can be obtained. Transport measurements by determining directly the solute redistribution can be quantitatively evaluated. In this part of the chapter these direct methods for measuring transport which admittedly can in some applications be quite laborious are described. The methods used in particular for measuring ADP/ATP transport in mitochondria were first described elsewhere (21–28).

The measurements of the metabolite transport require separation of mitochondria from the incubation medium. In general the intra-mitochondrial volume and therefore also the amount of metabolite within the mitochondria is 50–200 times smaller than that of the surrounding medium. Therefore, the disappearance of metabolites from the surrounding medium is insignificant, hence the appearance of metabolites in the mitochondria is used for following transport processes. Often the metabolism of the imported substrate can pose problems. Interconversion can occur during the incubation time and after mitochondria have been separated. The latter problem can be overcome by using appropriate inhibitors and/or by immediate quenching of the separated mitochondria in, for example, perchloric acid (PCA).

The relatively small volume on the one hand and the limited uptake capacity, in particular in the case of exchange, result in a rather short time span for following transport until equilibration is reached. Therefore, the time resolution of the sampling and separation steps must be high, as compared with transport measurements in cells, for example erythrocytes. High resolution techniques have been developed for determining the kinetics of metabolite transport in mitochondria (24–28). These methods have also been combined with rapid quenching procedures in order to compensate for the high intra-mitochondrial metabolic rates.

A survey of various metabolite transport systems in mitochondria has been given in several reviews (29–33) and for the ADP/ATP carrier (29, 34–37). These systems are characterized according to their metabolic function, unidirectional or exchange transport, the electrical balance and their inhibitors. Interestingly, in mitochondria metabolite transport occurs overwhelmingly by exchange. Net transport is only found for the uptake of phosphate, glutamate and pyruvate. The great importance of the exchange system is understandable in view of the intracellular location of the mitochondria and the function of these transport systems as linkers between cytosolic and intra-mitochondrial reactions. Moreover, exchange systems guarantee osmotic neutrality. As shown in this chapter, transport by exchange can be measured in a quite different manner from transport by net uptake. By radioactively labelling the intra- and extramitochondrial solutes, an exchange for the unlabelled solute can be measured. In general, exchange is at least one order of magnitude faster than net uptake.

Another feature important for practical applications is the electrical charge balance during transport. Some systems are electroneutral because the charges of the substrates are neutralized by H^+ co-transport, as is true for the citrate–malate exchange and also for phosphate uptake (38). In these cases the translocation rate and the net distribution depends on the (ΔpH) across the membrane. A higher pH value will facilitate co-transport of H^+ (39).

Two transport systems of key significance, the ADP/ADP exchange and the aspartate/glutamate exchange, are electrically imbalanced, since one negative charge is translocated during the 1:1 exchange. Therefore the rate of exchange can be greatly influenced by the membrane potential. This is, however, only true for the metabolically significant hetero-exchange. In the case of a homo-exchange, for example ADP/ADP or glutamate/glutamate, the rate is, of course, not influenced by membrane potential.

The inhibitors for the various transport systems may originate from natural sources as antibiotics, or be synthetic substrate analogues, or amino acid reagents. The most specific inhibitors exist for the ADP/ATP carrier and have obviously been developed by nature as highly specific defence substances. The inhibitors have great practical importance: first, for identifying and segregating specific transport systems, second, for identifying translocators and, third, for application as rapid quenchers of the translocation rate in the inhibitor-stop methods. Provided that the inhibitors interact rapidly with the carrier, the 'inhibitor-stop' method gives the highest time resolution yet available when used in combination with the appropriate equipment described in the following sections.

4.2 **Pre-treatment of mitochondria**

The isolated mitochondria used for the principal studies should be as intact as possible. This is important for three main reasons: firstly, the inner membrane should be intact, secondly, the mitochondria should be able to build up a good electrochemical H^+ potential as a driving force for transport, thirdly, the mitochondria should contain a high solute content. The last condition is important for exchange systems which comprise the majority of transport measurements in mitochondria. This means that the mitochondria should be isolated under isotonic conditions preferably with sucrose (see Chapter 1).

4.2.1 *Forward exchange experiments*

The mitochondria may be used for the exchange studies either directly with the endogenous solutes or may require pre-loading with solute in order to obtain maximum exchange rates. Phosphate transport is the best defined net uptake transport system in mitochondria. Glutamate uptake is only slow and much less clearly characterized. ADP or ATP transport occurs in well defined counter-exchange systems which rely primarily on the endogenous nucleotide content. The high intra-mitochondrial nucleotide content is very much dependent on the integrity of the isolated mitochondria. It is difficult to increase the intra-mitochondrial ADP or ATP content by incubation with external ADP or ATP. In mitochondria from etiolated corn a strong uptake has been reported (40) under certain conditions and also in liver mitochondria the content can be increased under extreme conditions (41). On the other hand, mitochondria tend to lose the endogenous nucleotides, often rapidly, depending on the incubation conditions.

The dicarboxylate and tricarboxylate transport systems are complicated by the intercalation of the tri- and dicarboxylate carrier with the phosphate transport through the dicarboxylate-phosphate carrier. Thus dicarboxylates can be taken up by exchange against the phosphate, coupled to the net release of phosphate via the phosphate transporter. The intra-mitochondrial phosphate may originate either from the endogenous phosphate or from loading the intra-mitochondrial phosphate with external phosphate. Loading of mitochondria with internal substrate is necessary for measuring transport in these metabolite exchange systems for which the isolated mitochondria do not provide endogenous counter-substrate. This requires pre-treatment of the mitochondria with high external concentration of the counter-substrate in order to build up an endogenous pool of the counter-solute. This is necessary for determining the transport in the case of the dicarboxylate carrier, and the ketoglutarate/malate, citrate/malate and glutamate/aspartate systems.

4.2.2 *Substrate loading procedure for forward transport experiments*

Incubate the mitochondria in the standard incubation medium (0.25 M sucrose, 10 mM Tris−HCl, pH 7.8) for 15 min at 0°C in the presence of 2−100 mM substrate. For the di- and tricarboxylates select concentrations in the range between 2 and 10 mM. For glutamate incubate mitochondria in 100 mM glutamate.

4.2.3 *Backward exchange experiments*

In forward transport or exchange the uptake of radioactively labelled substrate is measured. This method has the disadvantage that after separation of the mitochondria corrections must be made for adhering external media; especially when measuring small initial uptake values this error can be considerable. The backward exchange method avoids this error. Here the endogenous substrates, nucleotides, etc. are first radioactively labelled by pre-incubation of the mitochondria with labelled substrates of high specific activity. After removal of the external substrate these mitochondria are ready to be used in the 'backward exchange'. By adding unlabelled substrate the internal labelled substrate is released into the medium against a low background. Therefore, the appearance of even small amounts of substrate in the early stages of transport kinetics can be more accurately detected than in the forward exchange. Radioactivity is determined in the external medium after removal of mitochondria by centrifugation or filtration. This method of back exchange has been perfected in particular for the ADP/ATP exchange. In combination with the inhibitor-stop method it gives the highest time resolution and sensitivity.

4.2.4 *Radioactive substrate loading procedure*

(i) For labelling the intra-mitochondrial solutes, incubate mitochondria (either directly after isolation or after loading with substrates as described above) at high protein concentrations (20–60 mg/ml) in 0.25 M sucrose, 10 M Tris–HCl, pH 7.8 with substrates of high specific activity, that is 0.5–2 μCi (74 kBq) [^{14}C]substrate per 0.5 ml at 0°C.

(ii) After 30–60 min, dilute the mitochondria to 5 mg protein/ml with medium and centrifuge (26 000 *g* for 15 min at 4°C).

(iii) Resuspend the pellet in isotonic sucrose to a concentration of about 20–50 mg protein/ml. This solution is now ready for application in the back exchange.

4.3 **Transport measurements**

Transport measurements require to be initiated by mixing the substrate with the mitochondria and termination of the uptake or release. These steps have to be sufficiently rapid, with respect to the required time of resolution. The reaction can be started either by manual stirring or by rapid mixing devices. The termination of the reaction can coincide with the separation of the mitochondria by centrifugation or sieve-filtration, for example by silicone layer filtration or, in the case of sieve-filtration by pressure micropore filtration. A more convenient and faster termination is achieved by rapid addition of inhibitor (inhibitor-stop method) either manually or by mixing devices; after this the time requirements are less stringent.

4.3.1 *Starting and stopping the reactions*

For incubation of the mitochondria 1.5 ml plastic microcentrifuge tubes (Eppendorf cups) are most widely and conveniently used. For running a series of experiments in parallel, for example for studying concentration dependences, a procedure has been worked out where additions and stopping can be done in parallel (29). This experimental setup is shown in *Figure 8*.

(i) Place the tubes in aluminum holders which are immersed in a water bath for temperature control.

(ii) Incubate the mitochondria in these vessels ready for the transport experiment.

(iii) Add substrates on a fork which may hold about 5–20 μl.

(iv) At zero time start the reaction by placing these forks into the vessels and stirring rapidly.

(v) Termination of the reaction may now be achieved by the addition of inhibitors. For this purpose load a second fork with an excess of the transport inhibitor and at time '*t*' rapidly add the inhibitor to stop the transport.

(vi) Subsequently, separate the mitochondria by centrifugation.

Either the uptake or the release of the radioactively labelled substrates is measured in the sediment or supernatant.

4.3.2 *Centrifugal filtration*

The separation of mitochondria by centrifugal sedimentation is simple but has two disadvantages, firstly, the amount of adhering fluid may be relatively large and secondly

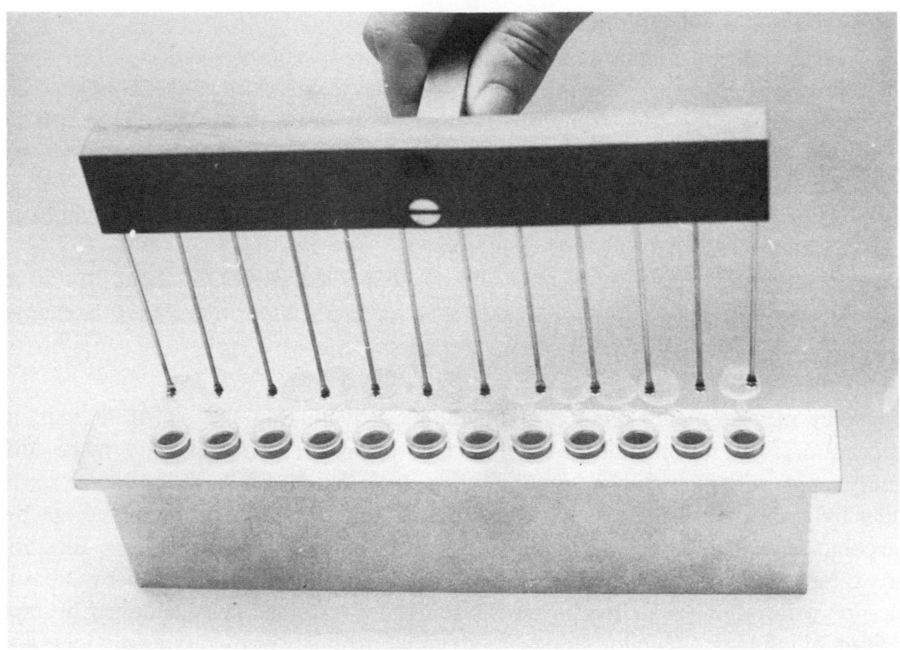

Figure 8. The fork system for starting and inhibiting a transport reaction in parallel samples. Several tubes are placed in a thermostatted aluminum block with the reaction suspension. The tongues of the fork are loaded with 5–20 μl of the starting or stopping reagent solution.

the transport metabolites may be altered because the mitochondrial pellet is anaerobic. In the centrifugal filtration procedure mitochondria are filtered through a silicone oil layer into acid (23, 42, 43). During the passage through silicone oil, the mitochondria stay aerobic and the immediate quenching and deproteinization by the acid retains the steady-state composition of the metabolites. These are immediately released into the extract as soluble constituents. The reaction can be started with the multiple fork system if several parallel samples are assayed. The reaction is terminated either directly by centrifugal filtration, if the time resolution required is not too high, or by the inhibitor-stop method.

Centrifugal filtration was first developed in the swinging bucket rotor using relatively large volumes up to 2 ml, which keeps the layers in a horizontal position. With larger volumes it is easier to measure accurately the metabolite composition in the acid extract by the various enzymatic assays. More practical and popular is the use of disposable 1.5 ml plastic tubes which in a microcentrifuge can also be used at a horizontal angle or, in the case of Eppendorf tubes, in a fixed-angle rotor, provided the silicone layer is sufficiently large to ensure separation of the incubation and the quenching layer; no traces of acid should reach the incubation layer.

Two procedures will be described; for Eppendorf tubes and for Beckman microcentrifuge tubes.

4.3.3 *The Eppendorf tube procedure*

(i) Layer into the cups the following solutions from the bottom to the top. 0.1−0.2 ml of 1.6 M PCA; 0.3 ml of silicone oil, 0.5−1 ml of incubation medium. For a mitochondrial suspension in 0.25 M sucrose (ϱ = 1.037 g/ml) the densities for the acid layer should be ϱ = 1.081 g/ml and for the silicone layer ϱ = 1.05 g/ml. The amount of mitochondria should not exceed 3 mg. The density of the silicone should be adjusted to the reaction temperature and to the different incubation media by mixtures, for example between type AR 100 (0−15°C) and 150 (>15°C) (Wacker Chemistry).

(ii) Finally add the mitochondrial suspension on top of the silicone layer and acid.

(iii) Initiate the reaction by addition of the labelled or unlabelled substrate using the fork system. For this purpose place the tubes in the aluminum block (*Figures 8* and *9*) and rapidly lower the forks, loaded with various substrates, into the tubes until a pre-determined stop, such that the forks penetrate only into the incubation layer.

(iv) Terminate the reaction either directly by centrifuging the tubes or by the addition of the transport inhibitor again by the fork system.

(v) For the silicone layer filtration procedure centrifuge the tubes at 12 000 *g* for 10 min until all the mitochondria have passed through the silicone layer into the acid.

(vi) Separate the layers by introducing a U-shaped metal wire into the upper layer and freezing the tubes in dry ice.

(vii) Remove the incubation layer completely. Rapidly decant silicone which remains liquid and wipe off residual silicone from the frozen acid layer with petrol ether.

(viii) Allow the separated supernatant and the acid to thaw and take aliquots for measuring radioactivity or determination by enzymatic methods.

4.3.4 *Microcentrifuge procedure*

With microcentrifuges (Beckman-, Coleman-type) the silicone layer filtration can be conveniently performed in still smaller volumes. However, in this case enzymatic assays of the mitochondria may be more difficult. For manipulating the start and termination in these tubes, an appropriate fork system is even more important in order to get reproducible results. A corresponding apparatus is shown in *Figure 9* where the microcentrifuge tubes are placed in centrifuge holders and where the additions for starting and termination of the reactions can be made by the appropriately dimensioned fork system. The block permits rapid change of the two sides, with one side for starting the reaction by addition of substrate and, after turning the holder, with the other fork side for terminating the reaction by addition of inhibitors. The tubes are thermostatted in an aluminum block.

(i) Layer into the plastic tubes the following solutions: 30 μl of 1.6 M PCA, 70 μl of silicone oil and on top of the silicone oil 100−150 μl of the mitochondrial suspension, containing 3−4 mg protein/ml.

(ii) Initiate and terminate the reactions as described above (Section 4.3.3) but using the apparatus depicted in *Figure 9* (*A* and *B*).

Figure 9. Fork systems for measuring transport in microcentrifuge tubes. In **A** the top section is rapidly pushed down into the centrifuge tubes. At the end of the rods the open cylindrical vessels take up the starting or stopping reagent solution. The shape of the vessel edge causes rapid mixing. This holder is turned by 180° for changing from the starting to the stopping additions. In **B** the bottom section is moved up and down for addition and mixing. The left rods contain the starting and the right rods the stopping solutions. At first the tube holder is placed in the left side for addition of the starting solution, then it is changed to the right side for addition of the stopping solution.

(iii) Centrifuge at 12 000 *g* for 10 min, that is until the mitochondria have passed into the acid phase.
(iv) Remove the upper supernatant, separate the bottom PCA extract by freezing and cutting the tubes at the lower end of the silicone layer and wipe away the remaining silicone with petrol ether.
(v) Thaw the extract, dilute it into 200 μl of water and remove the protein by centrifugation (10 000 *g*, 5 min).
(vi) Take 100 μl aliquots for radioactivity measurements.

4.3.5 *Micropore-filtration*

Separation of mitochondria by filtration through micropore filters (Sartorius or Millipore) can have a better time resolution than centrifugation. It is, however, inferior to the inhibitor-stop method. High resolution using a 'bypass' of mitochondria precipitated on these filters has been claimed using special equipment (29).

The following procedure can be used with any filter apparatus connected to a vaccuum line or by rapid filtration under pressure. The capacity of the filters for mitochondria is limited, since they easily become blocked. The capacity is increased to about 4 mg protein/filter of 20 mm diameter by combining two filters, one of larger pore size (0.8 μm) on top of one of a smaller pore size of 0.4 μm.

(i) Pipette a volume containing up to 4 mg of protein of the mitochondrial suspension, in which transport has been started by adding a substrate, onto the filter.
(ii) Terminate the reaction by applying suction at the required time.
(iii) Dissolve the filter with the accumulated mitochondria in the scintillation fluid for measurement of radioactivity.

Here the correction for the adherent external medium is particularly important and therefore the radioactivity of the filtrate must also be measured. The filtration can also be carried out to separate the mitochondria after the transport has been stopped by inhibitors. In this case the time resolution is defined by the rapidity of the inhibitor addition.

A disadvantage of the micropore-filtration is the relatively large proportion of adherent fluid in the filter. Therefore, a rather high background tends to obscure small amounts of radioactive uptake. The method is mostly used for back exchange and release of substrate. On the other hand, care must be taken that the mitochondria are not broken on the filters by suction for too long in which case they lose their endogenous substrates.

4.4 **Special equipment for higher time resolution in transport studies**

Higher time resolution can be achieved by using pressure instead of suction. An apparatus has been designed where a time resolution down to about 1 2 sec can be achieved by pressure filtration. In this case the reaction is started by addition of the substrate within the closed pressure chamber. Mitochondria are first pipetted onto the filter, the chamber is then closed and a stirring device within the pressure chamber is activated for starting the transport. A timing device is automatically started by the stirrer which, at a pre-determined time, releases a pressure of 20–30 p.s.i (1.5–2.0 bar). Filtration time is less than 0.5 sec.

For measuring more rapid transport kinetics, consecutive sampling techniques with high time resolution have been developed. This equipment however is relatively difficult to construct and not commercially available. In a rapid consecutive sampling method, not only the technicalities of rapid sampling but also the device for rapid mixing and termination of the reaction has to be provided. Two different types of equipment have been constructed, one of which is based on the inhibitor-stop method and subsequent separation of mitochondria by centrifugal sedimentation while the other is based on termination of the reaction either using the inhibitor-stop method and separation by rapid filtration or on termination directly with the pressure filtration.

The rapid mixing–quenching sampling apparatus (RAMQUESA) stops the reaction by withdrawing consecutive samples and rapidly mixing them with the inhibitor stop (24,26). The reaction is started by injecting the substrate into the mitochondrial suspension using the moving–mixing chamber with a time resolution of 100 msec. The reaction chamber is under pressure and with a special stepping motor-driven frictionless valve system precise samples are withdrawn. During the withdrawal, the fluid passes a mixing chamber into which the stopping inhibitor is injected and rapidly mixed with the sample. The stopped sample is then injected into the centrifuge tube of a multiple

Table 7. Some characteristics of the rapid mixing and sampling devices.

	RAMQUESA	*RAMPRESA*
Separation	Centrifugation	Filtration
Mixing time (msec)	10	10
Mixing ratio	20:50	20:100
Minimum sampling time (msec) variable for each sample up to 100 sec	80	250
Number of samples	12	12
Maximum sample volume (μl)	600	1000
Number of time steps/sample	5	8
maximum reaction volume (ml)	6	12
Temperature (°C)	−20 to 50	−20 to 50

sample holder. The movement of the sample holder is coordinated with the activation of the various valves by electronic time devices and stepping motors. The whole time sequence is pre-programmed.

In a rapid mixing and sampling pressure filtration apparatus (RAMPRESA) a similar mixing and sampling setup is employed (26, 44). After the reaction has been initiated samples are consecutively withdrawn and filtered under pressure at pre-set time intervals. Each filtered sample is injected into one of 12 filtration chambers containing the micropore filters. After injection the respective pressure chamber is tightly closed and the pressure valve is opened for rapid passage through the filter into a collecting vessel. Special rapid magnetic valves are used for the control of filtration pressure and the closing of the pressure chamber. Each time after pressure release and opening of the filtration chamber cover, the round table containing the 12 filtration chambers and collecting vessels moves on. The whole table is temperature controlled. The movement of this table is also controlled by a heavy-duty rapid-stepping motor through a pre-programmed time device. Some characteristics of these two rapid mixing and sampling devices are given in *Table 7*.

5. MEASURING METABOLITE TRANSPORT BY MITOCHONDRIAL CARRIER PROTEINS IN RECONSTITUTED SYSTEMS

5.1 **Introduction**

In the preceding section, elaborate methods for measuring metabolite transport in mitochondria have been described. What is the reason for extending these methods now to transport in reconstituted systems?

There are three main reasons for working with reconstituted systems instead of intact organelles:

(i) in general, it is the only way to identify a functional carrier protein;
(ii) very often, the transported substrates are further metabolized inside the mitochondria and measurement of pure carrier function is therefore difficult;
(iii) intact mitochondria represent a rather complicated system for the analysis of transport functions because of compartmentation, because of interfering enzyme activities and because of protein−protein interactions.

These and other reasons are responsible for the fact that several mitochondrial carrier systems have been functionally reconstituted. Reconstituted systems are much simpler than the intact organelle and provide some important advantages when measuring metabolite transport. Due to several specific properties connected with the reconstituted system, especially the small size of liposomes, the methods for measuring transport in reconstituted phospholipid vesicles differ from those applied to isolated mitochondria.

5.2 Reconstituted systems of mitochondrial carriers

Reconstitution of several mitochondrial transport proteins has been reported, either with purified carrier proteins (45–47) or with enriched fractions (48–53). The proteins to be reconstituted are solubilized by non-ionic detergents and incorporated into the membranes of liposomes. In general, the freeze/thaw/sonication method is used for reconstitution (45,54).

5.2.1 *Preparation and incorporation of ADP/ATP exchange carrier*

A detailed procedure for the reconstitution of the most important mitochondrial carrier protein, the ADP/ATP exchange carrier, has recently been published (55) and is given briefly.

(i) Solubilize beef heart mitochondria (10 mg protein/ml) at 0°C in 3% Triton X-100, 150 mM Na_2SO_4, 0.5 mM EDTA, 20 mM Tricine–NaOH, pH 8.0.

(ii) Centrifuge (10 000 *g* for 10 min) to remove debris.

(iii) Apply the supernatant to hydroxyapatite in a batch type procedure. Add about 7 mg of solubilized protein to 1 ml of wet (water-washed) hydroxyapatite.

(iv) Mix thoroughly and after 5 min centrifuge at 4°C briefly at 20 000 *g* to remove hydroxyapatite.

(v) Retain the supernatant containing the ADP/ATP carrier, which is about 75% pure. Since the carrier protein is very unstable in the solubilized state, it has immediately to be incorporated into liposomes.

(vi) Sonicate egg yolk phospholipids (50–80 mg lipid/ml) at 0°C in water (see also Chapter 5, Section 8) to form liposomes. All ions and substrates, especially nucleotides, which should afterwards be present in the internal volume of the reconstituted proteoliposomes have to be added at this stage.

(vii) Mix the solubilized carrier with the pre-formed liposomes.

Since in the basic method described here, a freeze/thaw/sonication procedure without removal of detergent is used, the amount of protein that can be incorporated is restricted by the fact that the solubilizing detergent is still present and the concentration of Triton X-100 has to be kept below 5% of total lipids (w/w).

5.2.2 *Freeze/thaw/sonication*

The freeze/thaw/sonication procedure of reconstitution starts by rapid freezing followed by slow thawing. This increases the liposome size making them suitable for kinetic measurements. The sonication step, leading to functionally active proteoliposomes, is very critical.

(i) Place the liposomes with incorporated ADP/ATP exchange carrier into a dry ice/acetone bath or liquid nitrogen.

(ii) Thaw the frozen liposome preparation in an ice/water bath allowing about 15 min for this process.
(iii) Place 1 ml of the thawed liposome preparation ($\sim$50 mg phospholipid per ml) in a 1.5 ml vessel and sonicate in a pulse mode.

The pulse mode sonication has to be thoroughly optimized and requires usually 8–12 sec net sonication time. The ADP/ATP carrier reconstituted into proteoliposomes is now relatively stable and is suitable for transport measurements (see also Section 4.1).

Reconstitution of the other mitochondrial carrier proteins can be carried out by very similar procedures (46–53).

5.3 Basic principles of transport measurements in proteoliposomes

In order to achieve kinetic resolution, the transport reaction initiated by addition of substrate has to be stopped after appropriate intervals. In principle, this can be done either by rapid separation of the vesicles from the surrounding medium or by an inhibitor-stop method followed by a separation step.

In the case of mitochondria, the former method, that is rapid separation, has been applied successfully in several transport systems (see Section 4.3). However, when using liposomes, this method has some drawbacks. On the one hand, it is impossible to sediment liposomes within a reasonable time. On the other hand, filtration of liposomes requires membrane filters with a very small pore size and is, therefore, by no means a rapid procedure, and in general does not lead to satisfactory results. Thus, inhibitor-stop kinetics is the method of choice to be applied in reconstituted systems, either with specific (45,47,48,50,51) or with non-specific inhibitors (46,49,53). After termination of the transport reaction by the appropriate inhibitors, the surrounding medium has to be separated from the vesicles. As mentioned above, centrifugation and filtration are not convenient means. Instead, for the separation step, advantage is taken of the ionic nature of the metabolites transported across the inner mitochondrial membrane, that is the external metabolites are bound to anion-exchange resins.

Another general question with respect to these measurements is whether to carry out the exchange reaction in the 'forward' or in the 'backward' direction (56 and Section 4.1). In mitochondria, the backward exchange experiment is more convenient and the data obtained by this method are easier to interpret than the results of forward exchange kinetics (22). In the case of the proteoliposomes the situation is just the opposite, as seen in the following section. It has to be mentioned, however, that although it is more complicated, backward exchange in proteoliposomes has one important experimental advantage. Phospholipid membranes are not indefinitely stable and some liposomes may be damaged during the kinetic experiment, the chromatographic procedures and the inhibitor treatment in the course of a transport experiment. If the carrier activity in an exchange experiment in the forward direction is found to be decreased, one cannot distinguish whether this is due to a decreased activity of the carrier protein or simply to a considerable portion of damaged (i.e. leaky) liposomes. This discrimination can, however, be achieved in backward exchange experiments where the efflux of radioactively labelled substrate from pre-labelled liposomes is recorded. The blank value, which of course represents the highest value in this type of transport measurement, decreases, if the liposomes have become leaky during the exchange experiment. The blank values are not decreased if the carrier itself has become inactivated.

5.4 **Procedures**

In the following section, the procedures for transport measurements are described in detail, using the ADP/ATP carrier as a representative example. Reconstituted transport catalysed by other mitochondrial carrier proteins can be analysed in a qualitatively similar manner; this will be discussed at the end of this chapter.

Start and stop of the transport reaction is effected by a method similar to that described for mitochondria in Section 4.3, using forks for the addition of substrates and inhibitors.

5.4.1 *Forward exchange of adenine nucleotides*

Since mitochondrial carriers predominantly function as exchange carriers, internal substrate has to be present during the preparation of proteoliposomes. After reconstitution (Section 5.2), the external adenine nucleotides have to be removed. The best way to do this is by gel filtration on Sephadex G-75 (Pharmacia), which is superior to removal by anion exchangers (55). The osmotic balance inside and outside the vesicles has to be considered during the gel chromatography.

For kinetic experiments with a series of samples, the liposomes are then distributed into 1.5 ml reaction vessels, which are placed in aluminum holders as described in Section 4.3. The exchange is started by addition of radioactively labelled adenine nucleotides on the first fork, followed by the appropriate inhibitor (usually carboxyatractylate) on the second fork. The time intervals between addition of substrate and inhibitor cannot be less than 5 sec, since this is a manual method. If a higher kinetic resolution is desired, the transport activity has to be slowed down by cooling the aluminum blocks in a water bath.

To separate internal and external labelled nucleotides, the samples are then applied to Dowex 1 × 8 anion-exchange columns. In order to minimize non-specific adsorption pre-equilibrate each column with a buffered solution of bovine serum albumin and egg yolk phospholipid liposomes (55). The radioactivity measured in the eluate represents the amount of labelled nucleotides in the internal volume of the proteoliposomes. Calculation of exchange kinetics in the forward direction has been described elsewhere (57).

5.4.2 *Backward exchange of adenine nucleotides*

The procedure for backward exchange is identical to that described above for the transport in the forward direction up to the Sephadex G-75 column. The internal nucleotides of the eluted proteoliposomes are then labelled by the addition of external [^{14}C]ADP or [^{14}C]ATP in 1–2 μM concentrations at room temperature. After 10 min the vesicles are applied to a second Sephadex G-75 column. After this column, the internally-labelled proteoliposomes are subjected to an exchange experiment in the way described above for the forward exchange. The backward exchange kinetics can be calculated essentially as described for mitochondria (57).

5.4.3 *Procedure for other carriers*

As stated before, the methods for measuring the activities of other mitochondrial carriers in reconstituted systems, such as the phosphate carrier (46), the oxoglutarate carrier (47), the aspartate/glutamate carrier (49), the tricarboxylate carrier (48), and the dicarboxylate carrier (50), are basically very similar to the method described here for the ADP/ATP carrier. When analysing transport kinetics, the different substrate affinities ranging from micromolar (ADP/ATP carrier) to millimolar (phosphate carrier) have to be considered. Furthermore, for the removal of the external substrate by anion-exchange chromatography, the binding affinity of the various anions to the resin has to be taken into account. That means, that Dowex–acetate has to be applied for the removal of oxoglutarate or asparate, whereas Dowex–Cl can be used for adenine nucleotides.

5.4.4 *Comments on the merits of forward and backward exchange experiments*

When using both the forward and the backward exchange technique, the high asymmetry of volume distribution between the internal and the external volume of the reconstituted proteoliposomes is the reason for several experimental restrictions. Since, in a typical exchange experiment with proteoliposomes reconstituted by the freeze/thaw/sonication procedure, the internal volume of the (active) liposomes is 100–500 times smaller than the external volume, restrictions have to be observed with respect to the internal and external substrate concentrations. Since an exchange carrier does not change the total concentration of substrates on the two sides of the membrane, the extreme volume asymmetry can partly be overcome by applying an opposite asymmetry of substrate concentrations. For instance, in an experiment in the forward direction (and, of course, also when labelling internal substrate in the course of a backward exchange experiment), the internal substrate concentration must be significantly higher than the external concentration. In typical experiments, the internal adenine nucleotides are present in 10–50 mM concentration, whereas the external nucleotides are applied in 5–500 μM concentration. Of course, this restriction does not hold true for backward exchange experiments. In this case it is useful to apply high external concentrations of substrate in order to avoid a backflow of radioactive nucleotides into the inside caused by an isotopic equilibrium between outside and inside.

In exchange experiments with proteoliposomes, always the amount of internal radioactive label is recorded, in contrast, for example, to backward exchange experiments with mitochondria. Thus the blank values in forward experiments are very low, that is less than 0.1% of the radioactivity which is applied onto the Dowex columns. Therefore already small amounts of labelled substrate, transported to the inside, can be detected, This makes the forward exchange method a suitable system for measuring initial velocities. On the other hand, the backward exchange starts with high amounts of internal label which makes it difficult to detect the small decrease of internal label at the very beginning of the transport kinetics. However, backward exchange has its advantages in detecting the intactness of the proteoliposomes, as mentioned above. Furthermore, backward exchange is the only feasible method when the experimental setup requires an application of very low internal or very high external substrate concentrations, due to the relatively small internal volume of the proteoliposomes (see above).

6. ACKNOWLEDGEMENTS

M.K. and R.K. acknowledge support from the Deutsche Forschungsgemeinschaft (Kl 134/23 and Kr 623/2).

7. REFERENCES

1. Denton,R.M. and McCormack,J.G. (1980) *FEBS Lett.*, **119**, 1.
2. Nicholls,D.G. (1982) In *Bioenergetics*. Academic Press, London.
3. Thayer,W.S. and Hinkle,P.C. (1973) *J. Biol. Chem.*, **248**, 5395.
4. Sorgato,M.C., Ferguson,S.J., Kell,D.B. and John,P. (1978) *Biochem. J.*, **174**, 237.
5. Casadio,R., Venturoli,G. and Melandri,B.A. (1981) *Photobiochem. Photobiophys.*, **2**, 245.
6. Nicholls,D.G. (1974) *Eur. J. Biochem.*, **50**, 305.
7. Rottenberg,H. (1979) In *Methods in Enzymology*. Fleischer,S. and Packer,L. (eds), Academic Press, New York, Vol. 55, p. 547.
8. Wilson,D.F. and Forman,N.G. (1982) *Biochemistry*, **21**, 1438.
9. Nicholls,D.G. (1978) *Biochem. J.*, **176**, 463.
10. Crompton,M., Moster,R., Ludi,H. and Carafoli,E. (1978) *Eur. J. Biochem.*, **82**, 25.
11. Scarpa,A. (1979) In *Methods in Enzymology*. Fleischer,S. and Packer,L. (eds), Academic Press, New York, Vol. 56, p. 301.
12. Scarpa,A. (1972) In *Methods in Enzymology*. San Pietro,A. (eds), Academic Press, New York, Vol. 24, p. 343.
13. Murphy,E., Coll,K., Rich,T.L. and Williamson,J,R. (1980) *J. Biol. Chem.*, **255**, 6600.
14. Simon,W., Ammann,D., Oehme,M. and Morf,W.E. (1978) *Ann. N.Y. Acad. Sci.*, **307**, 52.
15. Portzehl,H., Caldwell,P.C. and Ruegg,J.C. (1964) *Biochim. Biophys. Acta*, **79**, 58.
16. Sillen,L.S. and Martell,A.E. (1971) *Chem. Soc. (Lond.), Special Publ.*, **17**, 25.
17. Storer,A.C. and Cornish-Bowden,A. (1976) *Biochem. J.*, **189**, 1.
18. Perrin,D.D. and Sayce,I.G. (1967) *Talanta*, **14**, 833.
19. Crompton,M. and Carafoli,E. (1979) In *Methods in Enzymology*. Fleischer,S. and Packer,L. (eds), Academic Press, New York, Vol 56, p. 338.
20. Chappell,J.B. (1968) *Br. Med. Bull.*, **24**, 150.
21. Klingenberg,M. and Pfaff,E. (1967) In *Methods in Enzymology*. Estabrook,R.W. and Pullman,M.E. (eds), Academic Press, New York, Vol. 10, p. 680.
22. Pfaff,E., Heldt,H.W. and Klingenberg,M. (1969) *Eur. J. Biochem.*, **10**, 484.
23. Pfaff,E. (1967) Thesis, Marburg.
24. Klingenberg,M., Grebe,K. and Appel,M. (1982) *Eur. J. Biochem.*, **126**, 263.
25. Nohl,H. and Klingenberg,M. (1978) *Biochim. Biophys. Acta*, **503**, 155.
26. Palmieri,F. and Klingenberg,M. (1979) In *Methods in Enzymology*. Fleischer,S. and Packer,L. (eds), Academic Press, New York, Vol. 56, p. 279.
27. Duyckaerts,C., Sluse-Goffard,C.M., Fux,J.P., Sluse,F.E. and Leibecq,C. (1980) *Eur. J. Biochem.*, **106**, 1.
28. Dupont,Y. (1984) *Anal. Biochem.*, **142**, 504.
29. Klingenberg,M. (1970) In *Essays in Biochemistry*. Campbell,P.N. and Dickens,F. (eds), Academic Press, New York, Vol. 6, p. 119.
30. Scarpa,A. (1979) In *Membrane Transport in Biology*. Giebisch,G., Tosteson,D.C. and Ussing,H.H. (eds), Springer Verlag, Berlin, Vol. II, p. 263.
31. LaNoue,K.F. and Schoolwerth,A.C. (1979) *Annu. Rev. Biochem.*, **48**, 871.
32. Meijer,A.J. and van Dam,K. (1981) In *Membrane Transport*. Bonting,S.L. and dePont,J.J.H.H.M. (eds), Elsevier/North Holland Biomedical Press, p. 235.
33. LaNoue,K.F. and Schoolwerth,A.C. (1984) *New Comprehensive Biochemistry*. Ernster,L. (ed.), Elsevier Science Publishers B.V., Amsterdam, Vol. 9, p. 221.
34. Klingenberg,M. (1976) In *The Enzymes of Biological Membranes: Membrane Transport*. Martonosi,A.N. (ed.), Plenum Publ. Corp., New York, Vol. 3, p. 383.
35. Vignais,P.V. (1976) *Biochim. Biophys. Acta*, **456**, 1.
36. Klingenberg,M. (1982) In *Membranes and Transport*. Martonosi,A.N. (ed.), Plenum Publ. Corp., New York, Vol. 1, p. 203.
37. Vignais,P.V., Block,M.R., Boulay,F., Brandolin,G. and Lauquin,G.J.M. (1985) *Structure and Properties of Cell Membrane 2*. CRC Press, p. 139.
38. McGivan,J.D. and Klingenberg,M. (1971) *Eur. J. Biochem.*, **20**, 392.
39. Palmieri,F., Prezioso,G., Quagliariello,E. and Klingenberg,M. (1971) *Eur. J. Biochem.*, **22**, 66.
40. Abou-Khalil,S. and Hanson,J.B. (1979) *Plant Physiol.*, **64**, 281.
41. Aprille,J.-R. and Austin,J. (1981) *Arch. Biochem. Biophys.*, **212**, 689.
42. Pfaff,E. and Klingenberg,M. (1968) *Eur. J. Biochem.*, **6**, 66.
43. Werkheiser,W.C. and Bartley,W. (1957) *Biochem. J.*, **66**, 79.

44. Klingenberg,M. and Appel,M. (1980) *FEBS Lett.*, **119**, 195.
45. Krämer,R. and Klingenberg,M. (1977) *FEBS. Lett.*, **82**, 363.
46. Wohlrab,H. (1980) *J. Biol. Chem.*, **255**, 8170.
47. Bisaccia,F., Indiveri,C. and Palmieri,F. (1985) *Biochim. Biophys. Acta*, **810**, 362.
48. Stipani,I. and Palmieri,F. (1983) *FEBS Lett.*, **161**, 269.
49. Krämer,R. (1984) *FEBS Lett*, **176**, 351.
50. Kaplan,R.S. and Pedersen,P.L. (1985) *J. Biol. Chem.*, **260**, 10293.
51. Nalesz,M.I., Nalesz,K.A., Broger,C., Bolli,R., Wojtczak,L. and Azzi,A. (1986) *FEBS Lett.*, **196**, 331.
52. Hommes,F.A., Eller,A.G., Evans,B.A. and Carter,A.L. (1984) *FEBS Lett.*, **170**, 131.
53. Noel,A., Goswami,T. and Pande,S.V. (1985) *Biochemistry*, **26**, 4506.
54. Kasahara,M. and Hinkle,P.C. (1977) *J. Biol. Chem.*, **252**, 7384.
55. Krämer,R. (1986) In *Methods in Enzymology*. Fleischer,S. (ed.), Academic Press, New York, Vol. 127, p. 610.
56. Klingenberg,M. and Pfaff,E. (1967) In *Methods in Enzymology*. Estabrook,R.W. and Pullman,M.E. (eds), Academic Press, New York, Vol. 10, p. 680.
57. Krämer,R. and Klingenberg,M. (1982) *Biochemistry*, **21**, 1082.

CHAPTER 4

Sub-fractionation of mitochondria and isolation of the proteins of oxidative phosphorylation

C.I.RAGAN, M.T.WILSON, V.M.DARLEY-USMAR and P.N.LOWE

1. INTRODUCTION

In this chapter we turn from a consideration of the properties of intact mitochondria and focus our attention on some of the components comprising this organelle. This leads us to examine some of the techniques available for separating mitochondria into membrane, both inner and outer, fractions and matrix components. In the first section we describe such a fractionation process and give details of experiments which allow one to assign a particular enzyme activity to a given location within the mitochondrion. The remainder of the chapter deals with the preparation of purified inner membrane complexes which comprise the mitochondrial electron transfer chain. Discussion of the matrix enzymes is deferred to Chapter 6.

2. SUB-FRACTIONATION OF MITOCHONDRIA

The different compartments of the mitochondrion contain subsets of proteins contributing all or part of major metabolic pathways. For example the inner membrane contains most of the proteins of oxidative phosphorylation whereas the matrix contains the majority of the proteins taking part in the tricarboxylic acid (TCA) cycle. Large-scale preparation of these fractions for the purpose of enzyme purification will be dealt with in other chapters. It is often useful to separate the mitochondrial compartments on a smaller scale, for example for the purpose of assigning a newly discovered enzyme to its appropriate intramitochondrial compartment. On other occasions it may be necessary to separate fractions so that segments of the metabolic chain may be studied without interference. A case in point is the NADH cytochrome *c* reductase activity of the inner membrane electron transfer complexes I and III. The presence of an outer membrane NADH cytochrome *c* reductase enzyme which is rotenone insensitive makes accurate measurement of low levels of the inner membrane activity extremely difficult. This problem can be overcome if the outer membrane is first removed.

2.1 **Preparation of sub-mitochondrial particles (SMPs)**

SMPs can be produced by sonication which has the effect of inverting the orientation of the inner membrane. They are a convenient preparation for experiments concerned with electron transfer and are essentially devoid of matrix enzymes. However, they do contain variable amounts of outer membrane.

(i) Prepare mitochondria from rat liver as described in Section 2.1.1 of Chapter 1 and resuspend them at a concentration of 15 mg/ml in the medium of choice (e.g. 0.3 M sucrose, 5 mM Mops, 5 mM KH_2PO_4, 1 mM EGTA, pH 7.4).
(ii) Sonicate the preparation with a probe sonicator for six 5 sec bursts at the maximum energy setting interspersed with 30-sec cooling periods. Keep the mitochondria on ice to avoid heating. The mitochondria become more translucent during sonication because they scatter less light.
(iii) Dilute the sample with an equal volume of cold buffer and centrifuge at 15 000 *g* for 10 min at 5°C.
(iv) Resuspend the pellet in 10 times its volume of cold buffer and centrifuge again at 100 000 *g*. Repeat this process twice.
(v) Resuspend the pellet, after the final washing, in half the original volume used in step (i).
(vi) The activity of the matrix marker enzyme malate dehydrogenase (see Section 3.1) can be measured in the first washing compared with the final preparation to check the effectiveness of the sonication and subsequent washings.

2.2 Sub-fractionation of mitochondria using detergents

The inner and outer membranes of the mitochondrion can be isolated selectively by titration with detergents (1). Digitonin is used to dissolve the outer membrane and lubrol the inner membrane. The conditions given are a compromise which gives reasonably pure fractions for rat-liver mitochondria. If greater purity for a given fraction is required then this can be achieved by careful titration with the detergents.

2.2.1 *Preparation of mitochondria*

(i) Prepare mitochondria as described in Section 2.1.1 of Chapter 1. After the final wash, suspend the mitochondria in a minimum volume of buffer. Add only sufficient medium to suspend the mitochondria (i.e. about 1/10 volume of the pellet). A common mistake is to make the mitochondria too dilute, so be sure to suspend the mitochondria in the minimum volume.
(ii) This will give a protein concentration greater than 100 mg/ml. Measure the protein concentration by the Biuret method — other methods are too sensitive for such concentrated protein solutions. If necessary dilute the mitochondria to 100 mg/ml.

2.2.2 *Preparation of the outer membrane and mitoplasts*

(i) Dissolve digitonin (1.2% w/v) in the isolation buffer containing no bovine serum albumin (BSA) by gentle heating.
(ii) Slowly add 1.2 mg of digitonin for every 10 mg of mitochondrial protein, in ice and with stirring. This means that the volume of added digitonin is equal to the volume of mitochondrial suspension. Leave for 15 min, then dilute with 3 vols of isolation buffer and centrifuge at 15 000 *g* for 10 min at 5°C.
(iii) Gently resuspend the pellet (mitoplasts) in buffer.
(iv) For separation of the outer membrane, centrifuge the supernatant from step (iii) at 144 000 *g* for 20 min at 5°C.

(v) Resuspend the pellet in a small volume of buffer. This constitutes the outer membrane fraction and the supernatant the intermembrane fraction.

2.2.3 *Preparation of matrix and inner membrane*

(i) Prepare a stock solution of 20 mg/ml lubrol in the isolation medium containing no BSA.
(ii) Add 0.16 mg of lubrol per 1 mg of mitoplasts (30 mg/ml), leave on ice for 10 min, dilute with 3 vols of isolation medium and centrifuge at 144 000 *g* for 50 min at 5°C.
(iii) Resuspend the pellet in isolation medium. The supernatant is the matrix fraction and the pellet the inner membrane.

3. ASSAY OF FRACTIONS WITH MARKER ENZYMES

The relative purity of each fraction can be ascertained from the specific activity of marker enzymes (1). Cytochrome *c* oxidase is used as a marker for the inner membrane and its assay is described in Section 6.2 of Chapter 5. Malate dehydrogenase is used as a marker for the matrix enzymes and monoamine oxidase as a marker for the outer membrane. Adenylate kinase is used as a marker for the intermembrane fraction (see ref. 1 for conditions). A summary of these enzyme activities for fractions prepared according to the protocol given previously are shown in *Table 1*. The outer membrane and inner membrane fractions are not completely free of contamination with other fractions. Careful titration with the detergents digitonin and lubrol gives purer fractions of the respective membranes but a less pure matrix component.

3.1 Assay for malate dehydrogenase — a matrix marker

This assay is based on the reversal of the dehydrogenase reaction

$$\text{Oxaloacetate} + \text{NADH} + \text{H}^+ \rightleftharpoons \text{L-Malate} + \text{NAD}^+$$

and is monitored by following the oxidation of NADH at 340 nm.

(i) Make a stock solution of 5 mM NADH and 25 mM oxaloacetate in 50 mM Tris-HCl, pH 7.4.
(ii) Dilute the protein samples to 1 mg/ml in a buffer containing 50 mM Tris-HCl, pH 7.4, 0.1% Triton X-100.
(iii) To 0.8 ml of buffer in a 1 ml spectrophotometer cuvette add 10 μl of NADH and 10 μl of oxaloacetate to give final concentrations of 60 μM and 300 μM, respectively.

Table 1. Specific activity of marker enzymes in different sub-mitochondrial fractions.

Fraction	*Cytochrome c oxidase Cyt c^{2+} nmol/min/mg*	*Monamine oxidase nmol/min/mg*	*Malate dehydrogenase nmol/min/mg*
Mitochondria	1570	21	2530
Inner membrane	7500	25.3	2130
Matrix	0	0	3482
Outer membrane	3250	294	797
Intermembrane	0	13.6	1733

(iv) Record the baseline at 340 mm on a range of 1 absorbance unit on the spectrophotometer and a chart speed of 2 min/cm.

(v) Add 20 μl of the protein solution (about 0.02 mg/ml final concentration) and monitor the decrease in absorbance at 340 nm with time.

(vi) Measure the initial rate of absorbance decrease and using the extinction coefficient of 6.22/mM/cm for NADH$-$NAD$^+$ calculate the rate of NADH oxidation and divide by the protein concentration to give the specific activity.

3.2 Assay for monoamine oxidase — an outer membrane marker enzyme

This assay is based on the oxidation of benzylamine by monoamine oxidase to benzylaldehyde giving an increase in absorbance at 250 nm (1).

(i) Prepare a stock solution of 0.25 M benzylamine and dilute the protein solutions to 1 mg/ml in assay buffer (50 mM sodium phosphate, pH 7.5).

(ii) Add 0.8 ml of 50 mM sodium phosphate buffer, pH 7.5 to a quartz spectrophotometer cuvette and 10 μl of the benzylamine stock solution, record the baseline at 250 nm with a range of 0.3 A at a chart speed of 1 min/cm.

(iii) To initiate the reaction add protein to give a final concentration of 0.01$-$0.05 mg/ml and measure the increase in absorbance at 250 nm. The increase is linear with time.

(iv) Calculate the specific activity using an extinction coefficient of 13/mM/cm at 250 nm for benzaldehyde. Typical values for marker enzyme activities are given in *Table 1*.

3.3 Calculation of the purity of mitochondrial fractions

First calculate the proportion of total mitochondrial protein present in each submitochondrial fraction by using the following equation which makes the simplifying assumption that the marker enzyme is truly representative of the fraction and is only found in that fraction:

$$\frac{\text{specific activity of marker enzyme in mitochondria}}{\text{specific activity of marker enzyme in fraction}} \times 100$$
$$= \text{\% of total mitochondrial protein present in each fraction}$$

Results of such a calculation using the specific activity of marker enzymes from *Table 1* are given in *Table 2*. This information is useful when comparing the metabolic ac-

Table 2. Distribution of mitochondrial protein among sub-mitochondrial compartments.

	Mitochondria (activities: nmol/min/mg)	*Sub-fraction (activities: nmol/min/mg)*	*% Total mitochondrial protein*
Cytochrome *c* oxidase	1570	7500 (inner membrane)	21% inner membrane
Monoamine oxidase	21	294 (outer membrane)	7% outer membrane
Malate dehydrogenase	2530	3482 (matrix)	73% matrix

Table 3. Distribution of marker enzyme activities among sub-mitochondrial fractions.

	Marker enzyme: % total activity		
Sub-mitochondrial fraction	*Cytochrome c oxidase*	*Monamine oxidase*	*Malate dehydrogenase*
Inner membrane	92	12	15.0
Outer membrane	8	82	1
Matrix	0	0	80
Intermembrane	0	6.0	4

tivity of mitochondria from different tissues. The specific activity of the marker enzymes varies between mitochondria from heart, liver or lymphocytes presumably reflecting the relative importance of the metabolic pathways that the marker enzymes represent in these different types of cells.

Now we need to get the proportion of each marker enzyme activity in each fraction in order to assess the effectiveness of the sub-fractionation procedures. By comparing this data with that for the distribution of activity for another enzyme of unknown location then it too can be assigned to its sub-mitochondrial compartment.

The sum of specific activities for the sub-mitochondrial compartments normalized to the proportion of each fraction in terms of total mitochondrial protein gives the total activity for any marker. For the malate dehydrogenase activity shown in *Table 1* and the proportions of each fraction given elsewhere (1) that is, inner membrane = 21.3%, outer membrane = 4%, matrix = 67%, intermembrane = 6.3%. This gives:

$$\underbrace{(3.5 \times 10^3 \times 0.67)}_{\text{matrix}} + \underbrace{(2.1 \times 10^3 \times 0.213)}_{\text{inner membrane}} +$$
$$\underbrace{(0.8 \times 10^3 \times 0.04)}_{\text{outer membrane}} + \underbrace{(1.7 \times 10^3 \times 0.063)}_{\text{intermembrane}}$$
$$= 2.93 \times 10^3 \text{ (total activity)}$$

We can now calculate the percentage of the total activity present in each fraction. For example what percentage of the total malate dehydrogenase activity is present in the matrix?

$$\frac{3.5 \times 10^3 \times 0.67}{2.93 \times 10^3} \times 100 = 80\%$$

Table 3 reports the distribution of marker enzyme activities amongst sub-mitochondrial fractions calculated in the same way as described above for malate dehydrogenase. Under these conditions the matrix is effectively pure whereas other fractions are contaminated to varying degrees. However, the distribution of activities is such that assignment of an enzyme to its correct location would be unambiguous.

4. PREPARATION OF MITOCHONDRIA OR MITOCHONDRIAL FRAGMENTS FROM BEEF HEART

The procedures given in Sections 2 and 3 are useful for fractionating mitochondria into their component compartments and showing where marker enzymes occur. In the next sections procedures are described for the large-scale preparation of enzymes from the inner mitochondrial membrane. Here the more discriminating techniques of preparing mitochondrial subfractions are sacrificed to the need for obtaining larger quantities of material from which the electron transfer complexes can be purified.

Preparations of the mitochondrial complexes start with a preparation of mitochondria from beef heart. For preparations of complexes I–III intact mitochondria collected by relatively high-speed centrifugation are best used. For complex IV broken fragments of mitochondria suffice. For preparation of complex II some pre-treatment of the mitochondria is required, see Section 6.1.1. The quantity of tissue used largely depends on the available centrifuge capacity. The protocol(s) given are for 1 kg of minced tissue (about one beef heart ventricle) but can be scaled up or down.

4.1 Mitochondrial preparation suitable for complexes I, II, III and IV

(i) Free the ventricle of a beef heart from fat and internal connective tissue (valves, chordae, etc.) and cut it into 4 cm × 4 cm cubes and mince in the cold.

(ii) Suspend 200 g portions of the mince in 400 ml of 0.01 M Tris-HCl, pH 7.8 containing 0.25 M sucrose. Check the pH after stirring and adjust it to pH 7.8 with 2 M Tris base if necessary. Blend each portion for 3–4 min alternating 30 sec bursts of high speed with 30 sec periods of lower speed to avoid heating.

(iii) Combine homogenates and spin at 1200 *g* for 20 min at 4°C. Decant the supernatant through muslin cloth to remove lipid and loose precipitate and adjust it to pH 7.8 with 2 M Tris.

(iv) Centrifuge at 26 000 *g* for 15 min at 4°C. The pellet consists of a dark heavy layer of intact mitochondria and a loosely packed lighter layer of broken mitochondria. Both are suitable for preparations of complexes and thus the whole pellet can be used for the subsequent solubilization steps.

4.2 Additional mitochondrial preparation suitable for complex IV (cytochrome *c* oxidase)

(i) Free the ventricle of a beef heart from fat and internal connective tissue (valves, chordae, etc.) and cut it into 4 cm × 4 cm cubes and mince.

(ii) Wash the mince in cold water (~4°C) several times until the washing water is only faintly pink and then squeeze the mince in a muslin cloth.

(iii) Homogenize the washed mince in portions using approximately 1 vol of mince to 3 vols of cold buffer (0.02 M sodium phosphate, pH 7.4; about 4 litres in total) in a Waring Blendor. Blend each portion for 3–4 min, alternating 30 sec bursts of high speed with 30 sec periods at lower speed to avoid heating.

(iv) Combine the homogenates and centrifuge at low speed (~2500 *g* for 20 min at 4°C) to remove cell debris.

(v) Retain the supernatant and re-blend the soft pellet with its own volume of cold buffer (0.02 M sodium phosphate, pH 7.4). Centrifuge, as above, and combine the supernatant with that from the previous centrifugation.

This supernatant (typically several litres) can now be treated in one of two ways (Sections 4.2.1 or 4.2.2) to prepare mitochondrial fragments or Keilin–Hartree particles.

4.2.1 *Preparation by centrifugation*

Centrifuge the supernatant at 10 000–12 000 *g* for 60 min. This yields a rather loose, mitochondrial-rich pellet, therefore take care in decanting off the supernatant or aspirate using a water-driven suction pump. The resulting pellet (125–150 ml/kg of tissue) can be stored for several months at −20°C for use in cytochrome *c* oxidase preparations. The disadvantage of this method of preparing mitochondrial fragments is the time needed to centrifuge, at relatively high speed, the 4–5 litres of supernatant generated in steps (i) to (iv) above; an alternative possibility is to use a continuous-flow centrifuge such as a Sharples.

4.2.2 *Preparation by acidification*

An alternative method for the preparation of a mitochondrial-rich pellet, which is, however, contaminated to a greater degree with other cellular components, uses low-speed centrifugation, thus allowing large volumes of supernatant to be processed quickly, is as follows.

(i) Acidify, slowly and with stirring, the supernatant from Section 4.2 to pH 5.6 with 1 M acetic acid and centrifuge it at 2500 *g* for 20 min at 4°C.

(ii) Resuspend the resulting pellet in 1 litre of cold distilled water (*not* buffer) and re-blend for a few seconds to achieve a homogeneous suspension.

(iii) Centrifuge this suspension at 2500 *g* for 20 min. Resuspend the buff-coloured pellet, largely composed of sub-mitochondrial particles, in a minimal amount of 0.02 M sodium phosphate buffer pH 7.4 and store at −20°C or use immediately.

5. PREPARATION OF COMPLEX I: GENERAL STRATEGY

Only one procedure for purification of the mammalian enzyme has been devised (2). The method relies on selective solubilization by deoxycholate and fractionation by salts firstly to co-purify complexes I and III and secondly to separate the mixture into the individual enzymes. Even in experienced hands, the purification can prove difficult and the method described here is a little more flexible than that originally described and allows for greater variation between preparations.

5.1 **Purification procedure**

5.1.1 *Preparation of crude complex I–III*

Beef-heart mitochondria, prepared as described in Section 4 and frozen, are used as starting material, but they should not have been stored for more than 2 weeks before use. All steps are carried out at 0–4°C.

(i) Thaw the mitochondrial suspension in warm water and determine the protein content by the Biuret method [see Section 8.2.1 (i)]. A concentration of 60–100 mg of protein/ml is suitable. Dilute the suspension with 0.67 M sucrose, 50 mM Tris-HCl, pH 8.0 to 23 mg of protein/ml. The scale of the preparation is dependent on the availability of ultracentrifuges but between 5 and 10 g of mitochondrial protein is a convenient range for this step.

(ii) Selective solubilization of complexes I, II and III is now achieved by addition of deoxycholate. Earlier descriptions of this process stressed the need to purify commercially available deoxycholic (and cholic) acids. Purity is indeed essential for success, but it is now possible to buy bile acids of high enough grade (e.g. from Fluorochem, Glossop, Derbyshire). To make solutions of deoxycholate or cholate stir the bile acid in water and add concentrated NaOH in small volumes until the acid is completely dissolved. The final pH should be between 8 and 9. The process can be greatly speeded up by heating the solution.

(iii) To the stirred mitochondrial suspension, gradually add a solution containing 10% (w/v) deoxycholic acid to a final level of 2.2 ml/g of protein. Follow this immediately with solid KCl (3.3 g/g of protein). As soon as the KCl is dissolved, centrifuge the suspension at 75 000 *g* for 30 min. A clear red supernatant on a firmly packed greeny-brown pellet should be found. Dilute the red supernatant with 0.25 vol of cold water and re-centrifuge. This time only a very small pellet should be found. Dialyse the red supernatant against 10 mM Tris-HCl, pH 8.0. The volume of dialysis buffer is not critical as long as there is sufficient to allow stirring. Typically, the supernatant from 5 g of mitochondria should be distributed between four or five dialysis bags and stirred in 4 l of buffer. The optimum time for dialysis may vary considerably from one preparation to another although it is not that critical. A simple empirical rule is to invert the bags every 10 min or so to check for the onset of turbidity and to dialyse for 60 min after this time. Too long a dialysis will contaminate the final preparation with ATP synthetase. After dialysis, centrifuge the turbid suspension at 110 000 *g* for 75 min at 4°C. A red supernatant on a firmly packed red pellet should be obtained. If the pellet is not firm, centrifuge for a further 30 min, since subsequent success is critically dependent on a clear separation of pellet from supernatant. Roughly homogenize the pellets in sucrose–Tris-HCl buffer (no more than 3 ml/g of original protein) and freeze at −20°C.

For larger-scale complex I purification, it is convenient to process several batches of mitochondria through to this stage and to pool them before proceeding to the next.

The initial solubilization by deoxycholate is designed to leave cytochrome oxidase in the pellet, from which this enzyme may be purified. It is extremely important that no oxidase is solubilized at this stage and the recommended amount of deoxycholate is appreciably less than that used by Hatefi's group (2). Since the reason for this variability is not known, it is recommended that the precise amount of detergent to be used should be determined by titration. Treat 4 ml portions of the mitochondrial suspension with varying amounts of deoxycholate (say between 1 and 3 ml/g of protein). Add KCl to each portion, centrifuge and remove the supernatants. Determine the absorption spectrum of the supernatants between 500 and 650 nm before and after reduction with solid dithionite (see Section 8.1). The correct deoxycholate concentration to use is that which gives maximal solubilization of *b* and *c* cytochromes but no solubilization of cytochrome $a \cdot a_3$.

5.1.2 *Preparation of pure complex I–III*

(i) Thaw the batches of crude complex I–III, homogenize the suspension thoroughly in sucrose–Tris-HCl buffer and dilute to a protein concentration of 10 mg/ml (Biuret). The enzyme is insoluble at this point.

(ii) To the stirred suspension, slowly add a 10% (w/v) deoxycholic acid solution to a final level of 4.55 ml/g of protein. Some clarification occurs.

(iii) Fractionation is now carried out using ammonium acetate. Since ammonium acetate is very hygroscopic, an accurate 50% saturated solution is best made by dissolving the entire contents of a previously unopened bottle in water (1350 ml per kg). To the stirred suspension of complex I–III, gradually add 17.25 ml of 50% saturated ammonium acetate per 100 ml of suspension. Further clarification occurs until the last few millilitres of ammonium acetate are added which results in a return of turbidity.

(iv) After 10 min, centrifuge at 75 000 *g* for 30 min at 4°C. A red supernatant on top of a firmly packed light-brown pellet should be found. The pellets are discarded.

(v) The optimal amount of ammonium acetate to be added at the next stage varies from one preparation to another and should be determined by titration as follows. Take ten 5 ml portions of the supernatant. To the first add 0.2 ml of ammonium acetate solution. To the second add 0.225 ml and so forth, in 0.025 ml increments, up to 0.425 ml to the tenth tube. After 10 min, centrifuge at 75 000 *g* for 30 min at 4°C. Red supernatants overlaying light-brown pellets should be found, but examine the pellets carefully for the presence of a dark red 'button' or smear at their centre. This indicates the start of the precipitation of complex I–III. Supernatants from samples which have the red button should be pooled and saved for later.

(vi) To the main part of the supernatant from the previous stage, add a volume of ammonium acetate sufficient to precipitate the first traces of complex I–III. After 10 min stirring, centrifuge the suspension at 75 000 *g* for 30 min at 4°C and pool the red supernatant with those saved above.

(vii) To precipitate complex I–III, add 3.4 ml of ammonium acetate per 100 ml of supernatant, stir for 10 min and centrifuge at 75 000 *g* for 30 min at 4°C. It is advisable to check the amount of complex I activity which has been precipitated at this stage by assaying 10 μl volumes for NADH-ubiquinone reductase. Should 30% or more of the activity remain in the supernatant, add a further 1 ml of ammonium acetate per 100 ml of supernatant to precipitate most of the remainder.

(viii) The pellet of complex I–III should be a dark red-brown colour but it may be occasionally overlaid by a halo of contaminating light-brown material. To remove this, decant and save the supernatant, add a few millilitres of sucrose–Tris-HCl buffer to each tube and gently swirl liquid around. The contaminating material dissolves much more quickly than the complex I–III and can be discarded.

(ix) Dissolve the pure complex I–III in sucrose–Tris-HCl buffer (~10 ml/g of crude complex I–III) and freeze at −20°C. The supernatant from this stage contains a great deal of complex III and can be used as a source for purification of this enzyme (Section 7).

5.1.3 *Purification of complex I*

(i) Thaw the solution of complex I–III, measure the protein content (Biuret) and dilute to 10 mg of protein/ml with sucrose–Tris-HCl buffer.

(ii) Slowly add a solution containing 20% (w/v) cholic acid to a final level of 1.82 ml per 100 ml of complex I–III. Follow this with saturated, neutralized ammonium sulphate solution (65 ml per 100 ml of solution).

(iii) After 10 min stirring, centrifuge the suspension at 20 000 *g* for 15 min at 4°C. A light green–brown precipitate (complex I) and a pink supernatant should be

Table 4. A typical preparation of complex I.

A.	Preparation of crude complex I–III	
	Mitochondrial protein taken	9 g
	Volume after solubilization and centrifugation	470 ml
	Dialysis time	90 min
	The crude enzyme from three 9 g batches was pooled:	
	Volume of crude complex III	40 ml
	Protein concentration	66 mg/ml
	Total protein	2.64 g
B.	Preparation of pure complex I–III	
	Crude complex I–III protein taken	2.64 g
	Volume after first ammonium acetate step	298 ml
	Volume of ammonium acetate required in second step	6.5 ml/100 ml
	Volume of pure complex I–III	15 ml
	Protein concentration	48 mg/ml
	Total protein	720 mg
C.	Preparation of complex I	
	Complex I–III protein taken	720 mg
	Volume of complex I	12.6 ml
	Protein concentration	36 mg/ml
	Total protein	454 mg

obtained. The supernatant contains complex III from which this enzyme can be isolated (Section 7).

(iv) Dissolve the precipitate in two thirds of the original volume of sucrose–Tris-HCl buffer and re-precipitate the complex I by the addition of saturated ammonium sulphate (56 ml per 100 ml of solution).

(v) After stirring for 10 min, centrifuge at 20 000 *g* for 15 min at 4°C. This time the supernatant should be virtually colourless.

(vi) Dissolve the complex I in sucrose–Tris-HCl buffer (~10 ml/g of pure complex I–III) to give a clear yellow or green-brown solution. The colour depends on the protein concentration. The enzyme can be stored in small aliquots at −50°C or below indefinitely.

Occasionally, fractionation of the complex I–III with ammonium sulphate leads to a floating precipitate that still contains a considerable amount of complex III. The preparation can be rescued by dissolving the precipitate in sucrose–Tris-HCl buffer and re-precipitating with ammonium sulphate in the absence of added cholate. The enzyme should form a pellet and fractionation in the presence of cholate should then proceed normally. The specific activity towards ubiquinone analogues will almost definitely suffer, though, as a result of this treatment.

Table 4 summarizes the yields at various stages of a typical successful preparation. On average, the yields of crude complex I–III, pure complex I–III and complex I are, respectively, 12%, 2.6% and 1.3% of the original mitochondrial protein. Also shown in *Table 4* are the volumes encountered at various stages which may give some guidance as to the requirement for rotors and ultracentrifuges.

5.2 Properties of complex I

5.2.1 *Assay of complex I*

Complex I catalyses the reduction of many acceptors by NADH, but the assay which most faithfully reflects its physiological activity is the rotenone-sensitive reduction of ubiquinone analogues, of which ubiquinone-1 is most commonly used. A phospholipid supplement is included in the assay. This has the effect of stimulating the rotenone-sensitive rate by affecting the partition of ubiquinone-1 between the lipid and aqueous phases. The nature of the lipid is therefore not important and the simplest supplement is one prepared by sonication of crude soybean lipids in 2% (w/v) cholate. Soybean phosphatidylcholine works just as well and is much easier to disperse.

For comparison with most values in the literature, assays can be carried out at 30°C (5) or 38°C (2).

(i) To 0.95 ml of 10 mM potassium phosphate buffer, pH 8.0, contained in a 1 ml spectrophotometer cell, add 20 μl of 5 mM NADH, 20 μl of the phospholipid dispersion (20 mM by phosphorus analysis) and 5 μl of the enzyme or mitochondrial fraction appropriately diluted with phosphate buffer.

(ii) Allow equilibration for 1 min and start the reaction by addition of 5 μl of 10 mM ubiquinone-1 dissolved in ethanol or methanol.

(iii) Follow the reaction by the decrease in absorbance at 340 nm. Hatefi and co-workers use an extinction coefficient of 6.22/mM/cm but the contribution of ubiquinone-1 reduction to the absorption at 340 nm means that the true value is 6.81/mM/cm.

5.2.2 *Kinetic properties of the purified enzyme*

Hatefi and co-workers (2) have reported a specific activity towards ubiquinone-1 of 25 μmol of NADH oxidized/min/mg of protein at 38°C and at V_{max} for ubiquinone-1. In the authors' hands, typical activities at V_{max} range from 7 to 15 μmol/min/mg of protein at 38°C. In the assay of Section 5.2.1, at 50 μM ubiquinone-1 and 30°C, a specific activity of 3 μmol/min/mg of protein is acceptable. The apparent K_m for ubiquinone-1 is about 40 μM. Under the conditions of Section 5.2.1, activity should be at least 90% inhibited by rotenone (e.g. at a final concentration of 1 μg/ml).

Reduction of $K_3Fe(CN)_6$ by NADH is a very convenient assay because of the high specific activity of the enzyme towards this acceptor. It can be used as a criterion of homogeneity since the turnover number of the enzyme is identical in mitochondria and after isolation (4). However, it is not a good indicator of the physiological activity of the enzyme since it is unaffected by inhibitors such as rotenone and treatments which lead to the loss of ubiquinone-1 reductase such as delipidation with solvents, phospholipase A or detergents.

5.2.3 *Criteria of purity*

Two criteria are used, the prosthetic group content and the polypeptide composition.

(i) *Prosthetic group contents. Table 5* lists the composition of complex I. The non-covalently bound FMN content is the most easily measured quantity and values in ex-

Table 5. Composition of complex I.

	nmol/mg of protein
FMN	1.0–1.5
Non-haem iron	20–26
Acid-labile sulphur	20–26
Cytochrome $b + c_1$	0.1–0.2
Phospholipid phosphorus	200–240
Ubiquinone-10	2.0–4.5

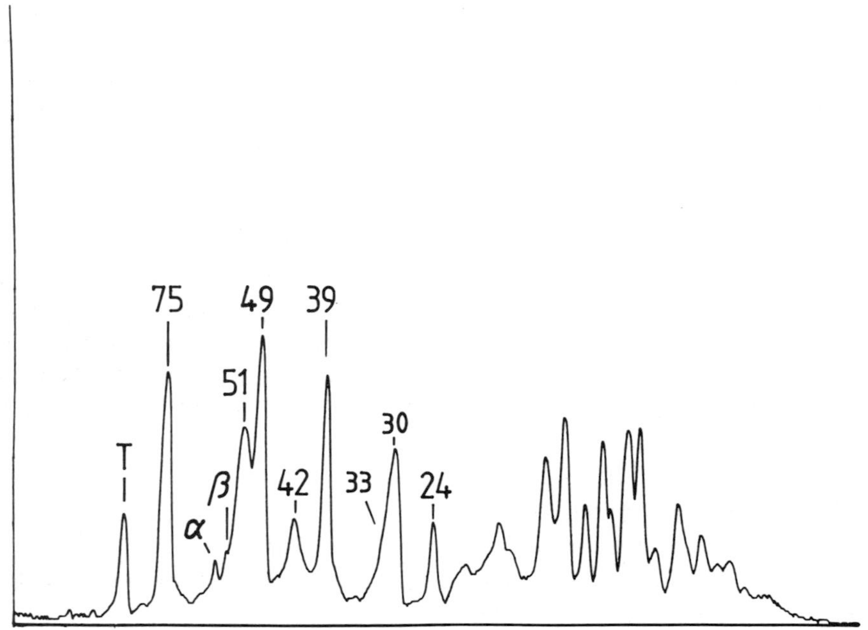

Figure 1. Polypeptide composition of complex I. Complex I was analysed on a 12–16% gradient SDS–polyacrylamide gel and stained with Coomassie blue. Major complex I subunits are indicated by their molecular weights (in kd). Impurities are T, transhydrogenase; α and β, the α and β subunits of the F_1-ATPase. Seven subunits of the mammalian enzyme are encoded by mitochondrial genes, but positive identification of these with subunits of isolated complex I has not yet been made. The ND-1 product co-migrates with the 33-kd protein, while the 42- and 39-kd subunits are in all probability identical with two of the ND-2, 4 or 5 gene products. The remaining ND-3, 4L and 6 products are to be found in the mass of poorly-resolved low molecular weight proteins.

cess of 1 nmol/mg of protein indicate a reasonable degree of purity. The higher values of 1.4–1.5 nmol/ml originally reported by Hatefi and co-workers have not been found by later workers, and the reason for this is not clear. Since virtually all the flavin is FMN, differential analysis for FMN and FAD is unnecessary. The following gives a very simple spectrophotometric assay suitable for purified complex I preparations.

(1) Dilute the complex I solution in a glass test tube to 1.5 ml with 50 mM Tris-HCl, pH 8.0, so that the final protein concentration is approximately 2 mg/ml.

(2) Wrap the tubes in aluminium foil and place them in a boiling water bath for 3 min.
(3) Cool the tubes on ice and sediment the protein in a bench centrifuge for 10 min.
(4) Remove the clear supernatant and measure the absorbancy change at 450–500 nm in a dual wavelength spectrophotometer following addition of a speck of solid sodium dithionite.

The flavin content can be determined using an extinction coefficient of 9.8/mM/cm. At the protein concentration stated, the absorbancy change is about 0.02. If a dual-wavelength machine is not available, the protein concentration should be raised or the flavin measured fluorimetrically (e.g. ref. 6).

(ii) *Polypeptide composition.* Complex I contains a very large number of constituent proteins, certainly in excess of 20. A typical profile is shown in *Figure 1*. Common identifiable impurities are also shown. The major ones are the nicotinamide nucleotide transhydrogenase, and F_1-ATPase whose α and β subunits can be discerned in most preparations. Contamination by cytochrome is better assayed spectrophotometrically but the core proteins of complex III can occasionally be seen on gels.

6. PREPARATION OF COMPLEX II: GENERAL STRATEGY

Although it is possible to isolate complex II from the same batch of mitochondria used for purification of complexes I and III (Sections 5 and 7), separation of complexes II and III can be difficult and higher yields of a purer product are obtained by the method described below. The unusual use of organic solvents is designed to bring about efficient separation of complex II and III, and strict adherence to the procedure, particularly the temperature, is essential.

6.1 **Purification procedure for complex II**

6.1.1 *Pre-treatment of mitochondria*

The method of preparing mitochondria is as described in Section 4.1 but 0.5 mM sodium succinate should be added to all the buffers.

(i) After sedimentation of the mitochondria (e.g. in a Sharples centrifuge), scrape out the paste without addition of further buffer and weigh it.
(ii) For each 100 g wet weight of paste, add 75 ml of post-mitochondrial supernatant.
(iii) Add 1 ml of 1 mM $CaCl_2$ to every 100 ml of suspension, homogenize the mixture thoroughly (e.g. in a small blender) and freeze at −20°C or below until needed.

As a rough guide, 200 g wet weight of mitochondria can be obtained from about 15 hearts, representing about 35 g of mitochondrial protein.

6.1.2 *Solubilization and ammonium sulphate fractionation*

(i) Thaw the mitochondrial suspension in cold water and add 40 ml of 1 M potassium phosphate, pH 7.4, (adjusted at room temperature) to each 100 ml of suspension.
(ii) Homogenize with a Potter–Elvehjem homogenizer and place on ice.
(iii) Determine the protein concentration (Biuret) and dilute the suspension to 70 mg of protein/ml with 0.25 M sucrose.
(iv) Place the suspension in a large beaker (e.g. 2 litre size) and warm in a 39°C

water bath for 15 min with continuous overhead stirring. The final temperature should be 34–36°C.

(v) Remove from the bath, and add, with stirring, a 20% (w/v) cholic acid solution to a final level of 3 ml/g of mitochondrial protein. The suspension clarifies and becomes less viscous.

(vi) Place the beaker in the water bath and continue stirring. After 10 min, the temperature should reach 37°C. Continue the incubation for a further 5 min.

(vii) Place the beaker in ice, continue stirring and gradually add solid ammonium sulphate (16.4 g per 100 ml of suspension).

(viii) When all the salt has dissolved and the temperature has lowered to 5°C, centrifuge the mixture at 75 000 *g* for 30 min. This and all subsequent steps are carried out at 0–4°C unless otherwise indicated. Note that, at this stage, the volume will be in excess of 600 ml for a preparation using 200 g wet weight of mitochondria, placing heavy demands on centrifuges.

(ix) Discard the light brown pellets, treat the supernatants with 5.6 g of ammonium sulphate per 100 ml of solution and centrifuge at 75 000 *g* for 30 min. Discard the supernatant and rinse the pellets with 0.25 M sucrose.

(x) Dissolve the precipitates in 0.25 M sucrose, 10 mM potassium phosphate, pH 7.4 (adjusted at room temperature). This is rather tedious because the pellets are sticky. Add a few millilitres of buffer to three or four tubes (no more than 5 ml each). Suspend the pellets with a glass rod, transfer the suspension to another tube and repeat. Finally, homogenize by hand with a Potter–Elvehjem homogenizer and determine the amount of protein using the Biuret test.

(xi) Dilute to 85–100 mg of protein/ml with sucrose–potassium phosphate and dialyse for 5 h with changes of buffer at 1 h and 3 h, using both internal and external stirring.

After dialysis, the solution can be frozen at −20°C or below.

6.1.3 *Solvent treatment*

(i) Thaw the dialysed material in cold or tepid water and add, with stirring, a 10% (w/v) deoxycholic acid solution to a final level of 7.5 ml/g of protein. Adjust the pH to 7.3 with 1 M HCl.

(ii) Stir the solution rapidly to ensure mixing and add cold 95% (v/v) ethanol from a separating funnel to the extent of 46 ml/100 ml of solution. The ethanol should be pre-cooled in an acetone–dry ice bath (−77°C) to offset the heating which occurs during the addition of ethanol. It is important to keep the temperature at 2°C or below.

(iii) Transfer the viscous, turbid solution to centrifuge tubes in a rotor pre-cooled to 0°C and centrifuge at 75 000 *g* for 30 min at 4°C.

(iv) Discard the large green-brown pellet and pool the pale brown supernatants. Store on ice.

(v) Mark a glass Potter–Elvehjem homogenizer at the 25 ml and 50 ml levels, and emulsify 25 ml volumes of the supernatant with 25 ml volumes of cold (4°C) cyclohexane by two rapid passes of a loosely fitting Teflon plunger.

(vi) Transfer the emulsion to centrifuge tubes and place them in rotors pre-cooled to 3°C. Do not put the tubes on ice as the cyclohexane will crystallize out below 3°C and ruin the preparation.

(vii) Centrifuge the mixture at 75 000 *g* for 45 min at 3°C. The emulsification and centrifugation should be carried out as quickly as possible.

(viii) After centrifugation, discard the supernatant, taking care that cyclohexane does not come into contact with the pellet.

(ix) Rinse the red-brown pellet five times with 0.25 M sucrose to remove all traces of solvent and resuspend in 0.25 M sucrose (no more than 10 ml per 200 g of original mitochondrial wet weight) by homogenization.

(x) Determine the protein concentration (Biuret) and dilute to 25 mg of protein/ml with 0.25 M sucrose.

6.1.4 *Deoxycholate–ammonium sulphate fractionation*

(i) To the suspension from the previous stage, slowly add a 10% (w/v) solution of deoxycholate to a level of 3 ml/g of protein, and follow this with saturated, neutralized ammonium sulphate to 10% of saturation. After stirring for 15 min, centrifuge at 75 000 *g* for 30 min at 4°C. Decant the red supernatant and discard the pellets. Since deoxycholate is inhibitory, it is necessary to remove this by gel filtration.

(ii) Apply the supernatant to a column (30 cm × 2.5 cm internal diameter) of Sephadex G-25 (coarse) equilibrated with 20 mM potassium phosphate, pH 7.4. The excluded material is largely free from deoxycholate and ammonium sulphate except for the trailing edge of the protein peak which can be discarded.

(iii) Collect the enzyme by centrifugation at 105 000 *g* for 90 min at 4°C.

Table 6. A typical preparation of complex II.

A.	Solubilization and ammonium sulphate fractionation	
	Mitochondrial wet weight taken	200 g
	Mitochondrial protein	35.5 g
	Volume after cholate addition	616 ml
	Volume of ammonium sulphate fraction	59 ml
	Protein concentration	89 mg/ml
	Total protein	5.25 g
B.	Solvent treatment	
	Protein taken from previous step	5.25 g
	Volume of supernatant after ethanol addition	118 ml
	Volume of fraction from cyclohexane treatment	14 ml
	Protein concentration	25 mg/ml
	Total protein	350 mg
C.	Deoxycholate–ammonium sulphate fractionation	
	Protein taken from previous step	350 mg
	Volume applied to column	15.7 ml
	Volume of complex II	8 ml
	Protein concentration	11.2 mg/ml
	Total protein	89.6 mg

(iv) Resuspend the dark-red pellet in 20 mM potassium phosphate, pH 7.4, by homogenization to a final volume of no more than 10 ml and freeze at −50°C or below in small aliquots.

Table 6 gives the yields and volumes for a successful preparation of complex II from 200 g wet weight of mitochondria. Hatefi and co-workers (7) report rather higher yields of complex II of between 0.3 and 0.4% of the original mitochondrial protein. They stress the importance of performing the ethanol step correctly in maintaining the high yield.

Unlike complex I, complex II from bovine heart has been purified by two other, quite different procedures. Comparisons between different laboratories are rather difficult but it appears as if the procedure described here may give rise to enzyme preparations with the highest turnover number while those of Yu and Yu (8) and Tushurashvili *et al.* (9) give rise to the highest purity as assessed by SDS−polyacrylamide gel electrophoresis although there may be little to choose between them in practice.

6.2 **Properties of complex II**

6.2.1 *Assay*

Complex II is most commonly assayed by the rate of reduction of ubiquinone-2 by succinate. To avoid possible problems with direct measurement of the quinone in the u.v., particularly with turbid samples, it is normal practice to follow the secondary reduction of a dye by the ubiquinol formed. The assay uses 2,6-dichlorophenolindophenol and is that described by Hatefi and Stiggall (7). For reasons described in Section 6.2.2, the inclusion of 0.004−0.01% Triton X-100 may be advantageous.

(i) To a 1 ml spectrophotometer cell, add 50 μl of 1 M potassium phosphate, pH 7.4, 20 μl of 1 M sodium succinate, pH 7.4, 10 μl of 10 mM EDTA, pH 7.3, and water to a final volume of 1 ml.

(ii) Equilibrate at the temperature of assay and add 16 μl of 4.65 mM 2, 6-dichlorophenolindophenol, 20 μl of 2.5 mM ubiquinone-2 in ethanol and approximately 1 μg of complex II protein in 2 μl of 20 mM potassium phospate, pH 7.4.

(iii) Follow the reaction by the decrease in absorbance at 600 nm and calculate specific activity using an extinction coefficient of 21/mM/cm.

6.2.2 *Kinetic properties of the purified complex II enzyme*

Hatefi and co-workers (7) have reported a specific activity with ubiquinone-2 as acceptor of between 50 and 55 μmol of succinate oxidized/min per mg of protein at 38°C. The preparation devised by Tushurashvili is comparable (specific activities in μmol of succinate/min per mg of protein of 24−28 at 25°C and up to 54 at 38°C). Other ubiquinone analogues can also be used as acceptors but these are less efficient and, in general, the activity is less sensitive to inhibition by 2-thenoyltrifluoroacetone or carboxin. Using ubiquinone-2, at least 90% inhibition by either 100 μM 2-thenoyltrifluoroacetone or 20 μM carboxin should be found. Tushurashvili *et al.* (9) dwell on complexities in the measurement of complex II activity apparently caused by dissociation of the enzyme. Whether this phenomenon depends on the type of complex II preparation is not known, but since the dissociation can be prevented by inclusion of 0.004% Triton X-100 in the assay, it may be a good idea to include this routinely. Indeed, the original complex II assay of Hatefi and co-workers contained 0.01% Triton X-100.

Phenazine methosulphate-mediated reduction of dichlorophenolindophenol can also be used to measure complex II activity. This assay is not affected by the dissociation phenomenon, it has a very similar turnover to the ubiquinone-2 assay but is only partially inhibited by 2-thenoyltrifluoroacetone. It is not, therefore, as good a guide to the 'intactness' of complex II as the assay described in Section 6.2.1.

6.2.3 *Criteria and purity of complex II*

The same criteria are used, namely (i) prosthetic group content and (ii) polypeptide composition.

(i) *Prosthetic group content. Table 7* lists the composition of complex II prepared by the method described here. The preparation contains covalently bound FAD and cytochrome *b* in equal amounts and either can be used as criteria of purity. Complex II prepared by other methods has a higher FAD content (~5 nmol/mg of protein) and lesser amounts of cytochrome *b* (1.2–2.5 nmol/mg of protein). The flavin content is, therefore, the more reliable indicator and a method for its determination is now given.

The covalently-bound flavin is released from the protein as a flavin-peptide by the action of proteases.

(1) In a foil-wrapped glass centrifuge tube, treat approximately 1 mg of complex II protein in 1 ml of phosphate buffer with 0.1 ml of 55% (w/v) trichloroacetic acid at 0°C.

(2) Centrifuge at 2000 *g* to sediment the precipitated protein, resuspend the precipitate in 2 ml of ice-cold acetone containing 0.016 ml of 6 M HCl and re-centrifuge as before.

(3) Resuspend the pellet in 0.2–0.3 ml of cold 1% (w/v) trichloroacetic acid, centrifuge again and repeat this washing step twice. Take care not to lose the loosely pelleted precipitate when decanting the supernatant in these latter steps.

(4) To the pellet add 10 μl each of solutions of trypsin and chymotrypsin (10 mg/ml in 0.1 M Tris base). Check that the pH is pH 8.0, and adjust if necessary.

(5) Incubate the mixture for 4 h at 38°C, cool on ice and dilute to 0.5 ml with phosphate buffer.

(6) Precipitate the protein by addition of 0.05 ml of 55% (w/v) trichloroacetic acid at 2000 *g*. Centrifuge and carefully remove the supernatant to a cold, foil-wrapped tube.

(7) Resuspend the pellet in 0.5 ml of cold 5% (w/v) trichloroacetic acid, and re-centrifuge.

Table 7. Composition of complex II.

	nmol/mg of protein
Covalently-bound FAD	4.6–5.0
Non-haem iron	36–38
Acid-labile sulphur	32–38
Cytochrome *b* haem	4.5–4.8
Phospholipid phosphorus	200
Ubiquinone-10	none

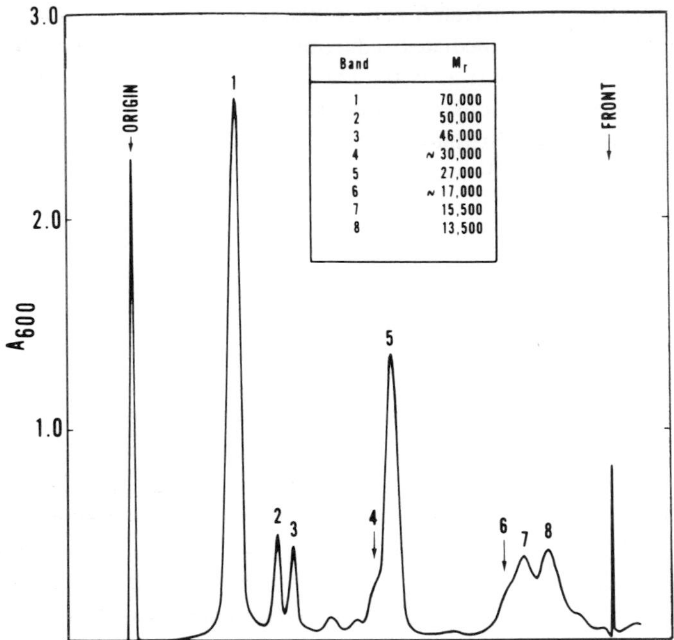

Band	M_r
1	70,000
2	50,000
3	46,000
4	~ 30,000
5	27,000
6	~ 17,000
7	15,500
8	13,500

Figure 2. Polypeptide composition of complex II. Complex II was analysed on a 10% SDS−polyacrylamide gel and stained with Coomassie blue. The four 'true' subunits are numbers 1, 5, 7 and 8. Other polypeptides are mostly contamination by complex III. Reproduced from (10) with permission.

(8) Pool the supernatant fractions and neutralize to pH 7 with KOH. Dilute the neutral solution to 2.0 ml with phosphate buffer.

(9) Determine the flavin in the extract spectrophotometrically, as described in Section 5.2.3.

(ii) *Polypeptide composition.* Complex II preparations contain four major polypeptides. The two larger ones (~ 70 and 25 kd) are the subunits of succinate dehydrogenase while the two smaller ones (~ 15.5 and 14 kd) are presumably involved in ubiquinone reductase activity. The preparation of complex II described here also contains much lesser amounts of complex III polypeptides (*Figure 2*) but reconstitution of succinate dehydrogenase with a purified preparation of the two small subunits gives rise to complex II containing only the four subunits. The preparations of Yu and Yu (8) and Tushurashvili *et al.* (9) appear to be largely free of complex III contamination, thus accounting for the slightly higher specific flavin contents. The role of the cytochrome *b* is, at present, highly contentious.

7. PREPARATION OF COMPLEX III: GENERAL STRATEGY

Two procedures are described here, both of which utilize fractions obtained from purification of complex I. The first makes use of the supernatant from purification of complex I−III (Section 5.1.2). This contains a lot of complex III and leads to isolation of relatively large amounts of the enzyme which, however, are slightly contaminated with complex II (11). The second procedure uses the supernatant from precipitation of complex I from complex I–III. The yield of enzyme is very much lower, but the product may be somewhat purer and free from complex II contamination (12).

7.1 **Purification procedure for complex III**

7.1.1 *Preparation of crude complex III*

(i) To the supernatant remaining after precipitation of complex I–III, slowly add solid ammonium acetate (450 g/l of supernatant) with continuous stirring.
(ii) When the ammonium acetate is completely dissolved, centrifuge at 75 000 *g* for 40 min at 4°C. The supernatant should be colourless.
(iii) Homogenize the red pellets in sucrose–Tris-HCl buffer (~10 ml/g of crude complex I–III used as starting material) and freeze the turbid suspension at −20°C.

7.1.2 *Preparation of pure complex III*

(i) Thaw the suspension of crude complex III, determine the protein content (Biuret) and dilute to 20 mg of protein/ml with sucrose–Tris-HCl buffer.
(ii) Slowly add a solution containing 20% (w/v) cholic acid to a final level of 1.14 ml per 100 ml of the suspension. Follow this with saturated, neutralized ammonium sulphate solution to 35% saturation (54 ml per 100 ml of suspension).
(iii) After 10 min stirring, centrifuge at 75 000 *g* for 15 min at 4°C. A red supernatant on top of a brown pellet should be found.
(iv) To the supernatant, add 8.35 ml of ammonium sulphate per 100 ml of solution (to 40% saturation), stir for 10 min and centrifuge at 75 000 *g* for 15 min at 4°C. A small red-brown precipitate should be present.
(v) To the red supernatant, add 3.45 ml of ammonium sulphate per 100 ml of solution (to 42% saturation), stir for 10 min and centrifuge at 75 000 *g* for 15 min at 4°C.
(vi) Discard the red pellet and precipitate pure complex III from the supernatant by addition of 11.5 ml of ammonium sulphate per 100 ml of solution (to 48% saturation).
(vii) Stir for 10 min, centrifuge at 75 000 *g* for 15 min and discard the supernatant.
(viii) Dissolve the pellets in sucrose–Tris-HCl buffer (~10 ml/g of crude complex III) and freeze at −50°C or below in small aliquots.

The ammonium sulphate fractionation is carried out in several stages since, as often as not, complex III begins to precipitate at lower salt concentration. Thus, a good deal of the red material may be found in the pellet following the second addition of ammonium sulphate (to 40% saturation). If this occurs, dissolve the precipitate in a few millilitres of sucrose–Tris-HCl buffer, measure the volumes of this solution and the supernatant and mix them. Add 20% (w/v) cholic acid, one half the amount previously used, and calculate the percentage saturation with ammonium sulphate. Continue the fractionation exactly as before, increasing the ammonium sulphate concentration to 42% and then 48% saturation. Most of the complex III should precipitate at the higher concentration.

Occasionally, precipitates obtained at higher salt concentration may float although in my experience this is very rare. If this occurs, the floating precipitates can be trap-

Table 8. A typical preparation of complex III.

A.	Preparation of crude complex III	
	Crude complex I−III protein taken	2.64 mg
	Volume after removal of complex I−III	292 ml
	Solid ammonium acetate added	131 g
	Volume of crude complex III	24 ml
	Protein concentration	22 mg/ml
	Total protein	528 mg
B.	Preparation of complex III	
	Crude complex III protein taken	528 mg
	Volume of complex III	12 ml
	Protein concentration	20.5 mg/ml
	Total protein	246 mg

ped by filtering the suspension through glass wool. Activity is unaffected, but the lipid:protein ratio is higher in such preparations (see the next section).

7.1.3 *Preparation of complex III from complex I−III*

(i) Precipitate complex I from a solution of complex I−III as described earlier (Section 5.1.3) and retain the supernatant.
(ii) Slowly add a further 6.5 ml of ammonium sulphate per 100 ml of solution, stir for 10 min and centrifuge at 75 000 *g* for 15 min at 4°C. A red supernatant overlaying a brown precipitate should be found. If the precipitate floats, remove it by filtration through glass wool.
(iii) To the supernatant, add 8 ml of ammonium sulphate per 100 ml of solution, stir for 10 min and centrifuge at 75 000 *g* for 15 min. The precipitate of complex III floats and can be recovered by filtration through glass wool or by withdrawal of the supernatant with a pipette.
(iv) Dissolve the oily red precipitate in sucrose−Tris-HCl buffer (~5 ml/g of complex I−III) and store at −50°C or below in small aliquots.

Table 8 gives the yields and volumes for a typical complex III preparation made by the higher yield procedure. According to Hatefi (12), the yield of complex III from the other method is approximately 15% of the complex I−III protein used, that is, from *Table 4*, approximately 100 mg of complex III might have been expected from 27 g of mitochondrial protein as opposed to the 246 mg obtained by the method of *Table 8*. Further differences between the preparations are detailed in Section 7.2.3.

7.2 **Properties of complex III**

7.2.1 *Assay of complex III*

Complex III is assayed by the rate of reduction of cytochrome *c* by ubiquinol-2. Other ubiquinol analogues are less suitable. Longer chain analogues give lower activities while shorter chain analogues give unacceptably high non-enzymic rates of cytochome *c* reduction. The preparation of ubiquinol-2 and the assay of complex III are now described.

(i) Dissolve ubiquinone-2 (10 μmol) in 1 ml of ethanol and acidify to pH 2 with 6 M HCl.

(ii) Add a pinch of $NaBH_4$ and 1 ml of water to reduce the yellow quinone to the colourless quinol.
(iii) Vortex mix the solution with 3 ml of diethylether:cyclohexane (2:1, v/v).
(iv) Collect the upper phase and vortex mix with 1 ml of 2 M NaCl.
(v) Collect the upper phase again, evaporate to dryness under a stream of N_2 gas and dissolve the residue in 1 ml of ethanol acidified to pH 2 with HCl.

The pale yellow solution can be stored at −2°C for long periods without autoxidation.

(i) To 0.93 ml of 50 mM potassium phosphate buffer, pH 7.4, contained in a 1 ml spectrophotometer cell, add 10 μl of 0.1 M EDTA and 50 μl of 1 mM cytochrome *c*.
(ii) Equilibrate at 30°C or 38°C (Section 7.2.2) and add 5 μl of the enzyme appropriately diluted with sucrose−Tris-HCl buffer.
(iii) Start the reaction by addition of 5 μl of the ubiquinol-2 solution.
(iv) Follow the reaction by the increase in absorbance at 550 nm and calculate specific activity using an extinction coefficient of 20/mM/cm.
(v) To correct for non-enzymic reduction of cytochrome *c* repeat the measurement in the absence of enzyme. Alternatively, the non-enzymic rate can be followed for a short period of time after which the enzyme can be added.

7.2.2 *Kinetic properties of the purified enzyme*

At 38°C, specific activities as high as 1000 μmol/min/mg of protein have been reported for the slightly purer enzyme made from complex I−III. For the other preparation, rates of 300−600 μmol/min/mg of protein at 38°C have been found. In the authors' experience and at 30°C, specific activities approaching 100 μmol/min/mg of protein are typical using the assay as described in above. The enzymic rate of cytochrome *c* reduction is completely inhibited by antimycin and other inhibitors of ubiquinol oxidation in mitochondria.

7.2.3 *Criteria of purity*

Again the criteria are (i) prosthetic group content and (ii) polypeptide composition.

(i) *Prosthetic group content. Table 9* lists the composition of complex III made by the two procedures described. The lower yield method from complex I−III gives the higher

Table 9. Composition of complex III.

	nmol/mg of protein	
	(a)	*(b)*
Cytochrome *b*	8.0−8.5	6.4−7.2
Cytochrome c_1	4.0−4.2	3.2−3.6
Non-haem iron	8−10	5.7−6.7
Acid-labile sulphur	6−8	−
Phospholipid phosphorus	650−800	200−240
Ubiquinone-10	1−4	1−4

[a]Prepared from complex I−III.
[b]Prepared from the supernatant from complex I−III purification.
In the authors' experience, the cytochrome c_1 content is generally greater than 3.5 mol/mg of protein.

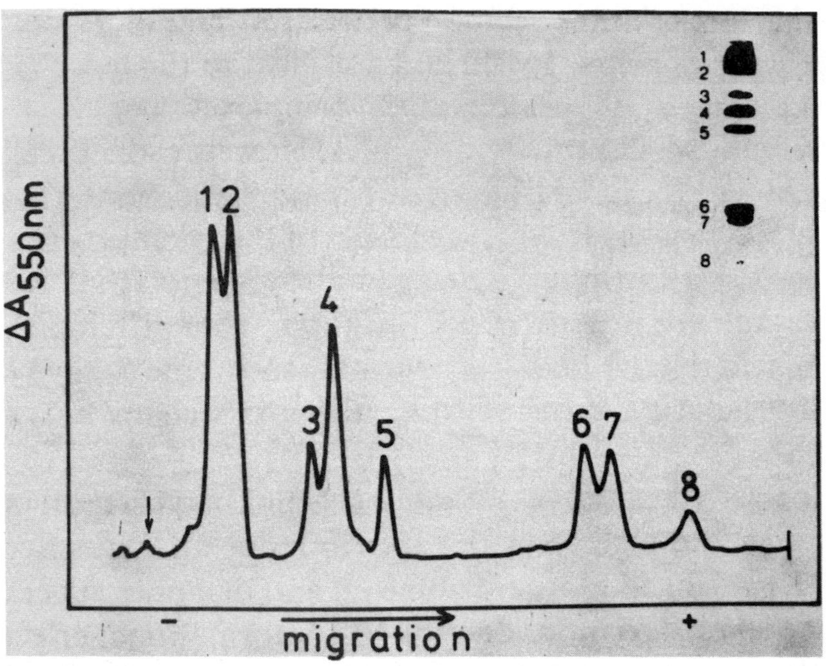

Figure 3. Polypeptide composition of complex III. Complex III was analysed on a 15% SDS–polyacrylamide gel and stained with Coomassie blue. The arrow indicates the position of the large subunit of succinate dehydrogenase whose concentration was very low in this preparation. Subunit identities are **1**, core protein I (49 kd); **2**, core protein II (45 kd); **3**, cytochrome *b* (34 kd); **4**, cytochrome *c*, (29 kd); **5**, Rieske iron-sulphur protein (24 kd). Reproduced from (14) with permission.

specific contents of cytochromes and non-haem iron. The phospholipid content depends not so much on the method of preparation of the enzyme but whether or not the precipitate of complex III floats following ammonium sulphate treatment. In the authors' experience, this rarely happens in the higher yield procedure and the phospholipid content is low, around 200 nmol of lipid P/mg of protein. Much higher levels are found in the enzyme prepared from complex I –III where floating of the precipitate invariably occurs.

The cytochrome c_1 content can be determined spectrophotometrically following reduction by ascorbate and this forms a simple assay of purity.

(1) Dilute the complex III with sucrose–Tris-HCl buffer to a final volume of 2.2 ml so that the final protein concentration is approximately 2 mg/ml. Divide the sample between two 1 ml spectrophotometer cells and scan a base line spectrum between 530 and 570 nm (0–0.2 absorbance full scale). Add a few grains of $K_3Fe(CN)_6$ to the reference cell and a few grains of ascorbic acid to the sample cell.

(2) After a few minutes, scan the spectrum and calculate the cytochrome c_1 content using extinction coefficients of 17.1/mM/cm for the reduced minus oxidized difference at 553 nm or 20.1/mM/cm for the reduced minus oxidized difference at 553–540 nm.

(ii) *Polypeptide composition.* Complex III from bovine heart consists maximally of 10 polypeptides when analysed by SDS–polyacrylamide gel electrophoresis. The five largest subunits are well characterized and their identities are indicated in *Figure 3*. Using simple gel systems, seven or eight polypeptides are normally resolved, depending on whether the cytochrome *b* protein, which migrates anomalously, is resolved from cytochrome c_1. Marres and Slater (13) deal with this topic at some length. The major obvious contaminant is succinate dehydrogenase, whose largest subunit is indicated in *Figure 3*. According to Hatefi (12), contamination is less with complex III prepared from complex I–III.

8. PREPARATION OF FERROCYTOCHROME *C*: OXYGEN OXIDOREDUCTASE EC 1.9.3.1 (COMPLEX IV): GENERAL STRATEGY

This enzyme is generally termed cytochrome *c* oxidase or cytochrome aa_3.

A number of ways to prepare cytochrome *c* oxidase in quantity from mammalian sources are now available. They generally have in common two main procedures. Firstly solubilization of membrane components using detergents and, secondly, fractionation of the resulting solution with ammonium sulphate. (For non-mammalian species other methods which include affinity and/or ion-exchange chromatography may be required.)

Although the methods for mammals are relatively simple, they require some practice to obtain good yields of pure material. This is largely because of the variations encountered from preparation to preparation. For example, there is a seasonal variation. Preparations in the summer months are generally more difficult than in autumn or winter. This problem is not so acute when beef heart is the starting material but with other species (e.g. pig) there can be considerable difficulty in separating the oxidase from contaminating cytochromes *b* and c_1. Also the fat content of the tissue can cause problems. Occasionally 'precipitates' float instead of forming a pellet on centrifugation. However, most preparations can be brought to a successful conclusion by carefully monitoring the spectral properties of the supernatants and precipitates generated in the fractionation steps. This ensures that the enzyme is not inadvertently discarded.

Two methods are described. The first uses the detergent sodium deoxycholate in the initial stages to dissolve the red cytochrome bc_1 complex from the inner mitochondrial membrane while leaving the green-brown oxidase in the lipid membranous phase. This method is termed the 'Red–Green' split. The second method dispenses with this step and proceeds directly to the complete solubilization of the membrane with sodium cholate.

The preparations generally take 2 days and may be stopped overnight (at 4°C) in the fractionation steps and at an ammonium sulphate saturation level of 20%. The shorter the time the enzyme is in 2% cholate the better for its eventual activity.

Both methods yield pure enzyme in good quantity; approximately 2–3 μmol oxidase functional units (0.5 g protein) per kg tissue. However, they generally differ in lipid content and thus the activity can be variable unless assayed in the presence of asolectin. Some workers have found that liposome reconstitution experiments are best performed with the enzyme prepared by the second method (see also Chapter 5, Section 8).

8.1 Assay of purification by absorbance spectroscopy

Cytochrome *c* oxidase activity is inhibited by ionic detergents used in the preparation and thus enzyme assays are only performed once the pure enzyme has been dispersed in non-ionic detergent in the final steps.

Purification is generally followed, therefore, via the characteristic absorption bands of the cytochromes in the 500–650 nm region of the optical absorption spectrum. Reduced minus oxidized difference spectra are the easiest to obtain and interpret for the turbid suspensions encountered in the initial steps.

(i) Dilute the mitochondrial suspension to approximately 1–10 mg/ml with 0.1 M potassium phosphate, pH 7.4.

(ii) Divide the suspension into a sample and reference cuvette for spectroscopic examination and record a baseline using a scanning spectrophotometer. Noisy spectra can result from scattering solutions such as these and steps should be taken to ensure that enough light passes through the suspension. If the absolute absorbance is much above unity the suspensions should be further diluted with buffer.

(iii) Add a few grains of solid dithionite ($Na_2S_2O_4$) to the sample cuvette (a large excess of dithionite will lower the pH and lead to precipitation), mix well and record the difference spectrum.

In the latter fractionation stages of preparation, where clear solutions are obtained, absolute spectra of the dithionite-reduced preparation may be collected from 400 nm to 650 nm by substituting a buffer blank for the reference. (Note: dithionite absorbs strongly below 370 nm).

Dithionite takes several minutes to reduce cytochrome a_3 and, as this chromophore makes a large spectral contribution between 400 nm and 470 nm, the reduced spectrum develops rather slowly in this region. This behaviour is much less evident in the visible region where rapidly reduced cytochrome *a* dominates the spectrum. Final, full, spectra of reduced enzyme preparations should therefore only be recorded some 10 min after dithionite addition.

8.2 The red/green split

Reagents: 50 mM Tris-HCl, pH 7.8 containing 0.66 M sucrose: 10% w/v sodium deoxycholate in water; cold saturated ammonium sulphate solution, neutralized to pH 7.4 with ammonia (at 0°C 390 g/l water).

8.2.1 *Procedure*

This method was introduced by Fowler *et al.* (15) and developed by van Buuren and the Amsterdam group (16) and is very similar to that already given in Section 5.1.1 (ii). The protein concentration in the suspension of mitochondrial fragments must be determined, as this is crucial to the initial steps. This is most quickly obtained using the Biuret method, as modified for haem-containing preparations by Yonetani (17).

(i) Add a minimum of cold distilled water to the mitochondrial pellet and briefly homogenize to give a uniform suspension. Take 0.1 ml of this suspension and mix with 0.1 ml of 10% (w/v) deoxycholate and 0.05 ml of 30% H_2O_2. Incubate at room temperature for 2 min and add water (0.75 ml) to 1 ml and 4 ml of Biuret reagent. Heat the solution at 80°C for 5 min and, after cooling, measure the absorbance at 540 nm. The protein concentration may be determined by reference to a bovine serum albumin calibration curve.

(ii) Dilute the mitochondrial suspension (Sections 4.1.1 or 4.1.2) to 23 mg protein/ml with 50 mM Tris-HCl, pH 7.8 containing 0.66 M sucrose. Take the reduced

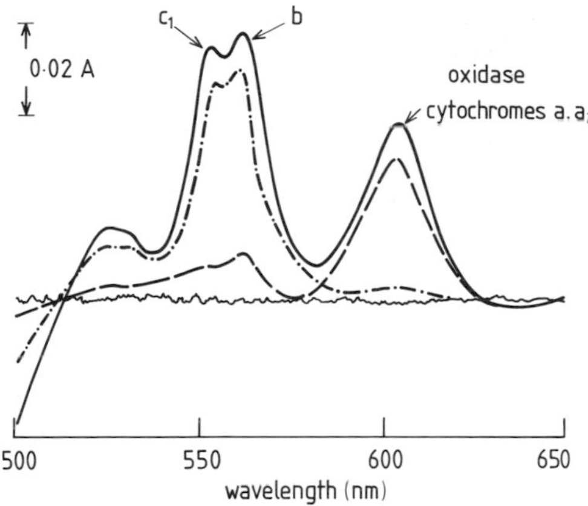

Figure 4. Difference spectra (reduced minus oxidized) obtained from the red/green split. —— sub-mitochondrial particles; —·—·— the supernatant following deoxycholate treatment and centrifugation; – – – the pellet from the centrifugation step.

minus oxidized difference spectrum of an aliquot diluted 10-fold in the same buffer (*Figure 4*).

(iii) To 30 ml of mitochondrial suspension add sodium deoxycholate (10% w/v in water) to a final concentration of 0.3 mg deoxycholate/mg protein (0.7% w/v deoxycholate). Add 2.1 g of solid KCl (70 g/l of suspension) and stir on ice for 20 min.

(iv) Centrifuge at 70 000 *g* for 60 min at 5°C. Take the difference spectra of the supernatant and the precipitate. These should appear as in *Figure 4*. If so, treat the bulk of the mitochondrial suspension with deoxycholate and KCl in the proportions given above and centrifuge, retaining the precipitate. If, on the other hand, the oxidase pellet still contains large amounts of cytochromes *b* and c_1, as indicated by prominent peaks in the 550–565 nm region of the spectrum, then, on a further sample, increase the deoxycholate concentration to 1% w/v, add KCl, centrifuge and check spectra.

(v) Resuspend the green precipitate in a minimal volume of sucrose–Tris-HCl buffer and, if available, freeze the suspension in liquid nitrogen overnight (or until needed). Freezing at −179°C seems to help disrupt the membrane.

(vi) Thaw and dilute the suspension to 25 mg protein/ml with sucrose–Tris-HCl buffer at 4°C and slowly add cold 10% (w/v) sodium cholate in the same buffer to a final concentration of 3% w/v. Stir the solution for 10 min to dissolve the membrane completely and then, while still stirring slowly, add saturated cold $(NH_4)_2SO_4$ and bring to 25% saturation (i.e. 1 ml of saturated solution for each 3 ml of solution). Stir for a further 10 min and centrifuge. Take the dark-brown supernatant to between 40 and 45% saturation with saturated $(NH_4)_2SO_4$ solution and collect the pellet of crude oxidase by centrifugation.

It is useful to know that to take *x* ml of solution at A% to B% saturation add *y* ml of 100% saturated ammonium solution where

$$y = \frac{x\,(B-A)}{100-B} \qquad \text{Equation 1}$$

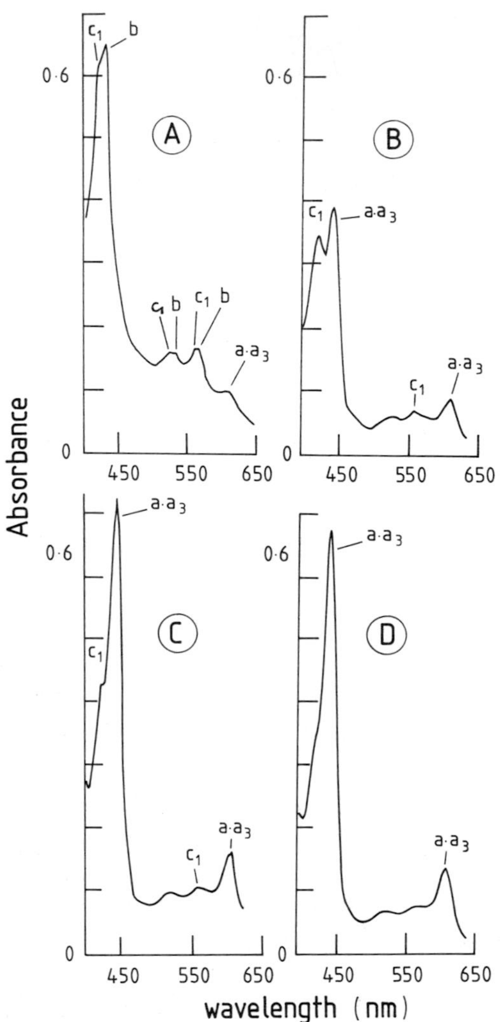

Figure 5. Spectra of the preparation of cytochrome *c* oxidase during fractionation. Small portions of the pellet, formed on centrifuging at the end of each fractionation cycle, were dissolved in 0.1 M potassium phosphate buffer, pH 7.4 (containing either 1% Tween 80 or 2% cholate) and reduced with sodium dithionite: (**A**) After solubilization of the mitochondrial cytochromes; (**B, C** and **D**) after one, two or three fractionation cycles, respectively. The peaks attributed to cytochromes *b*, *c* and *a* a_3 are indicated.

8.3 Ammonium sulphate fractionation

The final steps consist of a number of $(NH_4)_2SO_4$ fractionations monitored through the absolute spectrum of the dithionite-reduced preparation (*Figure 5*). The exact level of $(NH_4)_2SO_4$ saturation for each step must usually be judged from the spectra of previous steps. As a guide we may note that in this Tris buffer cytochrome *b* precipitates between 25 and 30% saturation and cytochrome *c* oxidase at around 40% saturation while cytochrome *c* remains in solution. The object is to obtain a precipitate which, when re-dissolved and reduced with dithionite, has a spectrum with no prominent peaks at

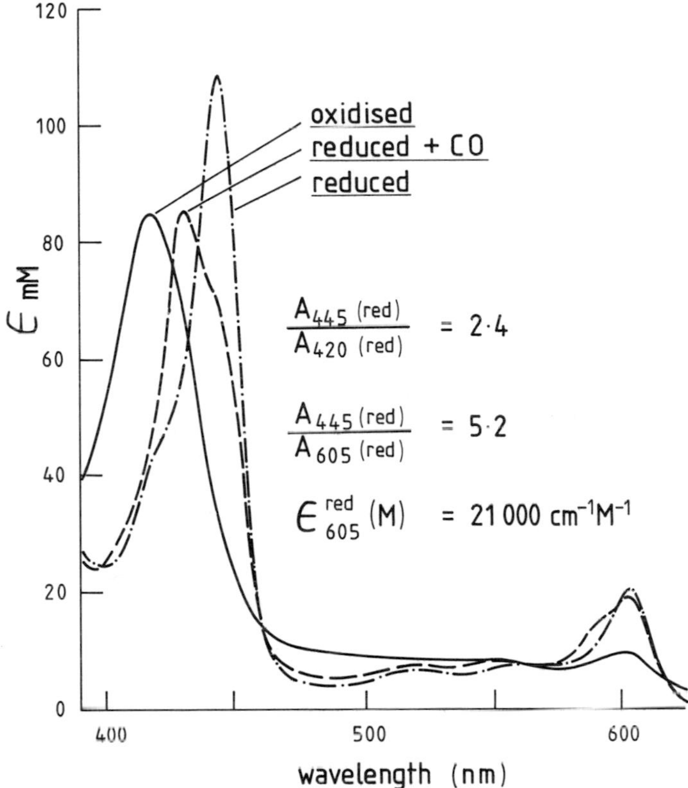

Figure 6. Spectrum of purified cytochrome *c* oxidase giving molar extinction coefficients and criteria of purity. The enzyme was dissolved in 0.1 M potassium phosphate buffer, pH 7.4, containing 1% Tween 80. Reduction was achieved by addition of a few grains of sodium dithionite and the carbonmonoxy derivative formed by bubbling CO gas through the dithionite-containing solution. The spectrum of the oxidized material may vary depending on the history of the enzyme and how it was prepared.

550–565 nm (see *Figure 5d*). A typical fractionation cycle is as follows.

(i) Dissolve the pellet of crude enzyme in 100 ml of cold 50 mM Tris-HCl buffer, pH 7.8 containing 2% w/v sodium cholate and bring to 29% saturation with $(NH_4)_2SO_4$. Centrifuge at 20 000 *g* for 20 min at 4°C and, after checking the spectrum, discard the pink precipitate of cytochrome *b*.

(ii) Bring the supernatant to 40% saturation and spin (as above). This precipitate should be enriched in oxidase. Dissolve a small portion in buffer and take the spectrum. This should show diminished bands at 550 and 560 nm compared with that at 605 nm.

Carry out several cycles of fractionation (usually about three) until material with the spectral characteristics shown in *Figure 6* is obtained.

8.4 Final steps

(i) Dissolve the final pellet in a small volume of buffer containing detergent, that is about one half of the volume of the pellet. This gives approximately 3–5 ml of 200 μM oxidase from each kilogram of starting tissue. This solution contains ammonium sulphate from the fractionation steps which may be removed by dialysis against the chosen buffer. The nature of the buffer depends on the experiments to be performed with the enzyme. If reconstitution experiments are planned then 50 mM Hepes–NaOH, pH 7.8 containing 0.2% cholate is appropriate. If the enzyme is to be used in free solution a wide choice of buffers at around pH 7.4 are available (e.g. phosphate, Tris-HCl, Hepes–NaOH, etc). Whatever buffer is selected it should also contain a non-ionic detergent, Tween 80 (0.5% v/v) or Emasol 1130 (0.5%) are suitable. Lauryl maltoside (0.1%) is also an excellent detergent for oxidase, giving monodisperse complexes and high activity. It is however expensive.

(ii) Centrifuge the final solution at 50 000–100 000 *g* for 30 min at 4°C to remove aggregates.

(iii) Divide the clear solution into small portions and store at −20°C. Alternatively drop 100 μl aliquots into liquid nitrogen. The small beads formed can be stored (best at liquid N_2 temperatures) and dispensed for use conveniently. The enzyme loses activity on repeated freezing and thawing.

8.5 **Yonetani method**

Reagents: 0.02 M potassium phosphate, pH 7.4, 10% w/v sodium cholate in 0.02 M potassium phosphate, pH 7.4, 0.1 M potassium phosphate, pH 7.4 containing 2% w/v sodium cholate; saturated ammonium sulphate solution (water, pH 7.4); solid ammonium sulphate.

This method is essentially that reported by Yonetani (18) but with slight modification. The method relies on dissolving the membrane with sodium cholate and, following precipitation of unwanted protein and lipid, fractionation of the resulting solution of the cytochromes.

Some steps involve the addition of solid ammonium sulphate. After each such addition the pH of the solution should be checked and restored to pH 7.4 with ammonia.

(i) Resuspend the Keilin–Hartree particles (Section 4.2) from 1 kg of tissue in 0.02 M potassium phosphate, pH 7.4 (4°C) to a final volume of 500 ml.

(ii) Add 10% (w/v) sodium cholate solution (125 ml), in the same buffer, to a final concentration of 2% and bring the $(NH_4)_2SO_4$ saturation to 25% by adding 90 g of the solid. Allow to stand in the cold with stirring for 2 h and then bring the saturation level to 35% by addition of 41 g of solid $(NH_4)_2SO_4$. Centrifuge at 7000 *g* for 20 min at 4°C.

(iii) Discard the precipitate and bring the supernatant to 50% saturation with solid $(NH_4)_2SO_4$ (10 g/100 ml supernatant). Centrifuge at 7000 *g* for 20 min and discard the supernatant.

(iv) Dissolve the precipitate (a mixture of the lipid-soluble cytochromes) in 200 ml of 0.1 M potassium phosphate, pH 7.4 containing 2% (w/v) sodium cholate and add 66 ml of saturated $(NH_4)_2SO_4$ solution (to 25% saturation). If the preparation is to be stopped and stored at 4°C overnight at this stage it is better to use only 20% saturation.

(v) Centrifuge at 7000 *g* for 20 min, discard the pinkish precipitate which comprises mainly cytochrome *b* and bring the supernatant to 35% saturation with 15.5 ml of $(NH_4)_2SO_4$ solution/100 ml of supernatant. Centrifuge at 7000 *g* for 20 min at 4°C to bring down the crude oxidase pellet.

As described for the red/green split a series of fractionation cycles, usually three, now completes the purification. In 0.1 M potassium phosphate, pH 7.4 containing 2% (w/v) cholate, cytochromes *b* and oxidase precipitate between 23 and 27% and beween 30 and 35% saturation with $(NH_4)_2SO_4$, respectively. Cytochrome c_1 remains in solution.

Care must be taken in monitoring the various precipitates and supernatants spectrophotometrically and judgement must be exercised as to the exact saturation levels needed. A final cycle is typically as follows.

(vi) Dissolve the oxidase pellet in 50–100 ml of 0.1 M potassium phosphate, pH 7.4, 2% cholate and bring to 27% saturation and centrifuge (7000 *g* for 20 min at 4°C). Bring the resulting supernatant to 33% saturation and centrifuge (7000 *g* for 20 min) once more. Dissolve the final pellet as described for the last stage of the 'red/green' split. The final enzyme preparations should display the spectral features as given in *Figure 6*.

8.6 Determination of concentration and criteria of purity

Dilute (1:30) the stock enzyme preparation into buffer, at pH 7.4, containing either cholate (at least 0.2% w/v) or a non-ionic detergent (at least 0.1% v/v). Record the oxidized and reduced spectra (see *Figure 6*).

The *total haem a* concentration may be determined by using the extinction coefficients in *Figure 6*. Generally the value of ϵ_{605}^{red} = 21 000/M/cm is used. If the enzyme concentration is required in functional units, the total haem concentration is divided by two as cytochrome *c* oxidase contains two haem *a* groups (associated with cytochromes *a* and a_3). The concentration of cytochrome *c* oxidase is given in terms of functional units in other sections dealing with this enzyme.

The haem a/protein ratio of the preparations are usually about 10 nmol/mg. A discussion of the polypeptide subunit composition of the enzyme is given in Chapter 5, Section 3.

8.7 Measurement of enzyme activity

The overall scalar reaction catalysed by ctyochrome *c* oxidase is as follows:

$$4 \text{ cyt } c^{2+} + 4 H^+ + O_2 \rightarrow 4 \text{ cyt } c^{3+} + 2 H_2O$$

The full reaction including vectoral proton translocation across the inner membrane is:

$$4 \text{ cyt } c^{2+} + 8 H^+ \text{ inside} + O_2 \rightarrow 4 \text{ cyt } c^{3+} + 2 H_2O + 4 H^+ \text{ outside}$$

The proton gradient generated is coupled to ATP synthesis (see Chapter 1).

The electron transfer activity may be measured either by monitoring the rate of oxidation of cytochrome *c* through absorbance changes or the rate of oxygen uptake with an oxygen electrode. It should of course be noted that turnover numbers (catalytic site activities) calculated from these methods may, depending on the definition of activity used, differ by a factor of four, reflecting the fact that cytochrome *c* is a single electron donor and oxygen is four electron acceptor.

Both spectroscopic and polarographic assays are described and discussed in Chapter 5, Section 6.2. Values of the catalytic activity (turnover number) depend strongly on solution conditions, the aggregation state of the enzyme and the presence or absence of lipids. For the isolated enzyme in 0.1 M phosphate buffer, pH 7.4, containing 0.5% Tween 80, typical values of the turnover number (expressed as moles cytochrome *c* oxidized/sec/mol oxidase) are in the region 50–200/sec. This value may increase by a factor of approximately five times in the presence of phospholipid. The steady-state kinetics of oxidase are however very complex and will not be further discussed here, (see also Chapter 5, Section 6.2).

9. PREPARATION OF ATPASE (ATP SYNTHETASE)

The mitochondrial ATP synthetase complex is bound to the inner mitochondrial membrane and catalyses the formation of ATP, coupled to H^+ translocation across the membrane. It can be divided into two sections, F_1 and F_0. The F_0 portion is membrane-bound, has no ATP hydrolase or synthetase activity, and is difficult to isolate in a purified form, because it consists of a mixture of intrinsic membrane proteins. The F_1 portion can be removed from the membrane by sonication (19), or by chloroform treatment as described in Section 9.1.2, is freely soluble in aqueous solutions and possesses ATP hydrolase (but in general not ATP synthesis) activity, and hence is often called 'soluble ATPase'.

Various detergent solubilized forms of the ATP synthetase, consisting of F_1, F_0 and some lipid have been described. Unlike the soluble ATPase, these preparations retain oligomycin sensitivity, a characteristic of the membrane-bound enzyme. For references to their preparation see ref. 20. Below we describe the purification of chloroform-released soluble ATPase.

9.1 **Purification procedure**

9.1.1 *Preparation of MgATP particles*

Beef-heart mitochondria, prepared as described in Section 4 and frozen are used as starting material. All steps are carried out at 0–4°C. Protein is determined by the Biuret method [Section 8.2.1(i), except H_2O_2 is omitted].

(i) Thaw the mitochondrial suspension and make it 4 mM in $MgSO_4$, 1 mM in sodium succinate and 1 mM in ATP. Freeze and thaw the suspension twice, and centrifuge it at 10 000 *g* for 10 min.

(ii) Resuspend the pellet in 0.25 M sucrose, 1 mM ATP, 1 mM sodium succinate, 15 mM $MgSO_4$, 20 mM Tris/H_2SO_4 pH 7.7 at a protein concentration of 10 mg/ml.

(iii) Sonicate, using a probe sonicator, in 40 ml aliquots for a total time of 3 min, in 30 sec bursts ensuring adequate cooling to maintain a temperature of less than 10°C.

(iv) Centrifuge the suspension at 10 000 *g* for 30 min to produce a shiny, brown, sticky pellet of submitochondrial particles. Wash the pellet by resuspension in 0.25 M sucrose, 10 mM Tris/H_2SO_4, pH 7.7, and recentrifugate. Finally, resuspend the pellet in the same buffer at a protein concentration of 30–40 mg/ml, and store at −20°C in aliquots of 1–2 ml.

9.1.2 *Preparation of chloroform-released ATPase (20–23)*

Carry out step (i) at 4°C. Carry out steps (ii–v) at room temperature (20–25°C) since removal from the membrane renders the ATPase cold-labile. Soluble protein is measured either by a modified Lowry procedure (23) or the Biuret method (8.2.1).

(i) Thaw MgATP particles, centrifuge at 100 000 *g* for 30 min, and discard supernatant. Resuspend the particles in 0.25 M sucrose, 10 mM Tris/H_2SO_4, 1 mM-EDTA, pH 7.6 at a protein concentration of about 5 mg/ml.

(ii) Transfer the suspension to a stoppered glass tube (a stoppered measuring cylinder is often suitable) and add a half volume of analytical grade chloroform. Immediately, shake vigorously for 20 sec which will result in the production of a creamy emulsion. The emulsion is broken by centrifugation in glass tubes in a bench centrifuge at about 5000 r.p.m. for 10 min, resulting in a clear chloroform layer at the bottom of the tube, a white interfacial layer containing denatured protein and membranes, and a slightly cloudy, aqueous layer containing all the ATPase activity. Remove the aqueous layer and clarify by centrifugation at 100 000 *g* for 30 min. (Care should be taken to check the centrifuge tubes for etching by residual chloroform before re-use.)

(iii) Concentrate the supernatant to a protein concentration of 0.5–2 mg/ml by ultrafiltration using a Diaflo XM100A membrane in a stirred Amicon ultrafiltration cell. By continuous ultrafiltration, change the buffer to 10 mM-Tris/H_2SO_4, pH 7.7. During this process pass a volume of buffer equivalent to 15 volumes of ATPase solution through the membrane. This not only removes sucrose, but also removes low molecular weight contaminants. Finally concentrate the ATPase solution in the ultrafiltration cell to between 2–4 mg/ml.

(iv) Freeze-dry small aliquots (e.g. 0.1 ml) which have been rapidly frozen by pipetting into 3 ml glass tubes, pre-cooled in solid CO_2/acetone. One may store the dry enzyme at −20°C over silica gel for several months. Reconstitute the ATPase by adding water or buffer at 20°C.

(v) This preparation is usually pure (see Section 9.2.2). However, the specific activity can be increased, by removal of some aggregated $\alpha\beta\lambda$ subcomplex (22). Load the concentrated enzyme onto a column (90 cm × 2.5 cm i.d.) of Ultrogel AcA34, equilibrated in 0.002 M Tris/HCl, pH 8.5, containing 0.05 M NaCl, 1 mM ATP, 0.5 mM EDTA and 0.1 mM dithiothreitol. Collect 5 ml fractions and pool the ATPase-active fractions (about fractions 35–45), which have a specific activity of 50–100 μmol/min/mg of protein.

9.2 **Properties of ATPase**

9.2.1 *Assay of ATPase activity*

ATP hydrolysis by the ATPase molecule liberates ADP which is reconverted to ATP by the action of pyruvate kinase, thus maintaining a constant concentration of ATP, and a low steady state concentration of ADP, a potent inhibitor of ATP hydrolysis. Pyruvate production catalysed by pyruvate kinase is monitored as a rate of oxidation of NADH by coupling to lactate dehydrogenase.

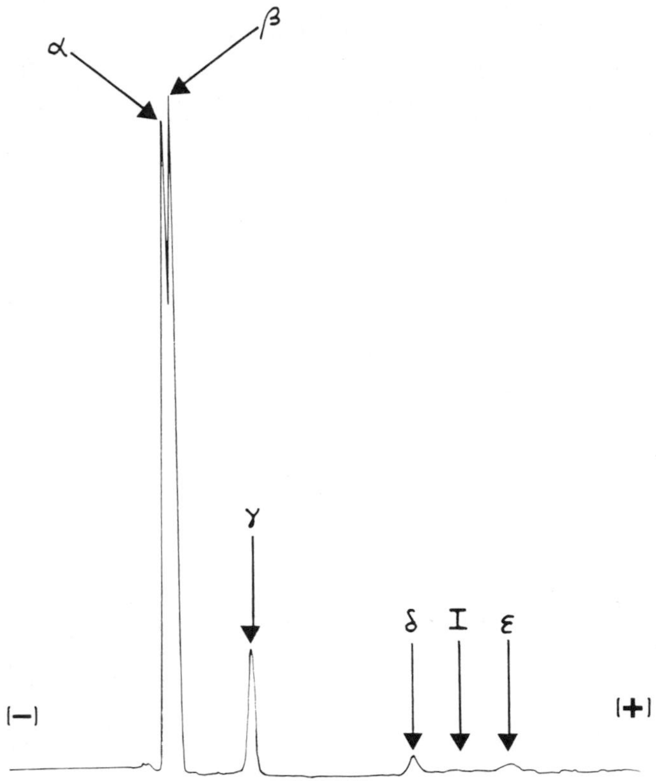

Figure 7. Polypeptide composition of purified soluble ATPase. ATPase was analysed on a 10% SDS-polyacrylamide gel. The inhibitor protein subunit (I) is present in very low amounts in this preparation [cf 9.2.2 (ii)].

(i) Prepare stock solutions of 0.5 M ATP, adjusted with NaOH to pH 7 and of 0.5 M sodium phosphoenolpyruvate. Store in aliquots at −20°C for not more than 2 months. Discard ATP if significant ADP has formed. This is noticed by a rapid decrease in A_{340} upon addition of ATP to the assay mix, as described in (iv) below except ATPase is omitted.

(ii) Store at 4°C a stock of 50 mM Hepes/KOH, pH 8.0, containing 5 mM $MgSO_4$.

(iii) As required for use on a single day only, dissolve NADH in the Hepes/KOH/$MgSO_4$ to a concentration of 0.35 mM.

(iv) To 0.95 ml of the NADH-containing buffer at 30°C, add 5 μl of phosphoenol pyruvate solution, 50 μg of pyruvate kinase (adenylate kinase free, Sigma Type II), 50 μg of lactate dehydrogenase (adenylate kinase free, Sigma Type II) and ATPase to give a final volume of 0.995 ml. Start the reaction by addition of 5 μl of ATP, and monitor oxidation of NADH by recording the decrease in A_{340}. Calculate rates assuming ϵ_{340} = 6.22/mM/cm. For assay of SMP ATPase, include in the assay mix 0.25 M sucrose to provide osmotic support and 2 μg of antimycin A per ml to prevent NADH oxidase activity catalysed by the electron transport system.

9.2.2 *Criteria of purity of ATPase preparations (20)*

(i) *Homogeneity under native conditions.* Homogeneity under non-dissociating conditions is usually checked by sedimentation velocity analysis and by polyacrylamide gel electrophoresis under non-denaturing conditions (23). On 5% disc gels the mobility of the ATPase is about 0.2 that of bromophenol blue tracking dye. The enzyme can be located by protein staining and by staining for ATP hydrolysis activity (23). The chloroform-released ATPase preparation is usually homogenous by these criteria, but it should be noted that heterogeneity can occur in pure soluble ATPase preparations since ATPase complexes of differing subunit composition and ATPase specific activities can be separated by gel electrophoresis.

(ii) *Subunit composition.* Polyacrylamide gel electrophoresis in the presence of SDS (for methodology see ref. 23 and Chapter 5, Section 2.2 for details of one system), reveals a pattern of 5 or 6 subunits which are common to most ATPase preparations (20). Chloroform-released ATPase clearly shows the presence of α (Mr 53 000), β (Mr 50 000) and γ (Mr 33 000) subunits (*Figure 7*), which comprise the bulk of the ATPase protein. In addition, δ (Mr 17 000), ϵ (Mr 7500) and inhibitor protein (Mr 10 000) subunits are present with much lower staining intensities. The amount of inhibitor protein present is variable and gel filtration [Section 9.1.2 (v)] removes it, so that in some preparations it is absent (22).

(iii) *Enzyme activity.* ATP hydrolysis activity *per se* is a poor guide to the purity of the ATPase since it can be modulated by heat treatment, anions, nucleotides, or by interaction with inhibitor protein (20). The chloroform-released ATPase preparation (step iv) has a specific activity of 30–40 μmol/min/mg of protein, but the activity can be increased to 70–90 μmol/min/mg by treatment at high ionic strength.

10. REFERENCES

1. Schnaitman,C. and Greenawalt,J.W. (1968) *J. Cell Biol.*, **38**, 158.
2. Hatefi,Y., Haavik,A.G. and Griffiths,D.E. (1962) *J. Biol. Chem.*, **237**, 1676.
3. Hatefi,Y. and Stiggal,D.L. (1978) In *Methods in Enzymology.* Fleischer,S. and Packer,L. (eds), Academic Press, New York, Vol. 53, p. 5.
4. Smith,S., Cottingham,I.R. and Ragan,C.I. (1980) *FEBS Lett.*, **110**, 279.
5. Ragan,C.I. (1985) *Coenzyme Q.* Lenaz,G. (ed.), Wiley & Sons, Chichester, p. 315.
6. Siegel,L.M. (1978) In *Methods in Enzymology.* Fleischer,S. and Packer,L. (eds), Academic Press, New York, Vol. 53, p. 419.
7. Hatefi,Y. and Stiggal,D.L. (1978) In *Methods in Enzymology.* Fleischer,S. and Packer,L. (eds), Academic Press, New York, Vol. 53, p. 21.
8. Yu,C.-A. and Yu,L. (1980) *Biochim. Biophys. Acta*, **593**, 24.
9. Tushurashvili,P.R., Gavrikova,E.V., Ledenev,A.N. and Vinogradov,A.D. (1985) *Biochim. Biophys. Acta*, **809**, 145.
10. Hatefi,Y. and Galante,Y.M. (1980) *J. Biol. Chem.*, **255**, 5530.
11. Rieske,J.S. (1967) In *Methods in Enzymology.* Estabrook,R.W. and Pullman,M.E. (eds), Academic Press, New York, Vol. 10, p. 239.
12. Hatefi,Y. (1978) In *Methods in Enzymology.* Fleischer,S. and Packer,L. (eds), Academic Press, New York, Vol. 53, p. 35.
13. Marres,C.A.M. and Slater,E.C. (1977) *Biochim. Biophys. Acta*, **462**, 531.
14. Marres,C.A.M. (1983) Ph.D. Thesis, University of Amsterdam, The Netherlands.
15. Fowler,L.R., Richardson,R.H. and Hatefi,Y. (1962) *Biochim. Biophys. Acta*, **64**, 170.
16. van Burren,K.J.H. (1972) Ph.D. Thesis, University of Amsterdam, The Netherlands.
17. Yonetani,T. (1961) *J. Biol. Chem.*, **236**, 1680.

18. Yonetani,T. (1960) *J. Biol. Chem.*, **235**, 845.
19. Knowles,A.F. and Penefsky,H.S. (1979) *J. Biol. Chem.*, **247**, 6617.
20. Lowe,P.N. (1984) in *Methods in Studying Cardiac Membranes*, Dhalla,N.S. (ed.) CRC Press Inc., Boca Raton, FL, Vol. 1, p. 111.
21. Linnett,P.E., Mitchell,A.D., Partis,M.D. and Beechey,R.B. (1979) in *Methods in Enzymology*, Fleischer,S. and Packer,L. (eds), Academic Press, New York, Vol. 55, p. 337.
22. Walker,J.E., Fearnley,I.M., Gay,N.J., Gibson,B.W., Northrop,F.D., Powell,S.J., Runswick,M.J., Saraste,M. and Tybulewicz,V.L.J. (1985) *J. Mol. Biol.*, **184**, 677.
23. Beechey,R.B., Hubbard,S.A., Linnett,P.E., Mitchell,A.D. and Munn,E.A. (1975) *Biochem. J.*, **148**, 533

CHAPTER 5

Reconstitution and molecular analysis of the respiratory chain

V.M.DARLEY-USMAR, R.A.CAPALDI, S.TAKAMIYA, F.MILLETT, M.T.WILSON, F.MALATESTA and P.SARTI

1. INTRODUCTION

The previous chapter demonstrated how sub-fractionation and purification of mitochondrial complexes can be achieved on a large scale. This chapter describes in detail how the complexes can be used in experiments designed to elucidate the mechanism of mitochondrial electron transfer. The techniques used cover a wide range of methods, from SDS−polyacrylamide gel electrophoresis to h.p.l.c. The inner mitochondrial membrane proteins have complex quaternary structures and analysis using SDS−polyacrylamide gels plays an important role in the interpretation of experiments concerning their structure as will become apparent. For the preparation of antibodies and in studies to determine topology of the subunits it is necessary to prepare subunits of the mitochondrial electron transfer complexes on a large scale. Cytochrome *c* oxidase is used as an example in this chapter. An example of the type of experiments that can be done using large-scale preparations of subunits is described in the next two sections starting with radiolabelling and antibody binding. Specific chemical modifications provide an invaluable approach in relating structure to function as illustrated here by cytochrome *c* and cytochrome *c* oxidase. Another functional aspect of the membrane-bound electron transfer proteins is their proton translocating activity which can only be observed in a closed membrane system. The reconstitution of cytochrome *c* oxidase and other membrane proteins into phospholipid vesicles and their assay is described using the more conventional approach and fast reaction methods.

2. POLYACRYLAMIDE GEL ELECTROPHORESIS OF INNER MITOCHONDRIAL MEMBRANE PROTEINS

Precise descriptions of the preparation of SDS gels is outside the scope of this book and the reader is referred to the volume in this series which deals specifically with electrophoresis (1). There are, however, some points which are pertinent to the study of the subunit composition of the inner membrane proteins which will be mentioned since they have led to confusion and misinterpretation in the past. These have largely arisen because of the hydrophobic nature of these proteins which cause anomalies in migration and artefacts during sample preparation.

2.1 Sample preparation

If the gel is to be stained with Coomassie Blue, 20 μg of purified cytochrome *c* oxidase or 100 μg of total membrane proteins is sufficient whereas if the silver stain is used then 2 μg and 10 μg, respectively give strong, easily detectable bands. For a 10 μM cytochrome *c* oxidase solution (~2 mg/ml) take 20 μl of a solution containing 0.25 M Tris-HCl, pH 6.2, 5% SDS, 8 M urea, 2% 2-mercaptoethanol and leave at room temperature for 20 min. It is most important that the sample is *not* heated since this will lead to irreversible aggregation of subunits I and III of cytochrome *c* oxidase and they will not then appear on the gel. Apply 20 μl of the sample to the gel.

2.2 Choice of SDS–PAGE system

A discontinuous slab gel is recommended and the recipe for one such system, which also contains 6 M urea, is given in *Table 1*. This gel can also be run without urea by making up the final volumes shown in *Table 1* with water. However, the presence of urea gives improved resolution of cytochrome *c* oxidase subunits II and III. The general principles detailed above for cytochrome *c* oxidase can also be applied to other inner membrane proteins. The Weber–Osborn system is not recommended as it is unable to resolve some of the inner membrane components. It is also important to realize that the order of migration of the polypeptides can vary according to the type of gel system. In the case of cytochrome *c* oxidase the amino acid sequences of all the constituent polypeptides are now known and so it is not necessary to rely only on migration order on SDS–PAGE for identification. Caution should be applied in using the apparent

Table 1. Recipe for 13% polyacrylamide–SDS–urea gel.

Stock solutions	*Stacking gel*[a]	*Main gel*
20% SDS (20 g/100 ml water)	62.5 μl (0.1%)	150.0 μl (0.1%)
30% acrylamide (30 g acrylamide 0.8 g bis-acrylamide/100 ml water)	3.0 ml (7.2%)	13.0 ml (13.0%)
urea	4.5 g (6 M)	10.8 g (6 M)
TEMED	10.0 μl	15.0 μl
ammonium persulphate[b] (100 mg/ml in water)	100.0 μl	300.0 μl
0.5 M Tris-HCl, pH 6.5	1.25 ml (0.05 M)	
3.75 M Tris-HCl, pH 8.6		3.0 ml (0.375 M)
make up to final volume with distilled water	12.5 ml	30.0 ml
Running buffer 10 × stock diluted before use. 1.92 M Glycine, 0.25 M Tris base, 1% SDS		

Final concentrations are shown in brackets.
[a]De-gas for 15 min before use.
[b]Make fresh and add last when ready to cast.

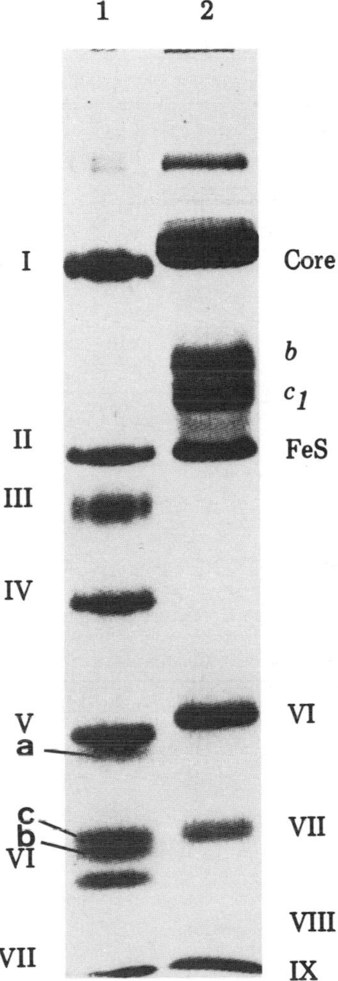

Figure 1. SDS–PAGE–urea gel of beef heart cytochrome *c* oxidase and ubiquinol cytochrome *c* reductase. **Channel 1:** 40 μg of purified cytochrome *c* oxidase. Subunits are numbered in order of decreasing molecular weight according to ref. 4. Other numbering systems do exist and some examples are shown in *Table 2* and generally reflect different researchers views on what constitutes a component of the complex rather than dramatic differences in the number of components or bands actually seen on the gel. **Channel 2:** 40 μg of purified ubiquinol cytochrome *c* reductase. The two core proteins co-migrate on this percentage gel. The polypeptide subunits are numbered after ref. 5. As with cytochrome *c* oxidase some controversy surrounds the assignment and numbering of the polypeptides. (See also Chapter 4, Section 7.2.3.)

molecular weights of inner membrane proteins calculated by reference to a standard curve based on the migration of water-soluble proteins of known molecular weight. The hydrophobic proteins tend to run at a lower molecular weight than predicted by the standard curve using water-soluble proteins (2). *Figure 1* shows an SDS gel of cytochrome *c* oxidase and ubiquinol cytochrome *c* reductase stained with Coomassie Blue. Subunits are numbered according to decreasing apparent molecular weight. More complex numbering systems are being developed by researchers active in this area as

Table 2. A comparison of different nomenclature systems of cytochrome *c* oxidase subunits.

Polypeptide		*Nomenclature*			
N terminus[a]	Mol. wt	Capaldi (6)[b]	Capaldi (4)[c]	Kadenbach (7)	Buse (8)
MFI	56 993	MtI	I	I	I
MAY	26 049	MtII	II	II	II
MTH	29 918	MtIII	III	III	
AHG	17 153	CIV	IV	IV	IV
SHG	12 434	CV	V	Va	V
ASG	10 670	CVI	a	Vb	VIa
FEN	6244	CVII	VIIs	VIIa	VIIIc
SHY	5541	CVIII	VIIs	VIIc	VIIIa
ITA	4962	CIX	VIIs	VIII	VIIIb
ASA	9418		b	VIa	VIb
AED	10 068		c	VIb	VII
STA	8480		VI	VIc	VIc

[a]See Appendix I for letter codes for amino acids.
[b]Used in Section 3.
[c]System used in other sections of Chapter 5.
Mt = mitochondrially coded.
C = cytoplasmically coded.

Table 3. Staining polyacrylamide gels with Coomassie brilliant blue.

Solution	*Conditions*
Stain	60 min with shaking
Water 750 ml, methanol 750 ml, acetic acid 150 ml, Coomassie Brilliant Blue R, 3.3 g	
De-stain	4–6 h with shaking
Water 1500 ml, ethanol 1500 ml, acetic acid 300 ml	change every hour
Storage	storage and
Water 900 ml, acetic acid 100 ml	overnight de-staining
Drying	30 min
Water 500 ml, ethanol 500 ml	

exemplified in Section 3 and *Figures 2* and *3*. This has naturally led to confusion when results from different laboratories using numbering systems based solely on order of migration have been compared. Several of these numbering systems are compared in *Table 2*.

2.2.1 *Staining and drying of the SDS–polyacrylamide gels*

Two staining methods are in common use; these are the Coomassie Blue and silver methods (3). Solutions and instructions for the Coomassie Blue and silver staining method are given in *Tables 3* and *4*. For drying, particularly with 16% gels, it is recommended

Table 4. Silver staining of acrylamide gels.

Solution	*Conditions*
Fix 1: 50% methanol 12.5% acetic acid in H_2O	1 × 200 ml for a minimum of 30 min
Fix 2: 5% acetic acid, 10% ethanol in H_2O	3 × 200 ml 10 min each
Reducing solution: 0.1% $K_2Cr_2O_7$, 0.014% HNO_3 (10 × stock, 5 g $K_2Cr_2O_7$ 1 ml 70% HNO_3 in 500 ml H_2O)	1 × 200 ml 5 min
Stain: 0.2% $AgNO_3$ (10 × stock, 10 g $AgNO_3$ in 500 ml H_2O)	1 × 200 ml for 5 min over light box (if yellow new 200 ml solution for 25 min under room light)
Rinse: water	2 × 200 ml 0.5 min each
Develop: Na_2CO_3, 0.02% formaldehyde: (4 × stock: 240 g Na_2CO_3 + 4.3 ml 37% formaldehyde in 2 litres H_2O)	2 × 200 ml 0.5 min each 1 × 200 ml - until bands appear
Stop and storage: 2% acetic acid	1 × 200 ml

Use gloves throughout all manipulations and touch the gel below the region of the marker dye if possible to avoid marks on the gel.

that they are partially dehydrated first in 50% ethanol for 30 min as this tends to reduce cracking.

3. LARGE-SCALE ISOLATION OF CYTOCHROME C OXIDASE SUBUNITS

Beef heart cytochrome *c* oxidase contains a number of different polypeptides (*Figure 1*). The numbering system used in this section is discussed in detail in reference (4) and is compared with others in *Table 2*.

No single step separation of the polypeptide components of the mammalian cytochrome *c* oxidase is available and purification of some of the smaller components relies on fractionation steps prior to gel filtration.

3.1 Preparation of subunits MtI, MtII, MtIII and C_{IV} of cytochrome *c* oxidase

The four largest subunits of beef heart cytochrome *c* oxidase can be obtained in pure form by gel filtration on a Bio-Gel P60 column in 2% SDS as first described by Steffens and Buse (8). For a rapid separation of subunits I → IV Bio-Gel P100 can be used (see Section 7.1 and *Figure 18*).

(i) It is important to use pure SDS and dissolve it in distilled water: salts in the elution buffer reduce the resolution of subunits MtI and MtII and also reduce the separation of subunit C_{IV} from smaller components.

(ii) The flow-rate of Bio-Gel columns depends critically on the careful preparation of the resin. Equilibrate the resin beads with 2% SDS overnight. Stir the suspen-

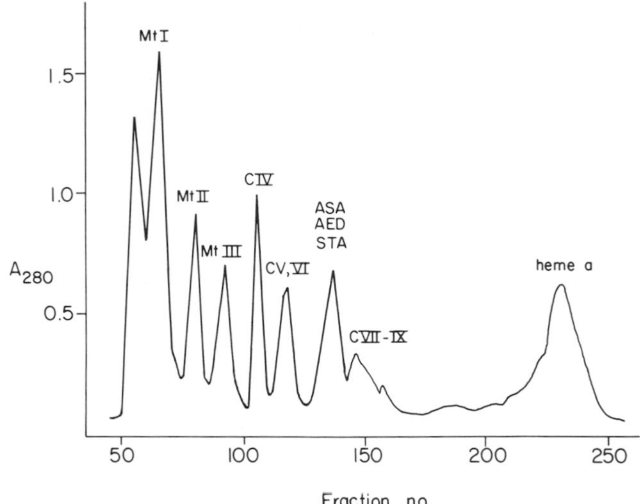

Figure 2. Elution profile of beef heart cytochrome *c* oxidase on Bio-Gel P60. Subunits are numbered according to ref. 6 MtI, II, III designates the mitochondrially-coded subunits and C the cytoplasmically-coded components. ASA, AED and STA identify the group of polypeptides migrating in the apparent molecular weight region 8000 – 10 000 by their three N-terminal amino acids using the single letter amino acid code. The enzyme (50 mg in 1 ml) is dissociated by addition of an equal volume 10% SDS, 5 mM 2-mercaptoethanol and 25% glycerol with incubation at room temperature for 60 min.

sion, allow to stand and carefully decant the fines. Failure to remove the small particles leads to slow flow-rates and even complete clogging of the column effluent filter with the inevitable loss of sample.

(iii) Establish a flow-rate of 2.5 ml/h under gravity (for a column of dimensions 2.6 × 90 cm).

(iv) Take 1 ml of cytochrome *c* oxidase (50 mg/ml) and dissociate by adding an equal volume of 10% SDS, 5 mM 2-mercaptoethanol and 25% glycerol and leave at room temperature for 60 min. Dithiothreitol (DTT, 10 μM) can be substituted for 2-mercaptoethanol and it has a considerably less offensive smell (see Section 7.1).

(v) Layer the sample under the 2% SDS solution, very carefully, using a syringe with an appropriate length of Teflon tube. As the sample is much denser than the 2% SDS it will form a distinct layer on top of the resin.

(vi) Collect fractions every 5 min and read the absorbance at 280 nm. A typical profile is shown in *Figure 2*.

3.2 Preparation of polypeptides C_V, C_{VI}, ASA and AED of cytochrome *c* oxidase

These four polypeptides of beef heart cytochrome *c* oxidase in the molecular weight range 12 500 – 8500 are not well separated on Bio-Gel P60 or on any other gel filtration medium used so far. Their purification depends on an initial fractionation in aqueous pyridine using a procedure modified from one developed by Yu *et al.* (9) to isolate C_V. This approach, the flow chart for which is given in *Scheme 1*, has six main steps.

(i) Dissolve cytochrome oxidase (200 mg or more can be used) in 0.1 M potassium

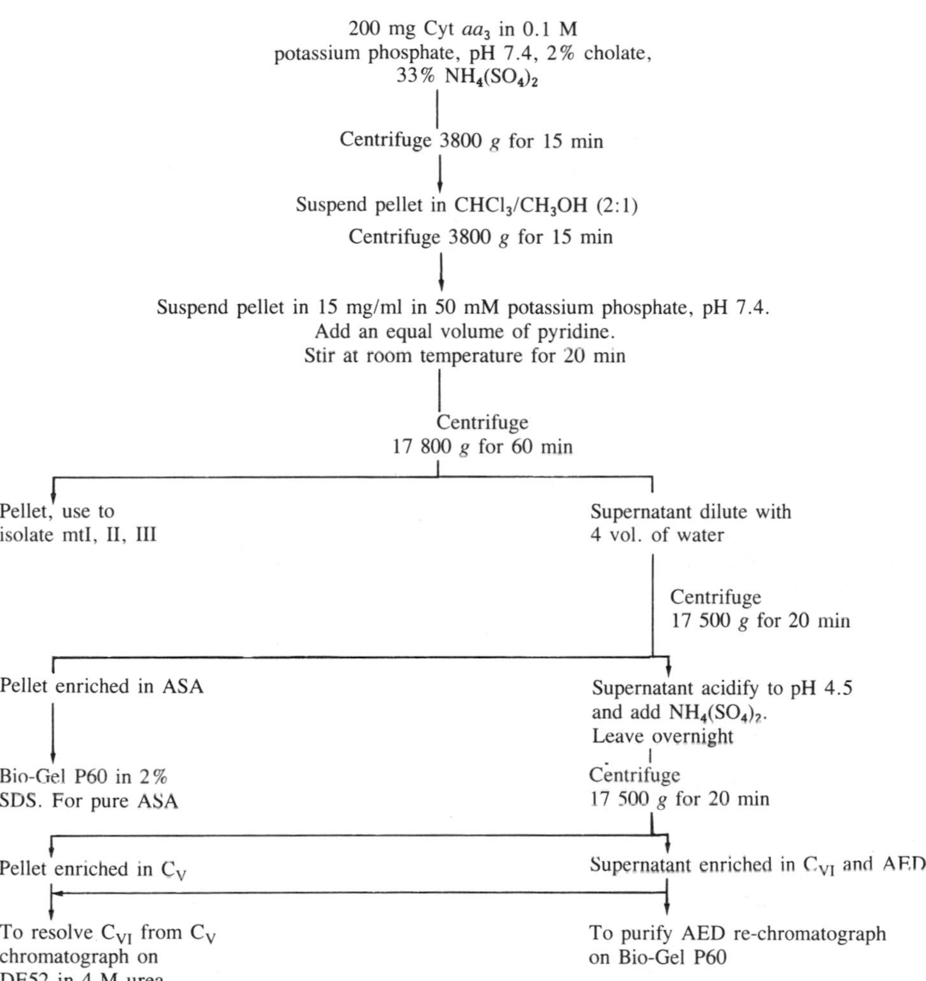

Scheme 1. The purification of polypeptides C_V, C_{VI}, ASA and AED of cytochrome oxidase (Section 3.2)

phosphate buffer, pH 7.4 by addition of 2% sodium cholate and then precipitate the protein by the addition of ammonium sulphate to 33% saturation. Collect the enzyme by centrifugation for 15 min at 3800 *g*.

(ii) Suspend the enzyme pellet in 10 volumes of chloroform–methanol (2:1 v/v), and homogenize in a Potter–Elvehjem homogenizer with a Teflon pestle (60 r.p.m., three strokes) and then collect the delipidated protein by centrifugation at 3800 *g* for 15 min using Pyrex tubes and Teflon stoppers.

(iii) Suspend the pellet of delipidated cytochrome *c* oxidase to 15 mg/ml in 50 mM potassium phosphate, pH 7.4, add an equal volume of pyridine to the suspension and stir at room temperature for 20 min. Centrifuge at 17 500 *g* for 60 min. The oily green pellet can be kept for isolation of larger polypeptides (see *Figure 3*). The supernatant contains polypeptides C_V, C_{VI}, ASA and AED.

(iv) To fractionate the soluble polypeptides further, dilute the pyridine supernatant with 4 volumes of distilled water at 4°C and collect the precipitate by centrifuga-

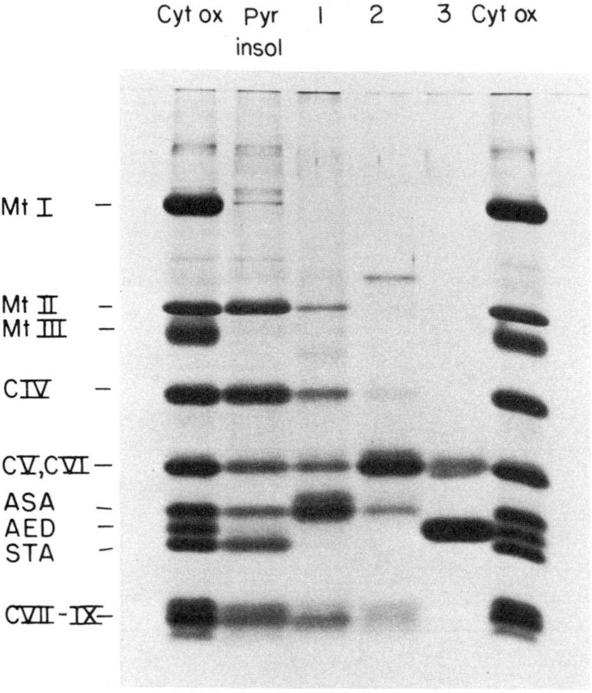

Figure 3. SDS–PAGE of fractions obtained by treating beef heart cytochrome *c* oxidase with aqueous pyridine. Fractions 1, 2 and 3 show steps in the purification of the cytoplasmic subunits [see Section 3.2 (iv)].

tion at 17 500 *g* for 20 min. This fraction (fraction 1 in *Figure 3*) is highly enriched in polypeptide ASA. Acidify the supernatant to pH 4.5 by addition of 12 M HCl, add amonium sulphate to 25% saturation and keep the solution overnight on ice. Collect the precipitate (fraction 2), which is highly enriched in C_V by centrifugation at 17 500 *g* for 20 min. Save the supernatant (fraction 3) which contains mainly C_{VI} and AED.

(v) To obtain polypeptides ASA and AED in pure form from fractions 1 and 3, rechromatograph on Bio-Gel P60 in 2% SDS as described already.

(vi) To separate C_V from the small amount of contaminating C_{VI} in fraction 2 and to resolve C_{VI} and C_V in fraction 3 requires ion-exchange chromatography on a DE-52 (Whatman) column. For small volumes of sample, pack the column in a Pasteur pipette (0.6 × 7 cm bed volume) and wash the resin with 100 ml of 5 mM potassium phosphate, pH 7.4 containing 4 M urea.

(vii) Load the protein dissolved in approximately 1 ml of the phosphate–urea buffer onto the column. It forms a sharp dark green band adsorbed at the top of the resin bed. Wash the column with 7 ml of the phosphate urea buffer and elute with a linear gradient of potassium chloride (0–0.3 M) in a total of 20 ml of the same buffer. Collect the eluent in approximately 0.5 ml aliquots. A broad elution profile is obtained with C_V eluting between 0.16 and 0.25 M KCl.

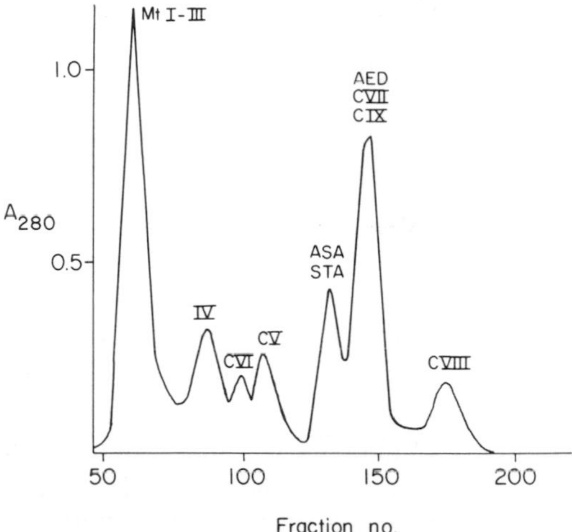

Figure 4. Elution profile of beef heart cytochrome *c* oxidase on Bio-Gel P10 in 10% acetic acid showing purification of C_{VIII} (see Section 3.3).

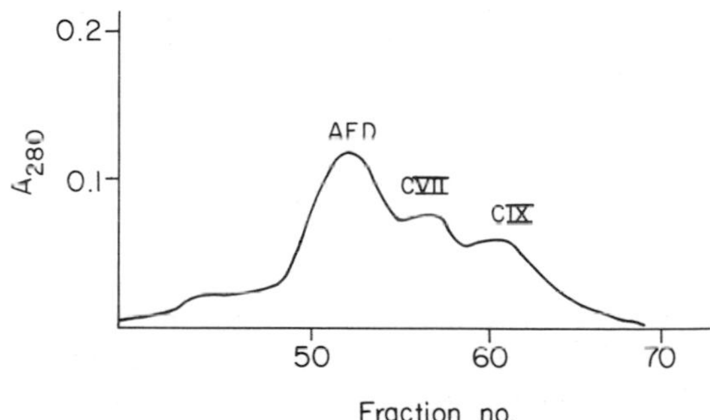

Figure 5. Elution profile of beef heart cytochrome *c* oxidase on Bio-Gel P10 in 70% acetic acid. The fraction containing C_{VII} and C_{IX} in *Figure 4* are re-chromatographed until pure.

3.3 Preparation of polypeptides $C_{VII}-C_{IX}$ of cytochrome *c* oxidase

The three smallest polypeptides of beef heart cytochrome *c* oxidase can be purified by gel filtration on Bio-Gel P10 (200–400 mesh) in 10% acetic acid as first described by Meineche *et al.* (10).

(i) Delipidate the enzyme by repeated cycles of cholate solubilization (1% detergent) and ammonium sulphate precipitation (50% saturation).

(ii) Dissolve the protein from the last precipitation step in 0.1% Tween 80 and 0.01% ammonium hydrogen carbonate, lyophilize, and then re-dissolve in 10% acetic acid.

(iii) Load the sample (100 mg in 2 ml) onto a Bio-Gel P10 column (5 × 130 cm, 200–400 mesh) and elute with 10% acetic acid. Polypeptide C_{VIII} can be obtained essentially pure by this single gel filtration step (*Figure 4*).

(iv) C_{VII} and C_{IX} elute together in a peak which also contains polypeptide AED. Resolve by repeated chromatography on a Bio-Gel P10 column (2.5 × 120 cm, 200−400 mesh) in 70% acetic acid (*Figure 5*).

4. ANALYSIS OF MOLECULAR STRUCTURE BY RADIOLABELLING AND ANTIBODY BINDING

Having discussed the preparative and analytical separation of subunits now the type of detailed analysis that can be undertaken using these preparations will be discussed.

4.1 **Detection of radiolabelled proteins after SDS−PAGE**

For detection of radiolabelled proteins after SDS−gel electrophoresis it is often necessary to provide accurate distance measurements in order to assign a radiolabelled polypeptide to a distinct band or spot on the same or similar gel stained for protein. For this the most reliable approach is to use autoradiography. The sensitivity of this method can be greatly enhanced by fluorography, that is by the use of scintillants to impregnate the gel. For a large number of samples it is the simplest and least time-consuming approach. Alternatively if quantitation of the amount of bound label is required the most reliable method is simply to slice the gel into small sections, dissolve them in a H_2O_2 solution, add scintillant and measure the amount of radioactivity. Since this can be done with stained gels, then fairly accurate assignment of radiolabel with specific bands is possible particularly with samples containing less than 20 distinct components.

4.1.1 *Staining and counting of the polyacrylamide gel*

After SDS gel electrophoresis stain and de-stain the gel and then slice into 1 mm sections. Incubate each 1 mm slice, in scintillation vials, with 0.5 ml of 15% H_2O_2 in an oven at 75°C for 8−12 h. Mix this solution with 3 ml of scintillation fluid (2 litres of toluene, 1 litre of Triton X-100, 8 g of Omnifluor) and measure the radioactivity of the samples using a liquid scintillation counter; take care to check that the samples are devoid of chemiluminescence. An example of such an experiment is shown in *Figure 6* in which native beef heart cytochrome *c* oxidase has been labelled with [^{14}C]iodoacetamide. The protocol described in the legend of *Figure 6* allows a distinction to be made between those sulphydryl groups which are reactive and those which are non-reactive and therefore putative ligands to redox centres (11).

4.1.2 *Autoradiography and fluorography of gels*

The staining method described in the previous section is time consuming when large samples have to be processed or not applicable when the location of protein spots on a two-dimensional gel relative to a stained gel is required. In these cases autoradiography is the most reliable and economical method. In the case of ^{3}H- or ^{14}C-labelled gels the sensitivity can be increased by permeating the gel with scintillant, this is termed fluorography.

(i) Soak the gel stained with Coomassie Blue or silver, in Enlightening (New England Nuclear) or equivalent product diluted with 3 volumes of water and 2 volumes of ethanol, for 30 min at room temperature.

(ii) Dry the gel on filter paper which gives support, if weak beta emitters such as

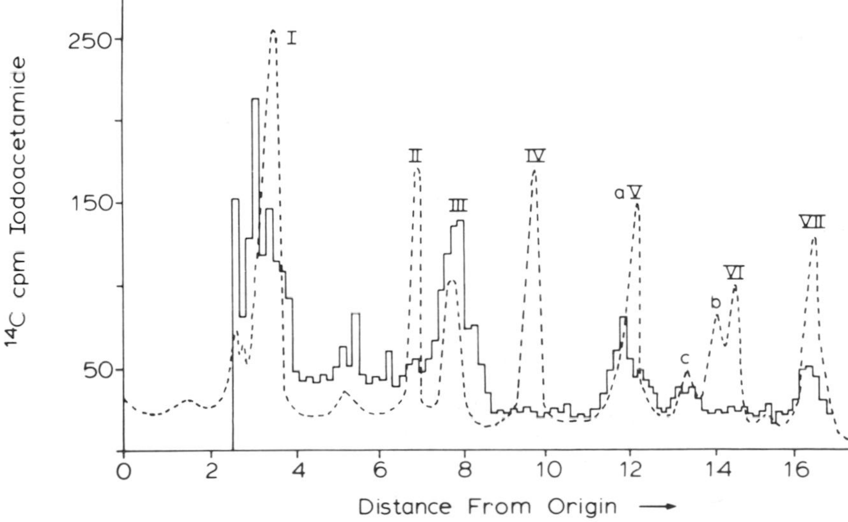

Figure 6. Labelling of native beef heart cytochrome *c* oxidase with [^{14}C]iodoacetamide. Cytochrome *c* oxidase was diluted to 20 μM in 25 mM Tris-HCl, pH 8.5 containing 1% Tween 80 and was mixed with 660 μM [^{14}C]iodoacetamide (2.27 Ci/mol) for 60 min at room temperature. The unreacted [^{14}C]iodoacetamide was separated by elution through a small Sephadex G-25 column. 100 μg was taken for SDS–PAGE–urea. Subunits are numbered according to ref. 4.

^{14}C or ^{3}H isotopes have been used then use non-permeable plastic on the upper side which will not adhere to the gel. This will ensure a completely flat surface on the gel after drying. If any material is placed over the gel during exposure to the X-ray film then the efficiency is greatly decreased with the low energy emitters.

(iii) Place the dried gel in a cassette with intensifier screens and X-ray film and store at −70°C before developing. As a rough guide a band with 1000 d.p.m. left for 5 days will leave a reasonably strong image on the X-ray film.

Figure 7 shows an example of an autoradiogram of mitochondrially-coded proteins labelled with [^{35}S]methionine which have been separated on a one-dimensional SDS–polyacrylamide gel (see Chapter 9). *Figure 8* shows a two-dimensional gel in which labelled proteins have first been separated by charge and then by size. The proteins were prepared as a total cell lysate from cultured lymphoblasts grown with [^{35}S]methionine and in the presence of the ionophore Nonactin. In the presence of this antibiotic, nuclear-coded mitochondrial proteins do not enter the mitochondria and consequently do not appear on the total cell protein map in their mature form but as precursors (13, 14). Three such components are labelled in *Figure 8*. The labelled proteins can be compared with their counterparts on the silver-stained gel (*Figure 8*).

4.2 Western blotting of mitochondrial proteins

The recent development of SDS–polyacrylamide gel electrophoresis (SDS–PAGE) systems capable of the high resolution necessary to separate the polypeptide subunits together with the ability to visualize these polypeptides by binding subunit-specific anti-

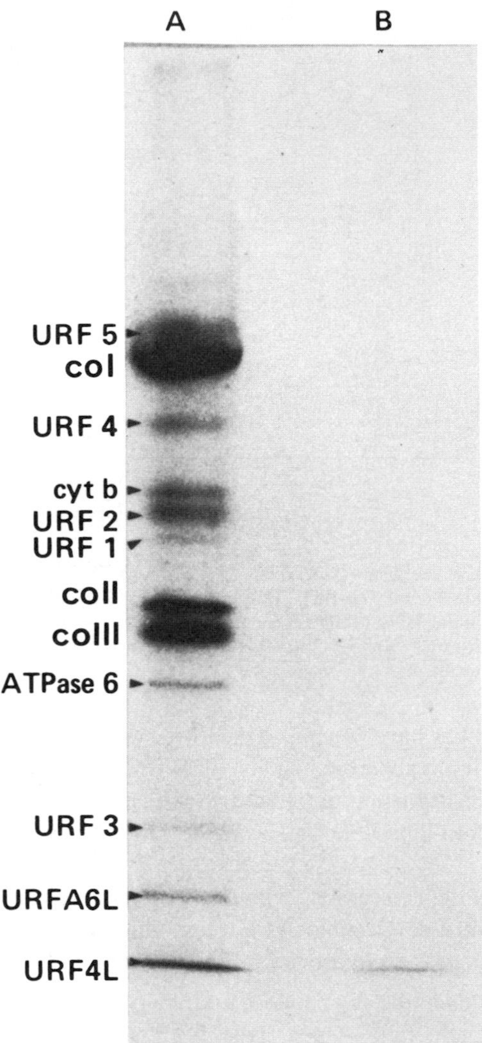

Figure 7. Autoradiography of radiolabelled mitochondrially-coded proteins. Raji lymphoblast proteins (10^7 cells/ml) were suspended in RPMI 1640 medium deficient in methionine and supplemented with 2% fetal calf serum. Cytoplasmic protein synthesis was inhibited by emetine (200 μg/ml) and mitochondrially-coded proteins labelled by adding 60 μCi/ml (2.2 MBq/ml) [^{35}S]methionine (1 mCi/mmol, 37 MBq/mmol). To a control flask was also added chloramphenicol 200 μg/ml which inhibits mitochondrial protein synthesis. After 3 h the cells were harvested, washed and the pellets dissolved in 2% SDS prior to electrophoresis. Electrophoresis and autoradiography were then performed as described in the text. **A:** mitochondrially-coded proteins numbered as in ref. 12. For a more detailed discussion of the identity of these components see Chapter 9. URF, unidentified reading frame. **B:** cells containing both chloramphenicol and emetine.

bodies has allowed investigators to extend their analysis of mitochondrial proteins to the complexes as they are situated in the mitochondrion itself. This is particularly important when material is very limited, as is the case when mitochondria are isolated from human biopsy samples or cultured cells, or when the mitochondrial proteins are

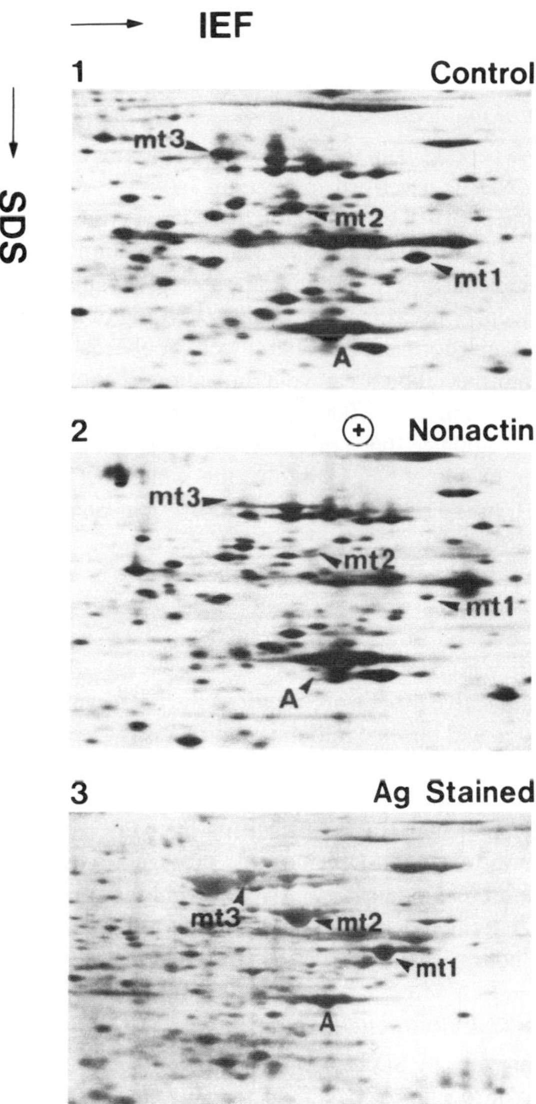

Figure 8. Two-dimensional gel electrophoresis of radiolabelled lymphoblast proteins and the 'mitcon' subset. Raji cells were suspended at a concentration of 2×10^6 cells per ml in methionine-deficient RPMI 1640 medium supplemented with [^{35}S]methionine (1 mCi/mmol, 37 MBq/mmol) and incubated for 16 h at 37°C. One batch of cells was grown in the presence of the K^+ ionophore Nonactin (50 μM). Cells were harvested and washed before dissolving in 2% SDS prior to electrophoresis. Proteins were separated first by charge in an isoelectrofocussing gel and then by size in an SDS gel. After electrophoresis the gel was stained and then impregnated with diluted Enlightening before drying. 2×10^6 d.p.m. were applied to the gel for autoradiography of labelled proteins and approximately 100 μg of cold proteins for staining with silver. **Gel 1** shows the pattern of spots in an autoradiogram of a section of the two-dimensional gel of labelled proteins. Compare this with the same region of a gel labelled under identical conditions (**gel 2**) but in the presence of the K^+ ionophore Nonactin. Several spots labelled MtI, MtII, and MtIII are greatly diminished under these conditions and form part of the 'mitcon' subset (13, 14). 'A' shows the protein Actin. **Gel 3** shows cold Raji lymphoblast proteins stained with silver. By superimposing the autoradiogram with the silver-stained gel the mitochondrial proteins mt1, mt2 and mt3 have been identified. N.B. These proteins have not yet been characterized fully but are *not* cytochrome *c* oxidase subunits (14).

to be studied in intact, that is orientated systems, for example isolated mitochondria or electron transport particles.

4.2.1 *Preparation of antibodies*

(i) Prepare polypeptide subunits of the respiratory complexes by column chromatography as described in Section 3.
(ii) Determine the purity of each fraction by SDS−PAGE and then suspend 0.5 mg in a volume of 0.5 ml with an equal volume of complete Freund's adjuvant and inject subcutaneously into the hindquarters of a rabbit.
(iii) Give subsequent booster injections using 0.25 mg of antigen at 2 week intervals and after two injections and a further 2 weeks collect a small sample (10−20 ml) of blood from the rabbit's ear vein and allow it to clot.
(iv) Centrifuge the sample at 2 000 *g* and decant the serum. Make the serum 40% in ammonium sulphate [66.7 ml saturated $(NH_4)_2SO_4$/100 ml] and centrifuge at 10 000 *g* for 15 min at 5°C. Dissolve the white precipitate in half the original volume in 10 mM Tris-HCl, pH 7.4 and use it as an enriched immunoglobulin fraction.

The polypeptide subunits prepared for this method are not in their native conformation since SDS is used to dissociate the proteins prior to gel chromatography. Because of this it is quite common to find that the antibodies raised to these proteins do not cross-react with the native protein.

If antibodies to the native protein are required the detergent-solubilized complex can be used as the antigen and injected into the rabbit. Assays for detection of the antibody which rely on diffusion and the formation and precipitation of an antigen−antibody complex ought to be conducted on the conformation of the protein used to raise the initial antibody. The Ouchterlony double diffusion assay can then be used for native proteins but not those raised in SDS since this detergent would denature the antibody as well as the antigen. Consequently the Western blotting method, described below, is used for assaying antibodies raised to both native and denatured proteins. Why this system works for both native and SDS-denatured proteins is not clear.

4.2.2 *SDS−PAGE and electroblotting*

First separate the proteins by SDS−PAGE (see *Table 1* for an example of one system) and then transfer to nitrocellulose paper electrophoretically (see *Table 5*). The addition of 0.1% SDS to the electroblotting buffer greatly enhances the transfer of the more hydrophobic components of the mitochondrial inner membrane.

Block the parts of the paper which do not contain protein to prevent non-specific binding and incubate the subunit-specific antibody with the paper. After an appropriate period of time (4−16 h) wash the paper and incubate with a second antibody labelled with a readily detectable probe directed against the first antibody. After an appropriate period of incubation visualize the antibody binding and compare the labelled bands with a lane(s) from the same gel stained for protein.

Commercial apparatus is available from several sources and simply consists of a buffer container with platinum electrodes and a cassette for holding the paper and the gel.

Table 5. Solutions for Western blotting.

Solution	*Composition*	*Volume (ml)*
Blot buffer	0.19 M glycine, 20 mM Tris base 0.1% SDS, 20% methanol	3000
Blocking solution	2% polyvinylpyrrolidone, 2% Ficoll	200
Ca^{2+}/Mg^{2+}-free phosphate-buffered saline (PBS)	1.37 M NaCl, 27 mM KCl, 80 mM Na_2HPO_4 15 mM KH_2PO_4 (10 × concentrated-dilute before use)	500
Antibody carrier solution	5% bovine serum albumin in PBS	100
Horseradish peroxidase: stain solution	dissolve 18 mg of 4-chloro-1-naphthol in 6 ml of methanol, then add 94 ml of PBS + 25 μl of 39% H_2O_2	100

Table 5 gives the stock solutions necessary for the blotting and subsequent antibody binding. The gel is prepared for blotting as follows.

(i) Put on surgical gloves and cut the section of the gel to be blotted from the unstained gel. Trim the gel to the minimum size for blotting as this will decrease the amount of antibody used in subsequent steps. Careful measurements should be made of the gel before cutting to enable comparison with the section of the gel stained for protein.

(ii) Gently place the gel over two sheets of filter paper and cut the paper round the gel. Make a notch to mark the bottom right hand corner of the gel.

(iii) Cover the gel with cling film to prevent it drying while other manipulations are being carried out.

(iv) Use the filter paper as a template to cut out a piece of nitrocellulose paper (Biorad 162.0113 or equivalent product) of the same shape.

(v) Dampen the cassette with blotting buffer and make a sandwich (see *Figure 9*) of one layer of filter paper, the gel, nitrocellulose paper and a final layer of filter paper. Dampen all these sheets in blotting buffer and lay the nitrocellulose paper over the gel with ample buffer to avoid air bubbles between the paper and gel.

The cassette is then closed and placed in the electroblotting cell so that the proteins are electrophoretically transferred from the gel to the paper. With SDS bound all of the proteins have a net negative charge and will therefore migrate to the positive electrode. Typical electrophoresis conditions are 190–200 V with a current of 20–30 mA. over a period of 2 h on a constant voltage setting on the power pack.

4.2.3 *Staining the nitrocellulose paper for proteins*

To ensure that the proteins you are interested in have been transferred to the paper two approaches are possible.

(i) Staining the gel for protein after transfer and comparing it with the control portions of the stained gel which were not used for transfer tells you whether transfer

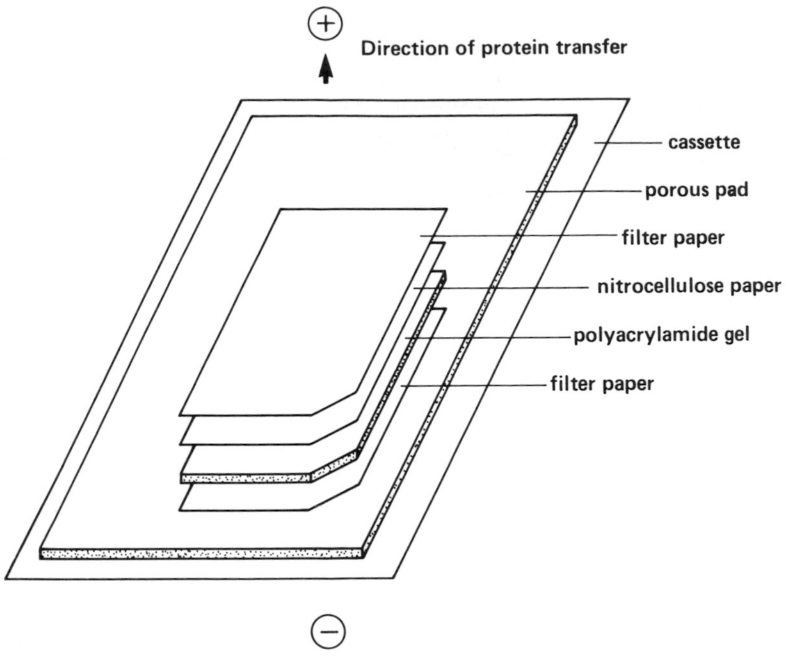

Figure 9. Electroblotting of proteins from an SDS–polyacrylamide gel onto nitrocellulose paper. The 'sandwich' of gel and nitrocellulose paper is shown. It is important to wet all the fillings of the sandwich thoroughly to avoid bubbles forming between the gel and nitrocellulose paper.

has occurred (see *Table 3* for the protein staining method).

(ii) Staining of the paper tells you whether the protein has remained adsorbed once transferred. A sensitive commercial stain is available from Janssen Life Sciences called Aurodye.

When removing the paper from the gel after electroblotting, again wearing gloves, take care that the side of the paper which was in direct contact with the gel is facing upwards. This side of the paper has most of the protein adsorbed unless the capacity has been exceeded.

4.2.4 *Antibody binding and visualization*

After electroblotting transfer the paper to a small plastic dish with the side originally in contact with the gel face up.

(i) Blocking. The vacant sites on the paper after electroblotting must be blocked to prevent non-specific binding and this can be achieved by incubating the paper with either 2% Ficoll, 2% polyvinylpyrrolidone in water or 5% bovine serum albumin (BSA) in Ca^{2+}/Mg^{2+}-free phosphate-buffered saline (PBS) for 30 min.

(ii) Incubation with polypeptide-specific antibody. The affinity of antibody preparations to their respective antigens shows great variability and because of this the optimum dilutions and length of incubations must be determined by the investigators for their samples. The times and dilutions given are those which were optimal for the author's experiment.

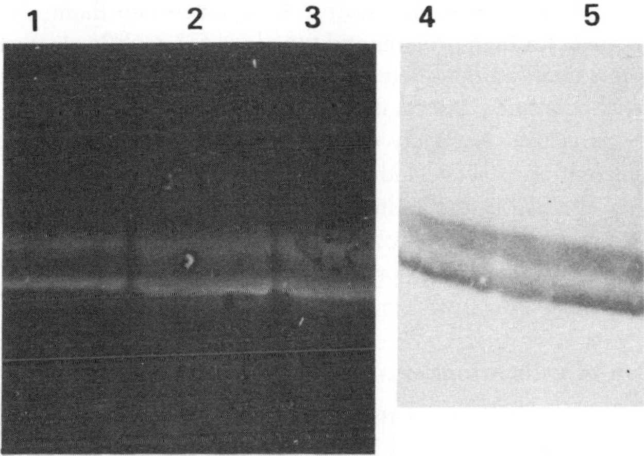

Figure 10. Western blotting of complex III core proteins. Beef heart mitochondrial protein was subjected to SDS–PAGE. After electrophoresis proteins were transferred to nitrocellulose paper and anti-core antibody bound according to the protocol described in the text. The bound antibody was visualized using a goat anti-rabbit antibody conjugated to fluorescein (**Lanes 1–3**, 2, 5 and 50 μg protein) and photographed under u.v light, or a goat anti-rabbit antibody conjugated to HRP (**Lanes 4, 5**; 5 and 50 μg protein).

Dilute 100 μl of antibody in 10 ml of 5% BSA in PBS and incubate with nitrocellulose paper for 2 h. After washing with PBS for 30 min the paper is ready for binding the second antibody which will allow visualization. Three methods are presently available. All involve visualization of the bound specific antibody by binding a second antibody which is either conjugated to an enzyme, a fluorescent label or a radioactive label. The former two of these methods will be described.

(iii) Goat anti-rabbit antibody conjugated to horseradish peroxidase (HRP) (available from Sigma). Dilute the antibody 1/500 and incubate for 2 h at room temperature and after washing add substrate solution (*Table 5*) and develop the stain until bands become visible.

(iv) Goat anti-rabbit antibody conjugated to fluorescein isothiocyanate (FITC) (available from Sigma). Dilute the antibody 1/100 and incubate for 2 h and after washing the bound antibody can be visualized under u.v. light. *Figure 10* shows an example of beef heart mitochondria after Western blotting using an anti-core proteins primary antibody stained with both HRP stain and FITC.

5. PREPARATION OF SELECTIVELY MODIFIED ANALOGUES OF CYTOCHROME C

Cytochrome *c* is still the only member of the mitochondrial electron transport chain for which a high resolution X-ray crystal structure has been determined (15). It is a small, highly basic haem protein with 19 lysines out of 104 amino acid residues. The interaction of cytochrome *c* with both the cytochrome bc_1 complex and cytochrome *c* oxidase is known to be largely electrostatic, but the mechanism by which it transports electrons between the two proteins is poorly understood. A central question is whether cytochrome *c* has separate binding domains for the cytochrome bc_1 complex and cytochrome *c* oxidase and forms an electron bridge between them, or instead has a single binding domain for both proteins and functions as a mobile electron shuttle. One way to address this question is to determine how specific modification of each lysine amino group on cytochrome *c* affects the interaction with the cytochrome bc_1 complex and cytochrome *c* oxidase. A variety of reagents have been used to accomplish this, including ethylthioltrifluoroacetate and *m*-trifluoromethylphenylisocyanate, which convert the positively charged lysine amino group to a neutral CF_3CO- or $CF_3PhNHCO$-lysine (16, 17) and 4-chloro-3,5-dinitrobenzoate, which converts the lysine to a negatively charged CDNP-lysine (18). The procedure below describes how to make singly-modified $CF_3PhNHCO$-cytochrome *c* derivatives.

5.1 Modification of cytochrome *c* with *m*-trifluoromethylphenylisocyanate

(i) Dissolve 250 mg of horse heart cytochrome *c* (Sigma, type VI) in 2.0 ml of 0.14 M ammonium acetate, pH 7.2, and place in a 13 × 100 ml test tube equipped with a small magnetic stirring bar.

(ii) Mount the test tube over a magnetic stirring motor with a small combination pH electrode inserted into the solution. The following steps should be carried out in a fume hood because of the toxic nature of the reagent.

(iii) Adjust the pH of the solution to pH 7.5 with 1 M NaOH dispensed from a syringe with fine polyethylene tubing.

(iv) To begin the reaction add 8 μl of *m*-trifluoromethylphenylisocyanate to the rapidly stirred solution at room temperature. (Use a clean, dry microlitre syringe for this step and avoid contaminating the stock reagent solution with water.) Maintain the pH at pH 7.5 continuously by addition of 1 M NaOH as needed.

(v) After 30 min, add a second 8 μl aliquot of the reagent and again maintain the pH at pH 7.5 for another 30 min.

(vi) To stop the reaction and remove excess reagent, apply the solution to a 1.0 × 20 cm Bio-Gel P-4 column equilibrated with 0.035 M ammonium phosphate, pH 7.2, and elute with the same buffer.

(vii) Collect the dark red cytochrome *c* fraction and dispose of the subsequent fractions containing the excess reagent, in the fume hood.

The type and amount of $CF_3PhNHCO$-cytochrome *c* derivatives obtained are dependent on the pH used in the reaction, with higher pHs generally leading to a greater yield of derivatives, but with less selectivity for specific lysine residues. It is useful to carry out the reaction on a smaller scale the first time to make sure that the amount of reagent used leads to the optimum distribution of singly-labelled cytochrome *c* derivatives.

5.2 Chromatographic separation of $CF_3PhNHCO$-cytochrome *c* derivatives

The chromatographic resolution of singly labelled $CF_3PhNHCO$-cytochrome *c* derivatives is generally carried out in two steps. The crude reaction mixture is first separated on a column of Bio-Rex 70, and then each peak is re-chromatographed on a column of Whatman CM32 carboxymethyl cellulose. The slightly different properties of these two columns lead to optimum resolution of the cytochrome *c* derivatives.

(i) To prepare a Bio-Rex 70 column equilibrate about 300 g of the dry Bio-Rex

70 resin (200–400 mesh) with distilled water in a large beaker, and remove the fines. Equilibrate the resin with 1 M HCl to convert it to the H^+ form, and then wash it several times with water.

(ii) To convert the resin to the desired NH_4^+ form equilibrate the resin with 0.14 M ammonium phosphate buffer, pH 7.2, made from phosphoric acid and ammonium hydroxide. Bring the pH of the resin slurry up to pH 7.2 by addition of 1 M ammonium hydroxide with gentle stirring. It takes several hours for the resin to equilibrate fully with the buffer, and after a stable pH 7.2 has been reached re-equilibrate the resin with fresh 0.14 ammonium phosphate buffer and measure the pH again.

(iii) Pour the resin into a 2 × 70 cm column, and elute with 0.14 M ammonium phosphate buffer for at least 10 h at 25 ml/h to equilibrate the column fully. It is very important to convert the resin fully to the ammonium form, as the elution of cytochrome *c* derivatives on the sodium form of the resin is quite different.

(iv) Oxidize the crude cytochrome *c* derivative mixture from Section 5.1 by the addition of several drops of 0.1 M potassium ferricyanide, and apply to the Bio-Rex 70 column. All of the cytochrome *c* should stick in a tight band at the top, and a little excess potassium ferricyanide should be observed as a yellow colour eluting down the column.

(v) Elute the column with 0.14 M ammonium phosphate, pH 7.2, at 25 ml/h and collect 4 ml fractions.

A typical chromatogram is shown in *Figure 11*. The peaks marked 1 and 2 are deaminated forms of cytochrome *c* present in the commercial Sigma type

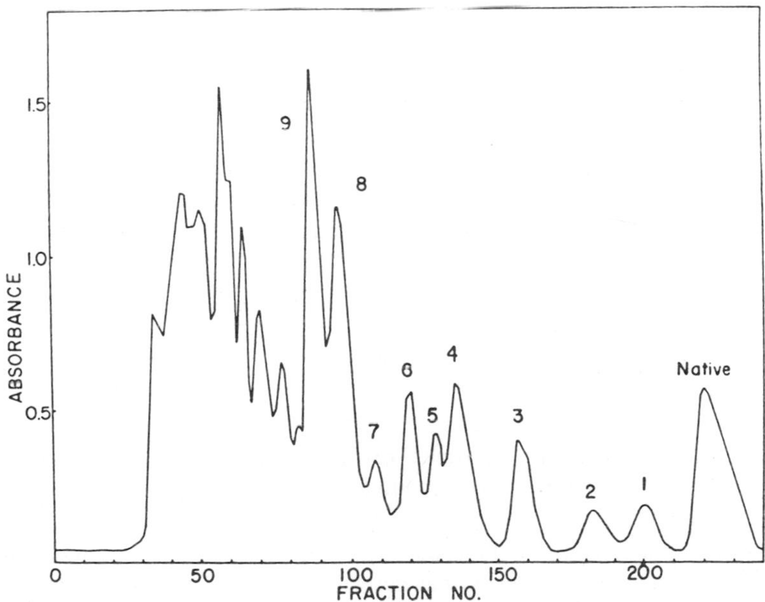

Figure 11. Chromatogram of 250 mg of a crude reaction mixture of $CF_3PhNHCO$-cytochromes *c* on a 2 × 70 cm Bio-Rex 70 column. The column was eluted with 0.14 M ammonium phosphate, pH 7.2, at a flow-rate of 25 ml/h, at room temperature. The fraction size was 4 ml. The absorbances were taken at 430 nm.

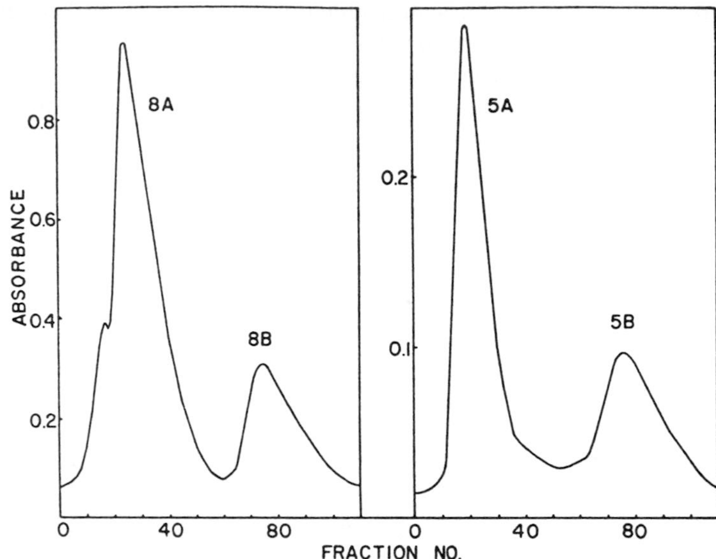

Figure 12. Chromatograms of $CF_3PhNHCO$-cytochrome *c* fractions 8 and 5 on 7 × 1.6 cm Whatman CM32 carboxymethyl-cellulose columns eluted with 0.08 M phosphate, pH 6.0, at 25 ml/h. Fraction collection was begun after 550 ml of buffer had eluted for fraction 8, and after 950 ml of buffer had eluted for fraction 5. The fraction size was 7 ml, and the absorbances were taken at 409 nm.

VI cytochrome *c*, and the batch to be used should be checked to make sure that these impurities are not present in very large amounts. The peaks marked 3−9 are mixtures of singly-labelled $CF_3PhNHCO$-cytochrome *c* derivatives.

(vi) Pool the fractions for each peak (from quarter-height to quarter-height) and concentrate as follows. Dilute the solution 3-fold with distilled water and apply to a 1 × 10 cm column of Bio-Rex 70 equilibrated with 0.03 M ammonium phosphate, pH 7.2. Elute the cytochrome *c* in a small volume (~0.5 ml) with 0.5 M ammonium phosphate, pH 7.2. De-salt this solution by elution through a 1 × 10 cm Bio-Gel P-4 column equilibrated with 0.02 M sodium phosphate, pH 6.0.

(vii) Re-chromatograph each of the peaks from the Bio-Rex 70 column on Whatman CM32 carboxymethyl cellulose. Treat the de-salted fraction with one drop of 0.1 M potassium ferricyanide to oxidize it fully, and apply to a 1.7 × 10 cm column of CM32 equilibrated with 0.08 M sodium phosphate, pH 6.0. Elute the column with 0.08 M sodium phosphate at a flow-rate of 25 ml/h, and collect 7 ml fractions. Peaks 3, 6 and 9 re-chromatograph on CM32 as single bands, while peaks 5 and 8 are resolved into two bands each, 5A and 5B, and 8A and 8B (*Figure 12*).

(viii) Pool the fractions under the peaks (from half-height to half-height), adjust to pH 7.2, concentrate and de-salt as described above. The final yields of purified derivatives from 250 mg of cytochrome *c* range from about 1 to 20 mg for derivatives 3−9.

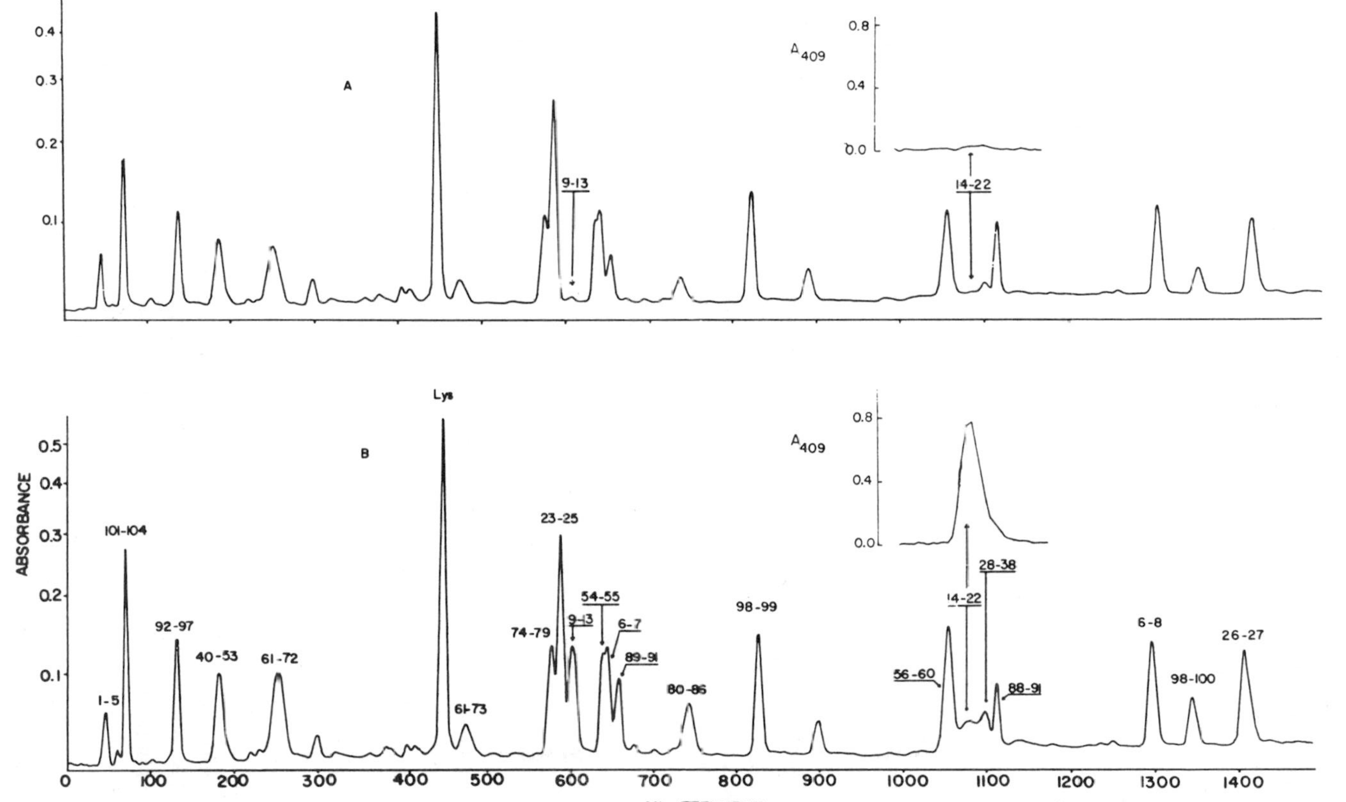

Figure 13: Elution profile of a 10 mg of tryptic hydrolysate of fraction 9 (**A**) and native horse heart cytochrome *c* (**B**) on a 0.9 × 23 cm Aminex A-5 column. The column was eluted first with a 1020 ml linear pH gradient from pH 3.3 (0.055 M pyridine–acetic acid buffer) to pH 4.9 (0.55 M pyridine–acetic buffer), and then with a 600 ml linear concentration gradient from the 0.55 M buffer to 3 M pyridine–acetic acid buffer (pH 5.3). The absorbances were automatically detected at 570 nm in a 3 mm flow cell after reaction with ninhydrin. The insets show the absorbances of the column eluent at 409 nm in the region where the haem peptide elutes.

5.3 Peptide mapping of CF_3PhNHCO-cytochrome *c* derivatives

The location of the CF_3PhNHCO-lysine and the purity of each derivative can be determined by running a tryptic peptide map by one of the following procedures: ion-exchange chromatography on a modified amino acid analyser (described here and in ref. 16), reverse phase h.p.l.c. (19) and two-dimensional thin-layer electrophoresis (18).

(i) Digest the cytochrome *c* derivative, at a concentration of 5 mg/ml in 50 mM Tris–HCl, pH 7.5, with Tos-PheCH_2C1-treated trypsin (5% by weight with respect to cytochrome *c*) at 37°C for 3 h. Lyophilize the sample, adjust to pH 3.5 and inject onto a 0.9 × 23 cm Aminex A-5 ion-exchange column operated at 52°C.

(ii) Elute the column with a pyridine acetate gradient as described in *Figure 13* at a flow-rate of 1 ml/min. The column eluent is stream-split, with 70% going to a fraction collector and 30% going through the amino acid analyser for ninhydrin detection and automatic plotting.

(iii) Subject each peak in the chromatogram (*Figure 13*) to amino acid analysis, and identify as a known peptide in the sequence of horse heart cytochrome *c* (18).

In the chromatogram of fraction 9, shown at the top of *Figure 13*, peptide 9–13 and the haem peptide 14–22 disappeared from the chromatogram and a new haem peptide remained at the top of the column. Since the peptide bond following a CF_3PhNHCO-lysine is not digested by trypsin, the lysine modified in derivative 9 is lysine 13. By the same procedure, fractions 3, 5A, 6, 8A and 8B were shown to contain singly labelled lysines at residues 100, 79, 27, 72 and 8, respectively. Each of the singly labelled derivatives was determined to be at least 95% pure by this peptide mapping approach.

6. REACTION OF CYTOCHROME C ANALOGUES WITH THE CYTOCHROME BC_1 COMPLEX AND CYTOCHROME C OXIDASE

A variety of assays have been described to study the reactions of cytochrome *c* with the cytochrome bc_1 complex and cytochrome oxidase. However, these assays often gave conflicting results when applied to cytochrome *c* analogues, and led to a certain amount of confusion in the early literature on the mechanism of cytochrome *c*. For example, in 1973 Margoliash *et al.* (21) reported that modification of lysine 13 at the top of the haem crevice increased the K_m for the reaction with cytochrome oxidase, but did not affect the reaction with the cytochrome bc_1 complex, suggesting that the two proteins had different binding domains on cytochrome *c*. Subsequently it was found that this was caused by the use of different ionic strengths in the two assays, and modification of lysine 13 affected both reactions equally when optimal assay conditions were used (22). It is important to recognize that the overall reaction monitored in these assays is very complex, and the rate-limiting step might not involve cytochrome *c*, but an internal reaction in the respiratory complex. Since this appears to be a general problem in studying the reactions of electron transport proteins, the criteria used to design an assay will be described in some detail.

6.1 Reaction of cytochrome *c* derivatives with complex III

The source of cytochrome c_1 used in these assays is beef heart succinate-cytochrome *c* reductase (see Chapter 4 for preparation).

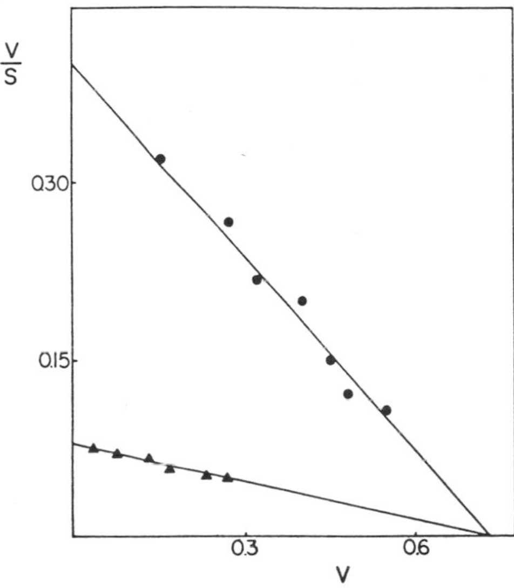

Figure 14: The rate of reduction of native cytochrome *c* (●) and CF_3NHCO-Lys-13 cytochrome *c* (▲) by succinate-cytochrome *c* reductase in Tris−HCl buffer pH 7.5 containing 0.2 M NaCl and 10 mM succinate at 25°C. The cytochrome *c* concentration, *S*, was in μM, and the initial velocity, *v*, c_1 concentration was 0.20 nM.

(i) Place cytochrome *c* in a cuvette at a concentration of 0.5 μM in 0.2 M Tris-HCl, pH 7.5, containing 10 mM sodium succinate. Re-crystallize the sodium succinate from ethanol, and prepare the buffer solutions fresh each day.

(ii) Record the spectrum from 500 to 600 nm to make sure that the cytochrome *c* derivative is fully oxidized.

(iii) Initiate the reaction by adding succinate-cytochrome *c* reductase to the cuvette to bring the final cytochrome c_1 concentration to 0.2 nM, and record the absorbance as a function of time at 550 nm.

It is important to make sure that at least 90% of the theoretical absorbance change is observed during the course of the reaction, based on a difference extinction coefficient of 18.6 mM/cm. The amount of succinate-cytochrome *c* reductase used in the assay can be adjusted to achieve this absorbance change in a convenient length of time. The function $\ln(A_\infty - A)$ plotted as a function of time should be linear at cytochrome *c* concentrations of 0.5 μM and below, indicating that the reaction is first order in cytochrome *c*. Repeat the assay at different cytochrome *c* concentrations (0.5−20 μM) for both the native protein and the cytochrome *c* derivatives. Plot the initial velocity *v* against *v*/S in an Eadie−Hofstee plot, and V_{max}/K_m is obtained as the intercept on the *v*/S axis, and V_{max} as the intercept on the *v* axis (*Figure 14*). Modification of lysine 13 did not affect the V_{max} of the reaction, but increased the K_m by a factor of 4. The effect of ionic strength is rather unusual, as shown in *Figure 15*. The V_{max}/K_m for native cytochrome *c* is nearly independent of ionic strength up to 0.1 M, but then decreases rapidly. At low ionic strength, the V_{max}/K_m for CF_3PhNHCO-lysine 13 cytochrome *c* is nearly the same as that for the native protein, but at higher ionic strengths it is about 4-fold smaller.

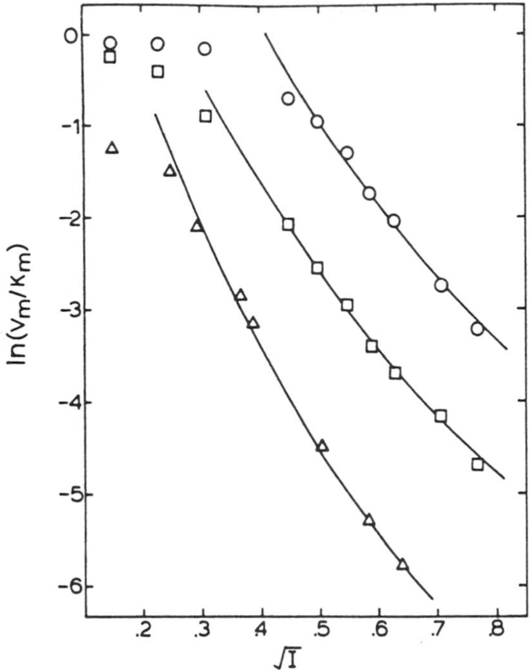

Figure 15. The effect of ionic strength on the spectrophotometric oxidase and reductase activities of cytochrome *c* at 25°C in 50 mM Tris−HCl, pH 7.5, containing 0−0.6 M NaCl. Δ, Cytochrome oxidase activity of native cytochrome *c* using 0.008 mg, of protein/ml of Keilin−Hartree particles. V_{max}/K_m is in min^{-1}. ○, Succinate-cytochrome *c* reductase activity of native cytochrome *c* using 10 mM succinate and succinate-cytochrome *c* reductase containing 0.08 nM final concentration of cytochrome *c*; □, succinate-cytochrome *c* reductase activity of CF_3PhNHCO-Lys 13 cytochrome *c*. The *solid lines* were calculated from equation 9 of reference 23, omitting the lysine 13 term in the case of the derivative.

To rationalize the contradictory effects of ionic strength and charge modifications, consider the following mechanism for the reaction between cytochrome *c* and the cytochrome bc_1 complex:

$$c_1^{2+} + c^{3+} \underset{k_2}{\overset{k_1}{\rightleftharpoons}} c_1^{2+}\ c^{3+} \underset{k_4}{\overset{k_3}{\rightleftharpoons}} c_1^{3+}\ \ c^{2+} \underset{k_6}{\overset{k_5}{\rightleftharpoons}} c_1^{3+} + c^{2+};\quad c_1^{3+} \overset{k_7}{\longrightarrow} c_1^{2+}$$

In this equation we have assumed that both oxidized and reduced cytochrome *c* can form a complex with cytochrome c_1, and that the oxidized cytochrome c_1^{3+} is reduced by an internal reaction in the cytochrome bc_1 complex at a rate k_7. If the cytochrome bc_1 complex is present in catalytic amounts, then the steady-state expression for the initial velocity is:

$V = V_{max}S/(K_m + S)$, where

$$V_{max} = k_3k_5k_7E_o/(k_3k_5 + k_3k_7 + k_5k_7 + k_4k_7)$$

$$K_m = (k_2k_5 + k_3k_5 + k_2k_4)k_7/(k_3k_5 + k_3k_7 + k_5k_7 + k_4k_7)k_1$$

Both V_{max} and K_m are highly complex kinetic parameters, and may not depend in a simple fashion on charge modifications. In particular, K_m is probably not equal to the equilibrium dissociation constant k_2/k_1. The parameter:

$$V_{max}/K_m = k_1k_3k_5E_o/(k_2k_5 + k_3k_5 + k_2k_4)$$

has a simpler form and is independent of the internal reduction rate k_7. One explanation for the data of *Figure 15* is that at low ionic strength the numerator and denominator terms of the equation for V_{max}/K_m are equally affected by charge interactions and cancel. As the ionic strength is increased, both dissociation rate constants k_2 and k_5 are expected to increase rapidly, and at some point the term k_2k_5 will become large compared with the other terms in the denominator, and V_{max}/K_m will become $k_1k_3E_o/k_2$. Under these conditions V_{max}/K_m will decrease rapidly as the ionic strength is increased, and the assay will be optimally sensitive to charge interactions. This phenomenon is also observed in the steady-state reactions of cytochrome *c* with many other proteins, including cytochrome oxidase (*Figure 15*).

6.2. Reaction of cytochrome *c* derivatives with complex IV

The reaction between cytochrome *c* and cytochrome oxidase can be assayed either by a spectroscopic method similar to that described above, or by an oxygen-electrode method developed by Ferguson-Miller *et al.* (18). The source of cytochrome oxidase can be either purified cytochrome oxidase, or Keilin–Hartree particles treated with deoxycholate.

6.2.1 *Preparation of reduced cytochrome* c *and assay*

(i) For the spectroscopic assay, reduce a cytochrome *c* stock solution of about 500 μM with 10 mM sodium ascorbate and then pass it through a 1 × 15 cm Bio-Gel P-4 column equilibrated with 0.2 M Tris-HCl, pH 7.5, to remove the excess ascorbate.

(ii) Place the reduced cytochrome *c* in a cuvette at a concentration of 0.5–10 μM in 0.2 M Tris-HCl, pH 7.5, containing 0.3% Tween 80, and record the spectrum from 500 to 600 nm to ensure that the cytochrome *c* is fully reduced.

(iii) Initiate the reaction by addition of about 10 nM cytochrome oxidase, and monitor the decrease in absorbance at 550 nm.

Since the reaction is rigorously first-order in cytochrome *c*, it is convenient to calculate the pseudo first-order rate constant *k* from a plot of $\ln(A - A_\infty)$ against time. Plot the rate constant *k* against *k*S on an Eadie–Hofstee plot. The V_{max}/K_m for native cytochrome *c* follows the same type of ionic strength dependence as observed for the cytochrome bc_1 complex (*Figure 15*), so it is important to run the assay at an optimal ionic strength.

The oxygen electrode assay is carried out in a Gilson Model KM Clark electrode cell (see Chapter 1 for description). The buffer contains 50 mM potassium morpholinopropane sulphonate, pH 7.5, 0.3% Tween 80, 7 mM sodium ascorbate (from a fresh stock solution of 0.5 M sodium ascorbate containing 1 mM EDTA), and 0.7 mM tetramethyl-*p*-phenylenediamine dihydrochloride (TMPD) (from a fresh stock solution of 50 mM TMPD). Addition of cytochrome oxidase (50 nM) to the buffer in the cell gives a low base-line rate of oxygen consumption which is subtracted from the rate of oxygen consumption after addition of cytochrome *c* (0.01–10 μM). The oxygen concentration of fully air-saturated buffer is assumed to be 250 μM. The Eadie–Hofstee

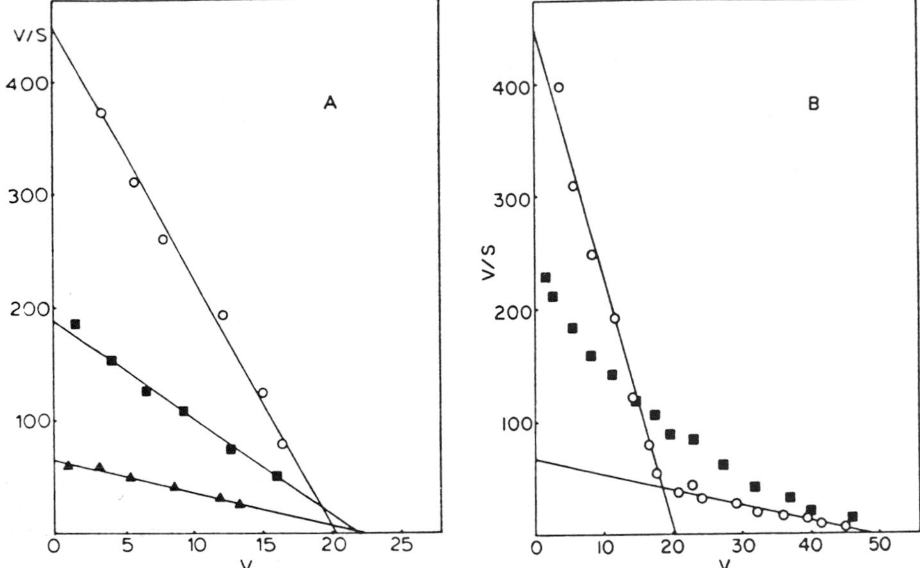

Figure 16. Steady-state cytochrome oxidase activities of $CF_3PhNHCO$-cytochrome *c* derivatives using the oxygen-electrode assay. (**A**) Activities of native cytochrome *c* (○), $CF_3PhNHCO$-lysine 8 (■) and $CF_3PhNHCO$-lysine 13 (▲) cytochrome *c* measured at cytochrome *c* concentrations of 0.01 – 0.50 μM. (**B**) Activities of native (○) and $CF_3PhNCHO$-lysine 79 cytochrome *c* (■) measured at cytochrome *c* concentrations from 0.01 – 6.0 μM. Velocities are in nanomoles O_2 reduced per min.

plot for native cytochrome *c* shows both a high-affinity and a low-affinity phase with apparent K_m values of about 0.05 and 0.7 μM, respectively (*Figure 16*). Modification of lysine 13 significantly increases the K_m of the high-affinity phase of the reaction indicating that this residue is important in binding to cytochrome oxidase.

6.3 **Interpretation of kinetic results**

Figure 17 summarizes the effect of specific modification of cytochrome *c* lysines on the reactions with cytochrome oxidase (A) and the cytochrome bc_1 complex (B). The contribution of each lysine to the total electrostatic interaction is estimated using the formula:

$$V_i = -\mathrm{RT}\ln Y_i$$

where Y_i is the ratio of V_{max}/K_m for native cytochrome *c* to that for a derivative modified at lysine *i*. The Y_i values obtained for a given lysine using reagents of different sizes were remarkably similar, indicating that the major effect of modification is to remove the electrostatic contribution of that lysine. *Figure 17* indicates that the lysines surrounding the haem crevice of cytochrome *c* are involved in binding both the cytochrome bc_1 complex and cytochrome oxidase. Since the binding domains for these two proteins are nearly identical, cytochrome *c* must function as a mobile electron shuttle during electron transport. A semi-empirical relationship for the electrostatic interaction of cytochrome *c* with its redox partners has been developed (23). It is assumed that each lysine surrounding the haem crevice of cytochrome *c* forms a charge-pair with

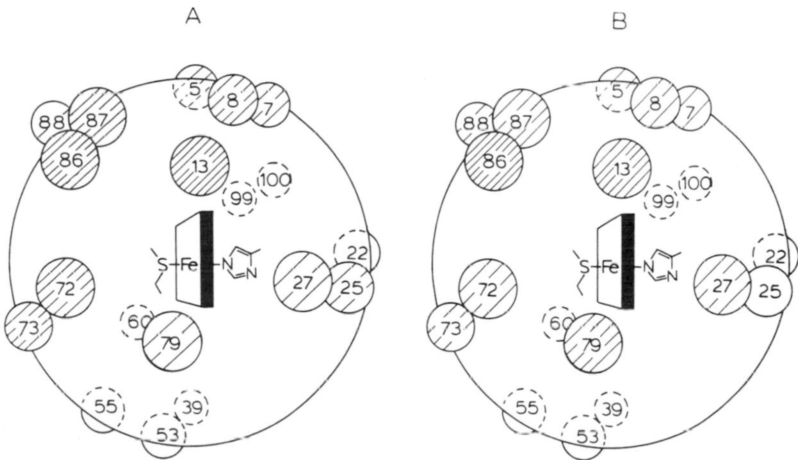

Figure 17. A schematic diagram of horse heart cytochrome *c* viewed from the front of haem crevice. The approximate positions of the β-carbon atoms of the lysine residues are indicated by *closed* and *dashed* circles for residues located toward the front and back of cytochrome *c*, respectively. The electrostatic free energy contribution of lysine *i*, v_i is indicated by the number of diagonal hatch marks in the circle, with −0.10 kcal/mol/hatch mark. (**A**) the interaction domain for cytochrome oxidase. (**B**) the interaction domain for cytochrome c_i.

a complementary carboxylate group on the redox partner, and the electrostatic energy of this charge pair is given by V_i. This relationship is in quantitative agreement with the experimental ionic strength dependence of the reactions of native cytochrome *c* with the cytochrome bc_1 complex and cytochrome oxidase (*Figure 15*).

7. THE USE OF CHEMICAL MODIFICATION AND PEPTIDE MAPPING TO LOCATE THE CYTOCHROME C BINDING SITE ON SUBUNIT II OF CYTOCHROME C OXIDASE

It has been shown above that the reaction between cytochrome *c* and cytochrome oxidase involves the formation of a complex stabilized by complementary charge pairs between lysine amino groups on cytochrome *c* and carboxylate groups on cytochrome *c* oxidase. This section describes a procedure to use a water-soluble carbodiimide, 1-ethyl-3-(-3-[^{14}C]trimethylaminopropyl)carbodiimide (ETC), to identify the specific carboxylate groups on cytochrome oxidase that are involved in binding cytochrome *c* (24). The methods used to digest subunit II of cytochrome oxidase with trypsin and separate the resulting peptides using the reverse-phase h.p.l.c. will be described in detail.

7.1 Reaction of cytochrome *c* oxidase with ETC

(i) Prepare duplicate samples of 100 μM beef heart cytochrome oxidase in 2 ml of 10 mM sodium phosphate, pH 7.0, containing 1% Triton X-100. Add cytochrome *c* (100 μM) to one of these samples to protect the carboxyl groups involved in binding from reaction with ETC.

(ii) Add 4 mM ETC [at a specific activity of 10 mCi/mmol (370 MBq/mmol) (24)] to each sample, and allow the reaction to proceed at 25°C.

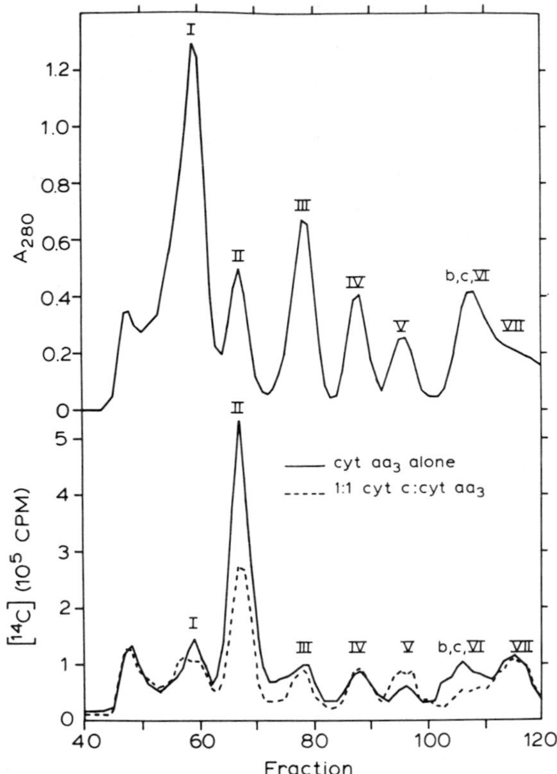

Figure 18. Effect of cytochrome *c* binding on the incorporation of [^{14}C]ETC into cytochrome *c* oxidase subunits. Cytochrome *c* oxidase (100 μM) with or without 100 μM cytochrome *c* was treated with 4 mM [^{14}C]ETC for 12 h in 10 mM phosphate, pH 7.0, and 1% Triton X-100. The oxidase subunits were separated on a 3 × 40 cm Bio-Rad P-100 column eluted with 3% SDS. The A_{280} trace at the top of the figure is for the sample treated in the absence of cytochrome *c*. The A_{280} trace for the sample with cytochrome *c* was essentially the same except for extra absorbance due to cytochrome *c* co-eluting with subunit V. The traces at the bottom of the figure show the incorporation of [^{14}C]ETC. The fraction size was 1.35 ml.

(iii) At intervals of 4 h remove 5 μl aliquots from each sample and assay for cytochrome *c* oxidase activity using the spectroscopic assay described in Section 6.2.1 with 10 μM cytochrome *c*.

(iv) When the activities of the samples with and without cytochrome *c* have been inhibited by about 20% and 80%, respectively, pass both samples through 2 × 10 cm Bio-Gel P-2 columns equilibrated with 50 mM sodium phosphate, pH 7.0, 1% Triton X-100, to remove excess ETC and stop the reaction. It should take about 12 h to achieve this level of inhibition.

(v) Dissociate each sample for 60 min in 5% SDS, 1% 2-mercaptoethanol, 5 mM EDTA and separate the subunits on a 3 × 40 cm Bio-Rad P-100 column (200–400 mesh) eluted with 3% SDS and 10 μM dithiothreitol as described in Section 3.

(vi) Analyse each fraction for protein at 280 nm, and for radioactivity by mixing a 5 μl aliquot with 2.5 ml of scintillation fluid and counting.

The chromatogram of *Figure 18* indicates that subunit II is the major site of radiolabelling, with about 1−2 mol of ETC incorporated per mol of subunit II. Equimolar concentrations of cytochrome *c* dramatically protect subunit II from modification by ETC, indicating that it is the major site of cytochrome *c* binding to cytochrome oxidase.

7.2 Digestion of subunit II and peptide mapping

To identify the specific carboxyl groups on subunit II that are modified by ETC, it is necessary to digest the subunit using trypsin and separate the resulting peptides by reverse-phase h.p.l.c. This is done as follows.

(i) Pool the fractions containing subunit II and pass them through a 5 × 20 cm Bio-Gel P-4 column equilibrated with 0.01% SDS to remove most of the SDS. The SDS elutes more slowly from the column than the peptide, and can be detected by a dramatic decrease in the drop size. Pool and lyophilize the fractions containing subunit II.

(ii) Dissolve subunit II (2 mg/ml) in 0.1% SDS, 1% octyl glucose and 20 mM bicine, pH 8.0, and add TPCK-trypsin (Worthington, Inc.) to a final concentration of 0.18 mg/ml and incubate the mixture at 37°C for 3 h. Octyl glucoside has been found to solubilize hydrophobic polypeptides such as subunit II for efficient proteolysis.

(iii) Inject the sample into a Brownlee RP-300 h.p.l.c. column and elute with a gradient from 5 mM sodium phosphate, pH 7.0 to 1-propanol as described in *Figure 19*. Monitor the eluent at 210 nm, collect 0.8 ml fractions and analyse 5% of each fraction for radioactivity.

(iv) Identify the seven major tryptic peptides from native subunit II by amino acid analysis, by reference to *Table 6*. These peptides elute in the order of increasing hydrophobicity, with the hydrophilic peptide $T_{172-178}$ eluting at a low concentration of propanol and the very hydrophobic membrane-spanning peptide T_{1-98} eluting at 46% propanol. The peak at 40% propanol is a solvent impurity.

Table 6. Primary sequence of subunit II from beef cytochrome *c* oxidase (8)

10 (Q) · * (E18) · 20 · 30
M A Y P M Q L G F Q D A T S P I M E* E L L H F H D H T L M I

40 · 50 · 60
V F L I S S L V L Y I I S L M L T T K L T H T S T M D A Q E

70 · 80 · 90
V E T I W T I L P A I L L I L I A L P S L R I L Y M M D E I

100 · 110 · * (D112) · * (E114) · 120
N N P S L T V K▼T M G H Q W Y W S Y E Y T D* Y E* D L S F D S

130 · 140 · 150
Y M I P T S E L K P G E L R▼L L E V D N R▼V V L P M E M T I

160 · 170 · 180
R▼M L V S S E D V L H S W A V P S L G L K▼T D A I P G R▼L N

190 · 200 · 210
Q T T L M S S R P G L Y Y G Q C S E I C G S N H S F M P I V

220 · * · 227
L E L V P L K Y▼F E K▼W S A S M L

The peptide bonds hydrolysed by trypsin are indicated ▼, and the residues labelled by [^{14}C]ETC are starred. See appendix I for 1 letter codes for amino acids.

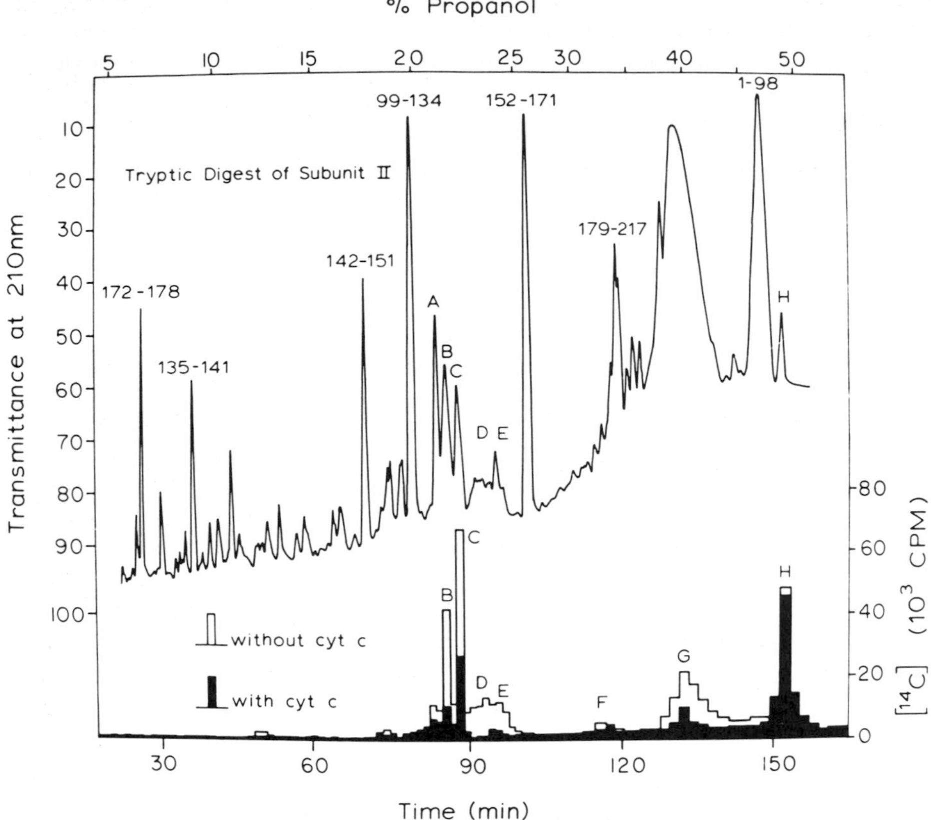

Figure 19. Separation of tryptic peptides of [^{14}C]ETC-labelled subunit II by reverse-phase h.p.l.c. Purified [^{14}C]ETC-labelled subunit II (0.6 mg) was digested with trypsin and eluted on a Brownlee RP-300 column with a linear gradient from 5 mM phosphate, pH 7.0, to 25% 1-propanol in 60 min. The flow-rate was 0.8 ml/min. The A_{210} trace at the top of the figure is for the sample treated in the absence of cytochrome *c*. The A_{210} trace of the sample treated in the presence of cytochrome *c* was the same except that peaks A–E were smaller. The histogram at the bottom of the figure shows [^{14}C]ETC incorporation in the presence and absence of cytochrome *c*.

The tryptic chromatogram of ETC-modified subunit II shown in *Figure 19* contains all the peptides found in native subunit II plus eight new peptides labelled A–H. Amino acid analysis indicated that peptides A–E are all modified forms of the native peptide T_{99-134}, peptide G is a modified form of $T_{197-217}$ and peptide H is a modified form of T_{1-98}. Each of the labelled tryptic peptides was then further digested with *Staphylococcus aureus* protease and h.p.l.c. analysis indicated that Glu-18, Asp-112, Glu-114 and Glu-198 are the major residues modified by ETC (24). Binding one molecule of cytochrome *c* to cytochrome oxidase protects Asp-112, Glu-114 and Glu-198 from labelling by ETC and inhibits the formation of peptides A–G. The negatively-charged carboxylates on these residues are therefore involved in binding cytochrome *c* to native cytochrome oxidase.

8. RECONSTITUTION OF MITOCHONDRIAL COMPLEXES INTO ARTIFICIAL PHOSPHOLIPID VESICLES

The procedure described below leads to insertion (reconstitution) of a membrane protein, into an artificial phospholipid bilayer, organized as a unilamellar, fairly spherical, liposome. This technology has been a key approach for the study of bioenergetic processes at the molecular level (25–27), although it should be borne in mind that up to now these artificial systems represent an oversimplified model, the properties of which cannot always be extrapolated to the situation *in vivo*.

8.1 Equipment and materials

An ultrasonic apparatus, of the probe or bath type, is the only special equipment required for reconstitution experiments. This apparatus, because of the large number of parameters controlling the disintegration efficiency by ultrasonic energy (probe or bath size, sample volume and density, actual power output of sonicator, etc.), gives variable results, so that quite often in the current literature it is written 'sonicate until clarity is reached'. In order to quantify the amount of energy transferred to the suspension to be sonicated, the acoustic energy, measured as vibrational amplitude (microns peak-to-peak) may be taken as a good index. The power output in Watts, of course, should be a more reliable index of sonication efficiency. However, this is true only under ideal conditions, when all sonication parameters are controlled. Once the best conditions to achieve clear suspensions of liposomes containing the correctly inserted functional protein have been found, they should be carefully recorded and reproduced from experiment to experiment.

8.1.1 *Re-crystallization of cholic acid*

(i) Dissolve cholic acid up to a concentration of 50 g/l (typically 1 litre) in 70% (v/v) ethanol containing 0.5 g/l of activated charcoal.
(ii) Stir the solution at room temperature for 1–2 h and then filter off the solution carefully.
(iii) Transfer the solution into a round-bottom flask and rotary evaporate it until the volume reaches approximately 1/3 of its initial value.
(iv) Add doubly-distilled water until the volume is restored to the original and let the crystals precipitate overnight in the coldroom.
(v) Filter off the material using a Buchner funnel connected to a water vacuum pump and rinse twice with 100 ml of diethyl ether.
(vi) Spread the white powder on a clean Petri dish (20 cm diameter) and dry in an oven at about 60°C.

8.1.2 *Purification of asolectin*

Reconstitution procedures commonly use soybean asolectin (28). This is a mixture of phospholipids (phosphatidylcholine, 40%; phosphatidylethanolammine, 40%; other phospholipids, 20%). Although being somewhat ill-defined it does give good results in protein reconstitution experiments and is thus often the phospholipid source of choice.

It is also inexpensive. The procedure is as follows.

(i) Suspend 25 g of asolectin (sold under the name of L-α-phosphatidylcholine, Type IV-S, from Sigma Chemical Co.) in 500 ml of acetone. Up to step (vi) all solvents should contain 25 μM phenyl-*p*-phenylenediamine as an anti-oxidant. Ideally, both acetone and diethyl ether should be dried over sodium metal before use, but this is not essential. Stir the yellowish suspension gently for 48 h at room temperature, under nitrogen.

(ii) Allow the phospholipid suspension to sediment and decant and discard the supernatant by use of a water vacuum pump.

(iii) Resuspend the insoluble fraction and stir for an additional 24 h as described in steps (i)–(ii).

(iv) Filter off the suspension by suction, using a Buchner funnel and rinse the resulting cake once with 200 ml of acetone.

(v) Dissolve the phospholipid mixture in 200 ml of diethyl ether and precipitate by slowly adding 400 ml (or more if necessary) of acetone with stirring: the phospholipid will appear as a sticky precipitate.

(vi) Decant the opalescent yellowish supernatant and repeat step (v) until the acetone is colourless (typically twice).

(vii) Dissolve the precipitate carefully in 100 ml of diethyl ether, under nitrogen, in a solvent-resistant bottle, if any sediment is present, centrifuge at 4000 *g* for 20 min.

(viii) Transfer the supernatant to a round-bottom flask and evaporate to dryness using a rotary evaporator.

(ix) Finally, scrape off the phospholipids from the flask walls and pulverize. Store the powder under nitrogen at −20°C in the dark.

8.2 Strategies for incorporation of membrane proteins into phospholipid vesicles

Thus far a number of reconstitution pathways have been proposed and more or less successfully used to study the vectorial properties of many membrane protein complexes from various biological sources (25). When dealing with mitochondrial membrane proteins, however, two strategies are more frequently followed for the reconstitution process.

(i) A suspension of phospholipids and protein are sonicated together so that while forming smaller and smaller vesicles the protein inserts into the liposome wall by making use of the disruptive ultrasonic energy. The major disadvantages of this approach are the topologically random insertion of the protein into the membrane and the inactivating effects of sonication on the protein.

(iii) Phospholipid and detergent mixed micelles are pre-formed by sonication and the protein, added and slowly exchanged with the micellar detergent during dialysis; thus the protein is incorporated into the phospholipid bilayer. The final equilibrium reached is in favour of a fairly homogeneous distribution of the protein molecules, almost unidirectionally oriented into the liposome wall. This approach can fail if the protein is affected by long exposure (24 h) to the detergent used. If this is the case other detergents or alternative reconstitution procedures (such as that described previously) can be used (25).

See Chapter 3, Section 5.2.1 for the reconstitution of the ADP/ATP exchange carrier and Chapter 4 for the preparation of inner-membrane proteins used in this chapter for incorporation.

8.2.1 *Reconstitution of cytochrome c oxidase*

Reconstitution can be carried out using purified oxidase (Chapter 4, Section 8.5) and a mixture of phospholipids from soybean (asolectin). Pure phospholipids or mixtures at different molar ratios have also been used, however the best respiratory control ratios (RCR, see Section 9.2 for determination) have been obtained using asolectin. The commercially available crude product tends to decrease the amount of protein properly inserted into the liposome (low RCR) so that the use of purified asolectin is recommended (see Section 8.1.2). The best reconstitution (RCR 7−14) is carried out by making use of the cholate dialysis method with minor modifications. The following procedure gives cytochrome oxidase vesicles (COVs) having the following properties: predominantly oriented (85%) with cytochrome *a* external to the liposome; a high rate of cytochrome *c* oxidation and redox-linked proton translocation; high oxygen uptake and RCR $\simeq 10$ (30). The RCR achieved following reconstitution is certainly dependent on the nature of the enzyme preparation. However reconstitution with different batches of enzyme prepared in the same way may give COVs with different RCR values. The reason for this variability is unclear at present. The procedure is as follows.

(i) Dissolve 50 mg/ml (typically 10−15 ml) of purified asolectin in 0.1 M Hepes-KOH, pH 7.3, containing 25 mM of re-crystallized cholic acid (see Section 8.1.1), re-adjust the pH with KOH and let the suspension stir gently under nitrogen for 60 min.
(ii) Sonicate the strongly opalescent suspension using a probe type sonicator setting the vibration power of the probe (9.5 mm diameter) to 12 μm amplitude, in a 25 ml cooled sonication vessel. Sonication must be carried out under nitrogen, avoiding cavitation and overheating (cycles of 30 sec duration with 1 min cooling intervals are recommended). The total sonication time is approximately 60 sec for every millilitre of suspension. The final suspension should appear slightly opalescent but definitely transparent and 'clarified'! (see *Figure 20*, test tube A, after sonication, as compared with B before sonication).
(iii) Add cytochrome *c* oxidase to a final concentration of 5 μM. The enzyme is best prepared for this usage following the Yonetani method (see Chapter 4) (as this gives better incorporation) and the final solubilization of the enzyme for reconstitution is generally in 0.2% cholate-containing buffer rather than in a non-ionic detergent.
(iv) Dialyse versus:
    (a) 100 volumes of 0.1 M Hepes-KOH pH 7.3 (4 h)
    (b) 200 volumes of 10 mM Hepes-KOH pH 7.3 containing 39.6 mM sucrose and 40.6 mM KCl (two changes lasting 4 and 16 h, respectively).
(v) Centrifuge at 8000 *g* for 30 min and discard the pellet. This contains large lipid/enzyme aggregates and titanium from the sonic probe.

The resulting COVs are unilamellar (i.e. contain only two phospholipid leaflets with

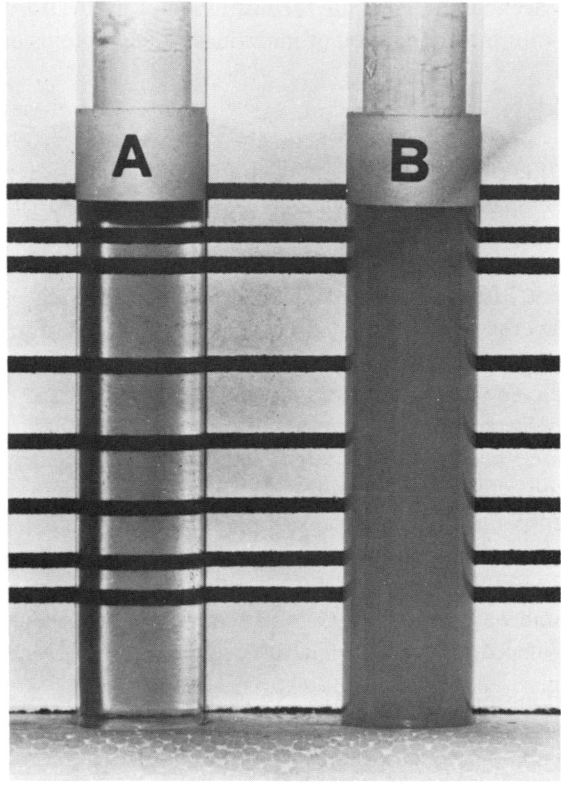

Figure 20. Effect of sonication on a phospholipid–detergent suspension. Test tube B is filled with a suspension of asolectin and K^+/cholate, in Hepes buffer, after 30 min stirring. Test tube A is after sonication. Tube dimensions: 1 × 9 cm.

hydrophobic tails inwards). Their average diameter is 30–40 nm and contain about 4000 molecules of phospholipid and one oxidase complex. If stored at 4°C they are fairly stable (in terms of RCR) for at least 1 week.

Notice that Hepes buffer is recommended although any kind of buffer, impermeable to the liposome membrane over a long time scale, could successfully be used. When dealing with cholate dialysis reconstitution methods a note of caution is in order. There is in fact some evidence that after 24 h dialysis a considerable amount of cholate (~10–100 molecules/liposome) is still present in the vesicle suspension (31). Keep in mind also that after step (i) the buffering power of the internal aqueous phase is fixed. It is therefore at this step that any desired changes in the internal buffer composition should be made. On the other hand, the final suspending medium of COVs is determined during the last dialysis or by further Sephadex G-25 gel chromatography. Proton pumping experiments, for instance, require a very low buffered suspending medium to allow protons to be measured, but a higher internal buffer (0.1 M Hepes) to provide protons. Typically a 50 μM Hepes-KOH buffer, pH 7.3 containing 46.1 mM sucrose and 43.6 mM KCl is used externally. The osmolarity and potassium concentration of the internal and external medium must be the same (30).

8.2.2 *Reconstitution of complex III*

The reconstitution of the bc_1 complex follows essentially the same procedure described for cytochrome *c* oxidase (32). However, the phospholipid and protein concentrations used are not uniquely defined, their optimal ratio depending on the characteristics of the complex III as purified during the mitochondrial respiratory chain fractionation. Typically 30–100 mg/ml of purified asolectin are sonicated to form liposomes and 2 mg/ml of bc_1 complex are then added [see Section 8.2.1 (ii) (iii)].

8.2.3 *Reconstitution of F_oF_1 ATP synthetase*

The strategy of reconstitution of mammalian F_0F_1 ATP synthetase is also essentially similar to the one described for cytochrome oxidase (33). Problems arise, however, from the intrinsic complexity of this protein. Mitochondrial ATP synthetase is composed of two structurally and functionally distinct units called F_0 (hydrophobic and membrane spanning) and F_1 (hydrophilic, protruding into the mitochondrial matrix space). When these two components are intimately associated, both proton translocation and ATP synthesis and hydrolysis can be studied. Reconstitution of ATP synthetase should include therefore both the hydrophobic and hydrophilic components. Reconstitution of the F_0 and F_1 proteins are however performed independently (the F_0 first) since when the whole crude complex is reconstituted in one step, partial or complete denaturation of F_1 and other polypeptides (e.g. oligomycin sensitivity-conferring protein and F_6) occurs. A typical reconstitution procedure is as follows.

(i) Dissolve 50 mg/ml of purified asolectin in 10 mM Tricine-NaOH buffer, pH 8.0 containing 45 mM cholate, 25 mM deoxycholate, 0.2 mM EDTA and 5 mM dithiothreitol (DTT).

(ii) Sonicate the suspension to clarity (see Section 8.2.1).

(iii) Add 4 mg of F_0F_1 to 1 ml of the liposome suspension.

(iv) Dialyse the proteoliposome mixture against 500 volumes of 10 mM Tricine-NaOH buffer, pH 8.0, 2.5 mM $MgSO_4$, 0.25 mM DTT and 0.2 mM EDTA for 24 h. At this stage F_0F_1 is reconstituted; however only F_0 is functionally competent whereas the F_1 is either partially or completely denatured.

(v) Add, therefore, an additional 1 mg of purified F_1 and incubate the mixture for 10 min at 30°C since F_1 readily denatures in the cold.

9. FUNCTIONAL ASSAYS OF RECONSTITUTED CYTOCHROME OXIDASE

9.1 **Proton pumping assay**

So far the proton pumping activity of mitochondria or reconstituted mitochondrial complexes has generally been investigated by making use of pH potentiometric measurements (34). More recently rapid mixing spectroscopic techniques have been employed as these allow study of protons released into a bulk aqueous phase by a pumping system to be measured in the millisecond time range (30). These have already been used to monitor the acidification of the external medium occurring during the oxidation of cytochrome *c* by COV in pre-steady-state experimental conditions.

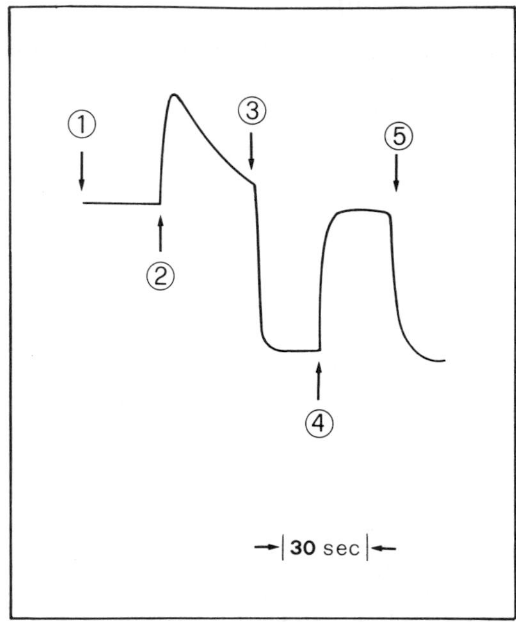

Figure 21. Potentiometric measurement of proton pumping in cytochrome *c* oxidase vesicles (COV).

9.1.1 *Potentiometric measurements*

The only special equipment needed is a thermostatted oxygen electrode cell with a fast-responding pH electrode and coupled to a millivolt chart recorder. The following method is the so called cytochrome *c* pulse technique and is typically carried out as follows.

(i) Dilute COVs to 0.5 μM (cytochrome *c* oxidase) with 50 μM Hepes-KOH, pH 7.3 43.6 mM KCl, 46.1 mM sucrose, add 5 μM valinomycin and establish a stable baseline ① in *Figure 21*).

(ii) Add 8 μM ferrocytochrome *c* (leading to each cytochrome *c* oxidase functional unit producing, on average, four molecules of water). At this stage proton extrusion is observed, provided the RCR is above 3 (② in *Figure 21*).

(iii) Add 10 μM carbonyl cyanide *m*-chlorophenyl hydrazone (CCCP). A very fast alkalinization is seen which levels off to the end point of the reaction. This is more alkaline with respect to the starting pH since protons are consumed by the scalar reaction inside the vesicle (③ in *Figure 21*).

(iv) Add 8 μM HCl (1 H^+/e^-). Since 8 μM H^+ have been consumed in the scalar reaction the addition of 8 μM H^+ should bring the pH to its initial value (④ in *Figure 21*).

(v) Add further a 8 μM ferrocytochrome *c*. Only the scalar reaction is observed. Therefore the baseline should return to the level at the end of step (iii). (⑤ in *Figure 21*).

9.1.2 *Rapid mixing methods*

Any rapid mixing device equipped with a rapid data capture system for example, oscilloscope or A/D converter plus memory, is suitable. A conventional stopped-flow

$$X = \frac{\Delta A_x \cdot 50}{\Delta A_T} = 10 \;\therefore\; 10\ \mu M\ c^{2+} \equiv 10\ \mu M\ H^+$$

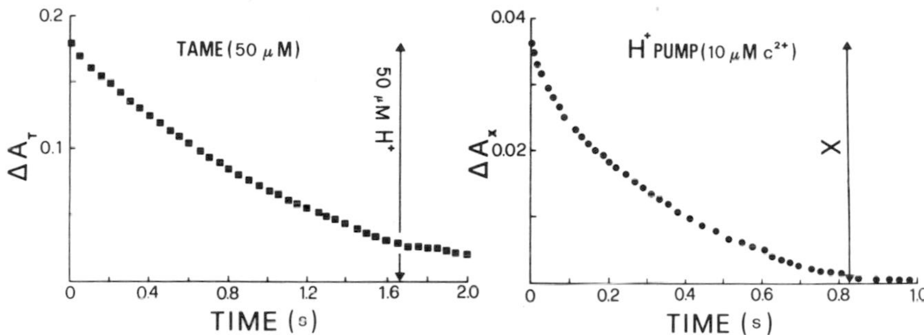

Figure 22. Photometric calibration of a proton pump signal in the stopped-flow apparatus. Both time courses refer to phenol red absorbance changes (556.7 nm) occurring when a known pulse of protons are released in the medium external to COV (**left trace**) and when protons are released during COV-catalysed cytochrome *c* oxidation (**right trace**). The simple proportionality relationship between the two signals and relative H^+ concentration is given (top). T = 20°C.

apparatus consisting of drive syringes, mixing chamber, an optical monitoring system, is ideal. However, mixing devices with dead times less than 20 msec, coupled to a conventional spectrophotometer with output to an oscilloscope, may also be used. The experiments outlined below were performed using a Durrum–Gibson stopped-flow apparatus. A typical experiment should be carried out as follows.

(i) Fill the drive syringe 1 of a stopped-flow apparatus with a COV preparation (see above) diluted 1:4 (to a final oxidase concentration of 1 μM) with a medium of a low buffer power at pH 7.3 (50 μM Hepes-KOH, pH 7.3, 46.1 mM sucrose, 43.6 mM KCl) containing 5 μM valinomycin.

(ii) Fill syringe 2 with 30 μM ferrocytochrome *c* (see Section 6.2.1) and 60 μM phenol red, both dissolved in the same buffer as contained in syringe 1 and at pH 7.3.

(iii) To calibrate the extent of acidificaton add 5 μM trypsin to syringe 1 and a known amount of *N*α-*p*-tofyl-L-arginine methyl ester (TAME; typically 100 μM) to syringe 2 (the final concentration of TAME will be 50 μM after mixing). The calibration is carried out substituting ferricytochrome *c* for its reduced form in order to eliminate any extra release of protons by the redox-linked pumping system (30).

Figure 22 shows the oscilloscope trace of the dye signal monitored at 556.7 nm. This wavelength is chosen to monitor the pH indicator as it is strictly isosbestic for the cytochrome c^{2+} to cytochrome c^{3+} transition. This is important in view of the possible large contribution of the cytochrome *c* reaction occurring synchronously with the proton release. This isosbestic wavelength should therefore be checked for the particular equipment used (thus allowing for any misalignment of the monochromator)

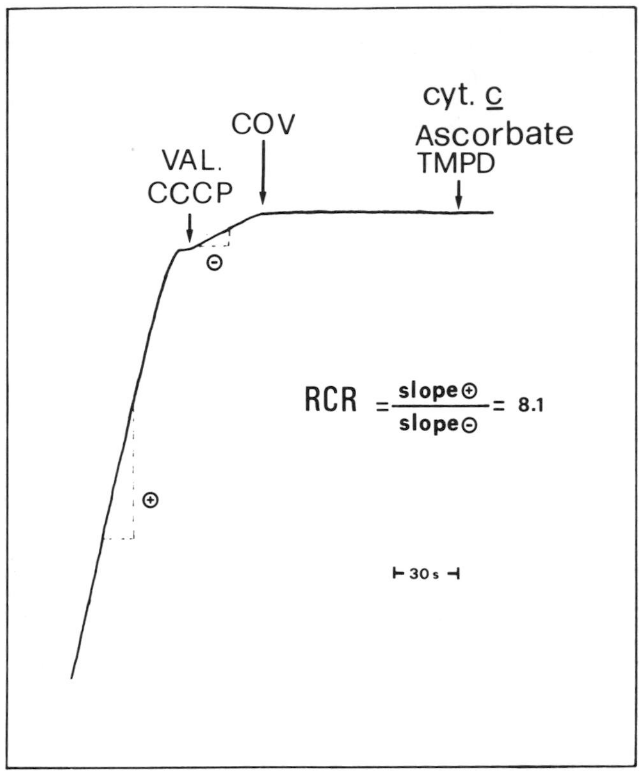

Figure 23. Oxygen electrode measurement of the RCR. Chart speed is 30 sec/cm from right to left. The arrows indicate when additions were made. T = 20°C. Each centimetre on the vertical scale represents 78 nmol 0_2.

as a preliminary to the proton pumping experiment.

The number of protons vectorially released per electron flowing through the enzyme is defined as the H^+/e^- stoichiometry. It is usually calculated by comparing the amplitude of the observed signal with the one obtained by adding a given amount of protons, for example, HCl, to the same system. The simple addition of acid, however, cannot be followed as a transient as proton re-equilibration in bulk water is diffusion limited. To overcome this problem the acidic pulse is obtained by hydrolysis of TAME catalysed by trypsin. This reaction releases one proton for each substrate molecule hydrolysed and at a rate which is determined by the enzyme (trypsin) concentration. It can be made to occur under the same experimental conditions used to follow the proton pumping reaction and therefore is easily detectable in the stopped-flow apparatus (35).

Note that when using COVs, or indeed any phospholipid vesicles as obtained using asolectin, a small external acidification always occurs on dilution. This occupies the first 50 – 100 msec, is unrelated to the proton pumping and can be subtracted from the experimental trace.

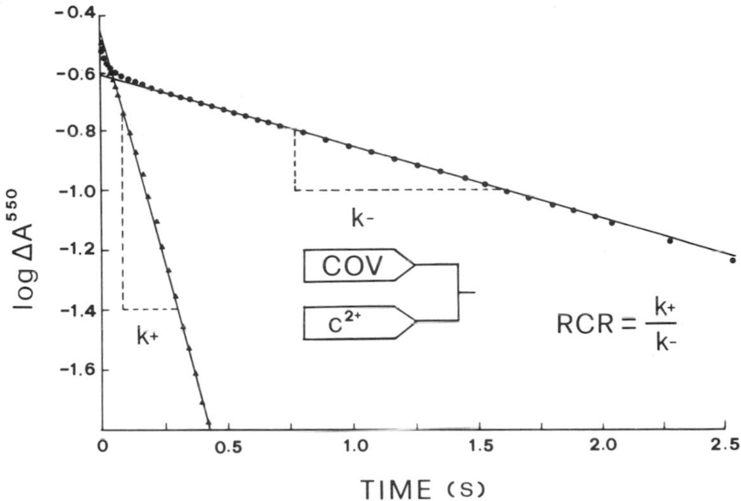

Figure 24. Semilogarithmic time courses of ferrocytochrome *c* oxidation catalysed by COV in the absence (●) and presence (▲) of ionophores as measured by stopped-flow spectroscopy. The two drive syringes **A** and **B** contain respectively: **A**: 1.25 μM COV in Hepes-KOH buffer, pH 7.3 **B**: 30 μM ferrocytochrome *c* in the same buffer. Ionophores are added to the COV-containing syringe. T = 20°C. The trace shows the extent and time course of pH changes following additions (indicated by the circled numbers). The addition sequence is identified by give steps as in Section 9.1.1. These numbers correspond to addition of (**1**) COV (**2**) ferrocytochrome *c* (**3**) CCCP (**4**) HCl (**5**) ferrocytochrome *c*. An upward or downward deflection of the pen represents an acidification or alkalinization, respectively.

9.2 Respiratory control ratio

The RCR is most commonly determined monitoring the rate of oxygen consumption by COVs in the presence of excess reducing power (see below), in the presence and absence of ionophores. Any Clark-type oxygen electrode is suitable. A typical assay is carried out as follows.

(i) Fill the electrode vessel (typically 3 ml) with 0.1 M Hepes-KOH buffer, pH 7.3 and wait for temperature equilibration (25°C).

(ii) Add 25–50 nM COVs (25–50 nM cytochrome *c* oxidase), add cytochrome *c* (20 μM), ascorbate (5 mM) and *N,N,N',N'*-tetramethyl-*p*-phenylene diamine (200 μM) and allow the system to turn over for approximately 30 sec, thereafter add 5 μM valinomycin and 10 μM CCCP. The ratio between the relative oxygen consumption rates in the presence and absence of ionophores (corrected for any drift detectable in the absence of COVs) gives the RCR (see *Figure 23*).

The RCR is also very easily measured photometrically in a stopped-flow apparatus by mixing ferrocytochrome *c* with COVs in the absence and presence of ionophores (36). The oxidation of cytochrome *c* can be monitored at 550 nm and the ratio between the pseudo first-order rate constants for oxidation in the presence and absence of ionophores gives directly the RCR (see *Figure 24*).

10. REFERENCES

1. Hames,B.D. and Rickwood,D., eds (1981) *Gel Electrophoresis of Proteins – A Practical Approach*. IRL Press, Oxford and Washington D.C.
2. Darley-Usmar,V.M. and Fuller,S.D. (1981) *FEBS Lett.*, **135**, 164.
3. Merril,C.R., Dunau,M.L. and Goldman,D. (1980) *Anal. Biochem.*, **110**, 201.
4. Ludwig,B., Downer,N. and Capaldi,R.A. (1979) *Biochemistry,* **18**, 1401.
5. Bell,R.L. and Capaldi,R.A. (1976) *Biochemistry*, **15**, 996.
6. Capaldi,R.A., Takamiya,S., Zhang,Y.-Z., Gonzales-Halphen,D. and Yanamura,W. (1986) In *Current Topics in Bioenergetics*. in press.
7. Kadenbach,B. and Merle,D. (1981) *FEBS Lett.*, **135**, 1.
8. Steffens,G.J. and Buse,G. (1976) *Hoppe-Seylers Z. Physiol. Chem.*, **357**, 1125.
9. Yu,C.A., Yu,L. and King,T.E. (1977) *Biochem. Biophys. Res. Commun.*, **74**, 670.
10. Meinecke,L., Steffens,G.J. and Buse,G. (1984) *Hoppe-Seylers Z. Physiol. Chem.*, **365**, 313.
11. Darley-Usmar,V.M., Capaldi,R.A. and Wilson,M.T. (1981) *Biochem. Biophys. Res. Commun.*, **103**, 1223.
12. Chomyn,A., Mariottini,P. and Cleeter,M.W.J. (1985) *Nature*, **314**, 592.
13. Anderson,L. (1981) *Proc. Natl. Acad. Sci. USA*, **78**, 2407.
14. Darley-Usmar,V.M. and Watanabe,M. (1985) *J. Biochem.*, **97**, 1767.
15. Swanson,R., Trus,B.L., Mandel,N., Mandel,G., Kallai,O. and Dickerson,R.P. (1977) *J. Biol. Chem.*, **252**, 759.
16. Smith,M.B., Stonehuerner,J., Ahmed,A.J., Staudenmeyer,N. and Millett,F. (1980) *Biochim. Biophys. Acta*, **592**, 303.
17. Smith,H.T., Staudenmeyer,N. and Millet,F. (1977) *Biochemistry*, **16**, 4971.
18. Ferguson-Miller,S., Brautigan,D.L. and Margoliash,E. (1978) *J. Biol. Chem.*, **253**, 149.
19. Yuan,P.M., Pande,H., Clark,B.R. and Shively,J.E. (1982) *Anal. Biochem.*, **120**, 289.
20. Canfield,R. and Liu,A.K. (1965) *J. Biol. Chem.*, **240**, 1997.
21. Margoliash,E., Ferguson-Miller,S., Tuolos,J., Kang,H.C., Feinberg,B.A., Brautigan,D.L. and Morrison,M. (1973) *Proc. Natl. Acad. Sci. USA*, **70**, 5234.
22. Ahmed,A.J., Smith,H.T., Smith,M.B. and Millett,F.S. (1978) *Biochemistry*, **17**, 2479.
23. Smith,H.T., Ahmed,A. and Millett,F. (1981) *J. Biol. Chem.*, **256**, 4984.
24. Millett,F., de Jong,C., Paulson,L. and Capaldi,R.A. (1983) *Biochemistry*, **22**, 546.
25. Nicholls,D.G. (1982) In *Bioenergetics: an Introduction to the Chemiosmotic Theory*. Academic Press, New York.
26. Casey,R.P. (1984) *Biochim. Biophys. Acta*, **768**, 319.
27. Racker,E. (1979) In *Methods in Enzymology*. Fleischer,S. and Packer,L. (eds), Academic Press, New York, Vol. 55, p. 699.
28. Kagawa,Y. and Racker,E. (1971) *J. Biol. Chem.*, **246**, 5477.
29. Wrigglesworth,J.M. (1978) *Proceedings of the 11th FEBS Meeting,* **45**, 95.
30. Sarti,P., Jones,M.G., Antonini,G., Malatesta,F., Colosimo,A., Wilson,M.T. and Brunori,M. (1985) *Proc. Natl. Acad. Sci. USA*, **82**, 4876.
31. Furth,A.J. (1980) *Anal. Biochem.*, **109**, 207.
32. Nalecz,M.J., Casey,R.P. and Azzi,A. (1983) *Biochim. Biophys. Acta*, **724**, 75.
33. Glaser,E., Norling,B. and Ernster,L. (1980) *Eur. J. Biochem.*, **110**, 225.
34. Wikstrom,M.K.F. and Saari,H.T. (1977) *Biochim. Biophys. Acta*, **462**, 347.
35. Sarti,P., Malatesta,F., Antonini,G., Colosimo,A. and Brunori,M. (1985) *Biochim. Biophys. Acta*, **809**, 39.
36. Brunori,M., Sarti,P., Colosimo,A., Antonini,G., Malatesta,F., Jones,M.G. and Wilson,M.T. (1985) *EMBO J.*, **4**, 2365.

CHAPTER 6

An enzymatic approach to the study of the Krebs tricarboxylic acid cycle

J.B.ROBINSON,JR, L.G.BRENT, B.SUMEGI and P.A.SRERE

1. INTRODUCTION

Measurements of enzyme activity in mitochondrial preparations and the use of mitochondria as sources of enzymes have been of great value in the study of metabolic pathways in mitochondria. This organelle is complex and its detailed structure incompletely understood, but recent studies on mitochondrial enzymes and their release has given us important information on mitochondrial structure and its relation to mitochondrial metabolism. The following sections will discuss mitochondrial disruption, enzyme assay and isolation of enzymes using some Krebs citric acid cycle enzymes as examples. These methods have been used by us to study interactions of Krebs cycle enzymes with each other and with the mitochondrial inner membrane (see ref. 1 for a review). Our results indicate that a complex of the Krebs tricarboxylic acid (TCA) cycle enzymes exists within the mitochondrion and that this complex (a metabolon) is an important part of the metabolic characteristics of oxidative metabolism.

2. DISRUPTION OF MITOCHONDRIA

The study of the metabolic pathways of mitochondria is usually begun by accurate measurements of the amounts of the various enzymes. The mitochondrion is a closed system with membrane barriers that are impermeable to many enzyme substrates so that the measurement of these enzymes cannot be done until the organelle is permeabilized. Additional information can be gained from knowledge of the release characteristics of the enzymes. There are a number of different methods available to enable this to be done.

Except where stated the procedures described here are for rat liver mitochondria. They have also been used for pig heart mitochondria.

2.1 Detergent treatments

The 'best' detergent would be one not deleterious to enzyme activity and not interfering with measurements of enzyme activity. The compound most often fulfilling these criteria is Triton X-100. Following closely in usefulness is octyl glucoside, which is not harmful to many enzyme activities and is removable to some extent by dialysis. Other non-ionic detergents which have been used are the Lubrol series (also trademarked as Brij), the Tergitol compounds, various sugar–alkane derivatives (e.g. dodecyl maltoside) and some newer compounds known by the acronyms of CHAPS and CHAPSO. Each has advantages and disadvantages which must be tested for each individual enzyme.

The anionic detergent deoxycholate is another compound in common use. Its use is limited since it tends to denature some enzymes, precipitate metal ions and form gels which are difficult to work with. If, however, it has been used successfully in the systems you study, you may also consider using the sulphonate derivative taurodeoxycholate, since it is usually as effective and avoids some of the undesirable properties of the parent compound.

It should be assumed that all detergents interfere with all enzyme assays until proven otherwise. Many spectrophotometric assays are done at wavelengths below 340 nm. The presence of detergents makes such assays difficult since detergent solutions scatter light and most detergents also absorb in the u.v. Enzyme assays must be linear with time and protein; this should be determined in the presence of detergent concentrations which bracket those used in your experiments.

Two more facts must be remembered in order to achieve reproducibility in detergent extractions. Firstly, aqueous solutions of detergents must be made fresh since in some cases peroxides and other harmful degradation products are formed on storage. Secondly, in designing experiments involving detergents the operative parameter is not detergent concentration, but the ratio of detergent to lipid. Since mitochondria have a constant protein to lipid ratio and protein is much more easily measured, one usually measures the ratio of detergent to protein. This is determined by measuring a marker enzyme activity and the activity of interest over a large range of protein to detergent ratios and looking for a range of ratios that give the same values, indicating that denaturation of enzyme activities or other interferences are not occurring. The best marker enzyme is citrate synthase (see below) although any relatively stable and easily assayed enzyme will do. It is best to remember that any change in temperature, ionic strength, pH or method of protein determination or enzyme assay, will change in a practical sense the ratio of detergent to protein and will also potentially change the validity of whatever standardizations you have already done.

2.1.1 *Triton treatment of mitochondria for complete disruption*

(i) Prepare a micropipette tip by cutting a 1 ml blue tip approximately 5 mm above the normal opening with scissors.

(ii) Using this tip, add 1.0 ml of Triton X-100 to 9 ml of water in a 12 × 75 mm glass test tube.

(iii) Cap the tube tightly with Parafilm and mix by gentle inversion avoiding foam formation as much as possible. The use of a rocking shaker is recommended. The final concentration of detergent is 10% (v/v). This solution should be prepared daily due to peroxide formation in dilute aqueous solutions of detergents.

(iv) Dilute mitochondria to 10 mg/ml with isotonic medium. If citrate synthase is used as a marker of disruption, use 10 μl of the mitochondrial suspension for the enzyme assay.

(v) Run a control assay (see Section 3.3) without detergent to determine damage to mitochondrial structure caused during preparative procedures.

(vi) Prepare an assay mix adding 100 μl detergent/ml, and assay the enzyme again.

The difference between the enzyme activities indicated by these two assays is the latent activity.

2.1.2 *Partial disruption of mitochondria*

(i) Dilute 1 ml of a freshly prepared solution of 10% (v/v) Triton X-100 with 9 ml of diluent. Mitochondria used in this experiment should be at a protein concentration of 100 mg/ml, and the diluent used by the authors is 220 mM mannitol, 70 mM sucrose and 20 mM Hepes–NaOH, pH 7.4, although any isotonic medium can be used.

(ii) Prepare in 1.5 ml microfuge tubes the following mixtures:

Mitochondria	Detergent	Diluent
100 μl	0 μl	900 μl
100 μl	5 μl	895 μl
100 μl	10 μl	890 μl
100 μl	20 μl	880 μl
100 μl	50 μl	850 μl
100 μl	100 μl	800 μl

which gives detergent concentrations of 0%, 0.005%, 0.01%, 0.02%, 0.05% and 0.1%, respectively. Mix by gentle inversion and centrifuge for 15 min at 30 000 *g* at 4°C.

(iii) After centrifugation, withdraw and save the supernatant solution and add 1 ml of diluent to the pellet.

(iv) Resuspend the pellet by drawing it through a no. 25 needle on a 1 ml syringe three times.

(v) Determine enzyme activities in all fractions.

The pellet fraction which usually contains 15 to 50% of total activity present (see above) is most appropriate for use in experiments concerning enzyme–enzyme and enzyme–membrane interactions. It may sometimes be necessary to prepare detergent concentrations intermediate to those above to obtain such a preparation and the detergent concentration which yields an appropriate preparation will vary slightly from preparation to preparation.

2.2 Sonication

Sonication consists of passing waves of ultrasound through the sample. **CAUTION:** during this procedure your ears are exposed to the sonic waves which can cause permanent loss of hearing. It is particularly insidious since it does not hurt and its effects are cumulative. Always wear hearing protection when using a sonicator.

Since all enzymes can be denatured irreversibly by sonication, one should begin with the lowest power setting for the shortest time of exposure possible, and measure the amount of enzyme exposed or enzyme loss that occurs in short increments of sonication. Sonication is terminated when no further increase in enzyme release occurs.

Sonication always heats a sample so the sample should be kept in an ice bath during sonication. The sonication should be applied in short bursts (no more than 30 sec) interspersed with cooling periods. Improper positioning of sonicator probes, or use of volumes too small for the machine can lead to foaming in the sample which destroys enzyme activity. Some enzymes, especially membrane-bound ones, may be destroyed by sonication no matter how carefully the procedure is carried out. If on successive treatments the enzyme activity reaches a peak and then declines rapidly, or if no activity is found, another method of disruption should be used.

2.2.1 *Disruption of mitochondria by sonication*

For complete disruption of mitochondria carry out the following procedure.

(i) Set up a Branson model 225 sonicator (max output 200 watts) with tapered microtip (probe tip diameter 3 mm), set the power to 4, and tune with the tip in 1 ml of water in a 1.5 ml microcentrifuge tube to maximum power output at this setting.

(ii) Place 1.0 ml of rat liver mitochondria in isolation medium into a 1.5 ml microfuge tube and place the tip of the sonicator into the solution.

(iii) Sonicate for 30 sec, and assay 2 μl of the suspension for citrate synthase activity (see Section 3.3.1).

(iv) Sonicate once more and assay again.

(v) Sonicate again for 15 sec and re-assay.

(vi) If the last two assays yield the same activity of citrate synthase ($\sim$0.15 change in absorbance per min), it is presumed the mitochondria are totally disrupted. If the two assays are not the same, sonicate again for 15 sec and continue the process until the last two values are equal.

2.2.2 *Preparation of easily sedimented exposed TCA cycle enzymes (2)*

To prepare mitochondria where there is still a possibility of residual native structure by sonication, the procedure is the same as in full disruption except that the sonicator (see previous section) is set at a power setting of 2 (the lowest setting in this system which gives any power output), and the initial time of sonication is reduced to 5 sec.

(i) Assay the preparation for exposed enzyme, and when half of the total enzyme (measured either by full disruption with sonication or with high levels of detergent) is obtained, stop the sonication (usually 15–30 sec total), and centrifuge the preparation at 33 000 *g* for 15 min at 4°C.

(ii) Suspend the pellet in isotonic medium by three passages through a no. 25 needle on a 1 ml syringe, determine the enzyme activities in supernatant solution and pellet.

In the case of citrate synthase, approximately equal parts of the enzyme are found in the pellet and supernatant.

2.3 **Osmotic shock**

This method is based on the principle that mitochondria, like all organelles with semi-permeable membranes, will increase in volume and then rupture when they are placed in a solution with low osmolarity. For the release of some enzymes, especially outer membrane and intermembrane space enzymes, disruption by osmotic shock usually works very well. The release of matrix enzymes by this method is not recommended since it is not always complete.

2.3.1 *Osmotic shock procedure*

Sediment the mitochondria and resuspend the resultant pellet into a large volume (to a final concentration of no more than 2–5 mg protein/ml) of cold water, or extremely dilute buffer (neutral 5 mM potassium phosphate or Hepes–KOH). The dilution of enzyme activity which is inherent to the method presents a problem, unless the enzyme assay is very sensitive or can tolerate relatively large volumes of enzyme samples. Also, some enzymes are not stable under these conditions so that assays should be performed immediately. The procedure can be repeated in cases where only a low percentage of enzyme is exposed the first time.

2.4 **Freeze–thawing and lyophilization**

These methods are treated together since they rely on the same general principle of disruption, that is, the formation of ice crystals in the organelle and concomitant disruption of membranes due to such crystallization. These are techniques which do not inactivate many enzymes.

In the freeze–thaw technique, plunge the tube containing the sample into a very cold bath, usually acetone mixed with solid carbon dioxide (although isopropanol can be used just as well), and swirl the tube in the bath until the sample is completely frozen, and then plunge the tube into water (room temperature) until it is completely thawed. The freeze–thaw cycle can be repeated several times until no further enzyme release is detected. The reason for rapid freezing and thawing is that some buffer substances tend to freeze out differentially, such that one form of the conjugate acid–base pair is frozen out first and the other later. This causes the enzyme to be subjected to extremes of pH so that unless the enzyme of interest is stable over very large pH ranges loss of activity will occur. The other common problem encountered is that some plastic laboratory ware dissolves in organic solvents while plunging glass containers into water when they are very cold can cause breaking of the tubes.

The major drawback of lyophilization is the length of time taken. The advantage is that if the enzyme of interest is stable in this procedure enzymes in very low concentration can be concentrated many-fold. Lyophilization can be carried out using the normal type of apparatus.

2.5 **Permeabilization of mitochondria using toluene**

A preparation of mitochondria permeable to small molecules (mol. wt of 1000 or less) can be made by treatment with toluene.

(i) Prepare a solution of 220 mM mannitol, 70 mM sucrose, 8.5% (w/v) polyethylene glycol (PEG, mol. wt 6000–8000) and 20 mM Hepes–KOH, pH 7.4.
(ii) Resuspend the mitochondria at 100 mg/ml in a glass container with either a glass stopper or Teflon closure. Do not use a Parafilm closure since the toluene will destroy it.
(iii) To the ice-cold suspension add 20 μl toluene/ml of suspension, close and shake gently for 10 min.

(iv) Centrifuge the suspension for 10 min at 12 500 *g* at 4°C and resuspend in the same volume of PEG-containing buffer. All enzyme assay buffers used in studying activities with this preparation should also be made 8.5% PEG since in the absence of this compound lysis of the mitochondria will occur instantly.

(v) Resuspend the pellet either by gentle homogenization using a glass–Teflon homogenizer or by repetitive pipetting through a wide tip 5 ml pipette.

2.6 **Release of enzyme activity by digitonin**

2.6.1 *Use of varying amounts of digitonin (procedure for yeast mitochondria)*

(i) Yeast spheroplasts at a protein concentration of about 20 mg/ml are used for the following steps.

(ii) Prepare a stock 1% digitonin solution in 1.35 M sorbitol, 10 mM EDTA and 10 mM Tris-HCl, pH 7.5 (buffer A) containing protease inhibitors (see Section 4.1). Add the appropriate amount of digitonin solution (to get final concentrations of 0–0.64%) to one aliquot (10 mg of spheroplast protein) and incubate for 5 min on ice. Unless you are able to obtain this specially purified, digitonin needs to be purified to make it completely water soluble (4).

(iii) Centrifuge at 3000 *g* for 10 min at 5°C. Separate the supernatant and assay for enzyme activities.

(iv) Also prepare a control sample which does not contain digitonin. Any activity measured in the control supernatant is exposed activity.

(v) The total enzyme activity is the activity assayed in the presence of 0.5% Triton X-100.

(vi) The percent of released enzyme is calculated as follows: activity assayed in supernatant/Triton X-100 activity–exposed activity.

2.6.2 *Release of enzyme activity by timed exposure to digitonin (5)*

(i) Into a 1.5 ml capacity polypropylene microcentrifuge tube add 0.4 ml of silicone oil.

(ii) Add 1 ml of buffer A containing protease inhibitors, 4 or 6.8 mg/ml digitonin and 0.2% Lubrol PX very gently on top of the oil layer so as not to disturb the oil (the digitonin solution is added so as to have 0.8 mg of spheroplasts protein).

(iii) For each experimental sample, add 0.2 ml of spheroplasts (4 mg) using an Eppendorf pipette. Rapid but gentle mixing is achieved by ejecting the spheroplasts right under the buffer layer and then pipetting up and down twice.

(iv) Incubate the sample on ice for the appropriate time (0–15 min) and then centrifuge at 8000 *g* for several seconds (8 sec) in a microcentrifuge.

(v) Carefully remove the supernatant with a Pasteur pipette. This supernatant contains the released enzyme activity.

(vi) Total activity of the enzymes is determined by adding 0.2 ml of spheroplasts to 1 ml of buffer A plus protease inhibitors containing 0.5% Triton X-100. Incubate this mixture on ice for 60 min and then centrifuge at 8000 *g* for 3 min. Assay the supernatant for total enzyme activities.

(vii) To ensure that whole cells, if present in the spheroplast preparation, do not contribute to the released enzyme activities measured, whole cells (prior to zymolyase treatment) are treated for 60 min with Triton X-100 and also for 15 min with digitonin as in the procedure above (in both cases no activity from whole cells was released).

3. ASSAY OF KREBS TRICARBOXYLIC ACID CYCLE ENZYMES

3.1 **Factors affecting the assay of enzymes**

3.1.1 *Turbidity*

Mitochondrial suspensions are quite turbid at concentrations useful for most spectrophotometric enzyme assays. This causes several problems in measurements of enzyme activities. The first is that addition of sufficient mitochondrial protein to achieve assayable amounts of the enzyme of interest will always add to the apparent absorbance due to light scattering by the organelles. Light scattering is inversely proportional to the wavelength of incident light, and increases as the fourth power of decreasing wavelength. This means that many enzyme assays which work perfectly well on purified enzymes cannot be used for measurements on mitochondria. The best protocol for spectrophotometric assays using mitochondrial preparations and fractions is to zero the machine against a water blank (or air) and measure the initial absorbance of the reaction mixture. If you are using a single-beam machine, any measurement where the absorbance of the reaction at any time goes over 1.0 will cause difficulty, and any such measurement over 1.5 is almost certainly wrong. It is better not to use any mechanical or electronic means to remove background absorbance artificially. If you are using a double-beam instrument, the limits of absorbance are 0.5 higher. If there is any doubt, add less mitochondria and re-assay the enzyme monitoring at a slower recorder speed.

There is also a problem in measuring a very slow enzyme rate which may void the use of an assay. Mitochondria will settle over a time course of minutes. This can be recognized by an apparent departure from linearity in the course of the enzyme reaction. Should you suspect this, prepare a blank omitting substrate and monitor the apparent change in absorbance (actually, of course, the change in light scattering as fewer mitochondria interact with the light beam) over the same amount of time as is required for the measurement of the enzyme. If this problem is present, it is sometimes possible to subtract this change either by use of a double-beam instrument's reference cell or arithmetically from separate measurements. There are spectrophotometers available with built-in stirrers which can be used, although in this case the possibility of mechanical disruption of the mitochondria during the assay must be considered.

3.1.2 *Interfering enzymes*

For some metabolites there are many enzymes which will act upon them. In the assay of mitochondrial enzymes there is a possibility that other reactions will utilize the substrates. Appropriate controls are needed to correct for these effects.

Mitochondrial preparations are often contaminated with lysosomes since these organelles tend to co-sediment with mitochondria. This means that the enzyme you wish to measure may be destroyed and the mitochondrial structure degraded by the action of hydrolytic enzymes released from the degraded lysosomes. Enzymes vary greatly in susceptibility to proteolytic attack and in extreme cases there are only a few minutes after disruption of mitochondrial fractions when some enzymes can be measured accurately. To estimate the degree of error introduced in your measurements, repetitive assays, spaced an hour or so apart, should be performed and if a progressive loss of activity is observed, then always measure the enzyme activity as quickly as possible. Addition of proteolytic enzyme inhibitors may be of use (see Section 4).

3.2 **General spectrophotometric techniques**

Spectrophotometric assays are usually performed in quartz cuvettes with a 1 cm light path. Care should be taken to ensure that the outsides of the cuvettes are clean and that you do not touch the optical faces. The cuvettes are kept in a thermostatted compartment at the required temperature. Mixing may be done either by inversion of the cuvette which is covered by Parafilm or by using a plastic spoon after the last addition; stirring is a less desirable option since it may damage any intact mitochondria present.

3.3 **Assays of individual mitochondrial enzymes**

3.3.1 *Citrate synthase (EC 4.1.3.7) (6)*

In all higher eukaryotic organisms, this is the best marker enzyme for mitochondria and, more specifically, the matrix fraction. The assay is rapid, specific, sensitive and easy, and the enzyme is stable for at least 24 h even in relatively impure preparations of mitochondria. Although the enzyme, like other Krebs citric acid cycle enzymes, is apparently bound to the inner membrane *in situ*, most disruption techniques now in use solubilize this enzyme. In cells which do not contain the glyoxylate shunt, the enzyme is located exclusively in mitochondria.

(i) *Reagents.* Prepare the following solutions: 13.2 mg of sodium oxaloacetate (OAA) in 10 ml of water (prepare daily); 4 mg of 5,5′-dithio-bis-(2-nitrobenzoic) acid (also known as DTNB, or Ellman's reagent) in 10 ml of 1.0 M Tris-HCl, pH 8.1 (prepare daily also); acetyl-coenzyme A 10 mg/ml in water acidified to pH 5. Alternatively, acetyl-CoA can be synthesized by dissolving the lithium salt of coenzyme A in 0.1 M Tris base to a concentration of 10 mg/ml and adding 13 μl of acetic anhydride. The solution is mixed gently and kept at 4°C overnight after which it may be used. Either acetyl-CoA preparation is stable for months when frozen and for several days when refrigerated.

(ii) *Assay.* Set a spectrophotometer with recorder at a full scale reading of 0−1 absorbance at 412 nm and set the recorder speed to 2 cm/min (or 1 inch/min). To a 1 ml cuvette (plastic, glass or quartz) add 810 μl of water, 100 μl of DTNB solution and 30 μl of acetyl-CoA solution. Then add 10 μl of mitochondrial suspension containing 0.02− 0.2 mg of mitochondrial protein, mix, put into the spectrophotometer and turn on the recorder. Most mitochondrial preparations contain acetyl-CoA hydrolase activity so this activity must be determined first for final calculations of citrate synthase activity. After 2 min, add 50 μl of the OAA solution and mix as before. Continue recording the absorbance (the absorbance should increase) for another 2−3 min or until the absorbance exceeds 1.0.

To calculate the enzyme activity, calculate the change in absorbance for the last minute of the reaction before addition of OAA and subtract this value (which is acetyl-CoA hydrolase activity) from the change in absorbance of the first minute of linear change in absorbance after addition of OAA (there may be a small lag of 15 sec or so before linearity is achieved). Divide this quantity by 13.6, and the resultant number represents the μmol coenzyme A produced per minute by the amount of enzyme added.

Reaction rates below 0.005 and above 0.15 change in absorbance/min are not desirable, and enzyme amounts should be chosen such that the rate falls in this range (numbers are after correction for hydrolase activity). For almost all mammalian mitochondria the osmolarity of the citrate synthase reaction mixture is sufficient to maintain mitochondrial integrity so the assay is useful in determination of damage done to mitochondria during preparation. No citrate synthase activity can be measured in a preparation of completely intact mitochondria.

3.3.2 *Aconitase (EC 4.2.1.3) (7)*

Aconitase exists both in the cytosol and mitochondria so that it is not a good marker for mitochondrial purity. The nature of the assay (measuring absorbance change at 240 nm) makes detergent addition impossible since detergent scatters light at this wavelength.

(i) *Reagents.* Prepare the following solutions: sodium chloride at 29 mg/ml (2.9 g/100 ml, 0.5 M stock); 17.4 mg of cis-aconitic acid in 10 ml of water, then adjusted to pH 7.2; 1.0 M Tris-acetate pH 7.2.

(ii) *Assay.* Set up a recording spectrophotometer at 240 nm, and set the recorder to a slow (0.5 cm/min) speed. Prepare in a quartz cuvette 755 μl of water, 20 μl of 1 M Tris-HCl buffer pH 7.4, 100 μl of sodium chloride solution and 100 μl of cis-aconitate. Mix by inversion and add 25 μl of mitochondria (~0.05−0.1 mg protein) and record the change in absorbance. The absorbance should decrease. The change in absorbance/min divided by 4.28 gives μmol cis-aconitate used/min.

3.3.3 *Isocitrate dehydrogenase (NAD^+-dependent) (EC 1.1.1.41) (8)*

(i) *Assay method.* Enzyme activity is determined by following the formation of NADH as measured by the increase in absorbance at 340 nm.

(ii) *Reagents.* The following reagents are used: 1.0 M Tris-acetate buffer, pH 7.2; 20 mM $MnCl_2$; 10 mM ADP; 10 mM NAD^+; and 20 mM threo-D_s-isocitrate.

(iii) *Procedure.* The enzyme activity is measured in a 1 ml volume at 340 nm. Add 0.7 ml of distilled water, 0.1 ml of Tris-acetate buffer, 0.05 ml of $MnCl_2$, 0.05 ml of ADP, 0.05 ml of NAD^+ and 0.05 ml of isocitrate into the cuvette and mix with a plastic spoon. Add the enzyme to the cuvette, mix and measure the absorbance change at 340 nm with a recording spectrophotometer. The enzyme activity is determined from the initial linear portion of the curve. Phosphate precipitates the manganese ions which cause an apparent increase in the absorbance at 340 nm, therefore phosphate buffer should not be added to the assay mixture. It is important to note that the yeast enzyme is activated with AMP, not with ADP, therefore the assay mixture of yeast enzyme should contain 0.5 mM AMP instead of ADP.

3.3.4 *Isocitrate dehydrogenase (NADP$^+$-specific) (EC 1.1.1.42) (9)*

(i) *Assay method.* The enzyme activity is determined by the formation of NADPH by detecting the increase in absorbance at 340 nm.

(ii) *Reagents.* The following reagents are used: 1.0 M Tris-HCl buffer, pH 7.4, containing 10 mM EDTA; 6 mM dithiothreitol (freshly made); 20 mM $MnSO_4$; 5 mM $NADP^+$; and 80 mM threo-D_sL_s-isocitrate.

(iii) *Procedure.* Put into a cuvette 780 µl of distilled water, 50 µl of Tris-HCl buffer, 50 µl of dithiothreitol, 50 µl of $MnSO_4$, 50 µl of $NADP^+$ and 20 µl of threo-D_sL_s-isocitrate. The reaction is started by the addition of enzyme. The enzyme activity is determined from the initial linear part of the curve. Remember: do not use phosphate buffer because it precipitates manganese ions causing an apparent increase in the absorbance.

3.3.5 *α-Ketoglutarate dehydrogenase complex (10)*

(i) *Assay method.* The α-ketoglutarate dehydrogenase complex catalyses the oxidative decarboxylation of α-ketoglutaric acid in the presence of NAD^+ and CoA-SH to succinyl-S-CoA and NADH. Since NADH is formed by the reaction, the enzyme activity can be determined from the change in absorbance at 340 nm.

(ii) *Reagents.* The following reagents are used: 0.5 M potassium phosphate buffer, pH 7.8; 2 mM CoA-SH; 20 mM cysteine-HCl; 4 mM thiamine pyrophosphate; 10 mM NAD^+; 40 mM α-ketoglutaric acid; and 20 mM $MgCl_2$.

(iii) *Procedure.* The enzyme activity is measured by the increase in absorption at 340 nm when incubated at 30°C. Into a cuvette put 600 µl of distilled water, 100 µl of phosphate buffer, 50 µl of CoA-SH, 50 µl of cysteine, 50 µl of thiamine pyrophosphate, 50 µl of NAD^+, 50 µl of α-ketoglutaric acid and 50 µl of $MgCl_2$. Mix and wait until thermal equilibration is reached. Start the reaction by addition of enzyme and determine the enzyme activity from the initial, linear part, of the curve. CoA-SH and α-ketoglutaric acid solutions always have to be fresh.

3.3.6 *Succinate thiokinase (mammalian) (EC 6.2.1.4) (11)*

(i) *Assay method.* The enzyme is generally assayed in the reverse direction because the stoichiometric formation of GDP from GTP makes it possible to couple this enzyme with pyruvate kinase and lactate dehydrogenase so the enzyme activity can be determined from the disappearance of NADH in the lactate dehydrogenase reaction (Equations 1–3).

$$\text{GDP} + \text{succinyl-S-CoA} + \text{P}_\text{i} \xleftarrow{\text{STK}} \text{GTP} + \text{CoA} + \text{succinate} \quad (1)$$
$$\text{GDP} + \text{phosphoenolpyruvate} \xrightarrow{\text{PK}} \text{pyruvate} + \text{GTP} \quad (2)$$
$$\text{Pyruvate} + \text{NADH} + \text{H}^+ \xrightarrow{\text{LDH}} \text{Lactate} + \text{NAD}^+ \quad (3)$$

(Arrows show the direction of the reaction under the assay condition, otherwise the reactions 1 and 3 are reversible.) The rate of formation of GDP in the first reaction is determined by the absorbance decrease at 340 nm due to the disappearance of NADH.

(ii) *Reagents.* The following reagents are used. 0.2 M $MgCl_2$ in 2.0 M KCl; 31 mM phosphoenolpyruvate; 4 mM NADH; 0.1 M Tris-succinate, pH 7.4; 1 mM GTP; 1 mM CoA; 2 mg/ml pyruvate kinase enzyme in 0.1 M Tris-acetate, pH 7.2; and 2 mg/ml lactate dehydrogenase in 0.1 M Tris-acetate, pH 7.2.

(iii) *Procedure.* Add to a cuvette 0.5 ml of Tris-acetate buffer, 0.1 ml each of CoA and GTP, 50 μl of $MgCl_2$, 50 μl of phosphoenolpyruvate, 50 μl of NADH, 50 μl of pyruvate kinase, 50 μl of lactate dehydrogenase in that order. Wait until thermal equilibrium at 30°C is reached and start the reaction by the addition of succinate thiokinase. Always use a blank cuvette that contains everything except CoA-SH so that you can determine the possible GTP hydrolysis and NADH oxidase which may be present. Sometimes there is transient phase in the beginning of the reaction, therefore wait until the linear rate is reached and determine the enzyme activity from this rate. Caution: all solutions have to be fresh except the $MgCl_2$ and buffer. GTP is extremely labile so use only fresh solutions.

3.3.7 *Fumarase (EC 4.2.1.2) (12)*

The assay for this enzyme is similar to aconitase in that the reaction is monitored directly by absorption in the u.v.

(i) *Reagents.* Prepare the following solutions: a stock solution of 0.5 M L-malic acid in water containing 65 mg/ml and adjusted to pH 7.4 with KOH; 0.1 M potassium phosphate buffer, pH 7.4. Both solutions are stable for several weeks if refrigerated.

(ii) *Assay.* Set up the spectrophotometer as for aconitase (see Section 3.3.2) except monitor the absorption at 250 nm. Add 890 μl of buffer, 100 μl of malate and 10 μl of the enzyme preparation and mix. Insert into the spectrophotometer and follow the reaction for 5 min. The absorbance should increase. The μmol of fumarate produced/min can be obtained for the sample added by dividing the change in absorbance/min by 4.88.

3.3.8 *Malate dehydrogenase (EC 1.1.1.37) (13)*

Although not as useful as a mitochondrial marker enzyme because cells contain a cytosolic isozyme, malate dehydrogenase is a very important and easily assayed mitochondrial enzyme. Its activity, measured by the conversion of OAA to malate, is high in disrupted mitochondria so that dilution is usually required to assay this enzyme accurately.

(ii) *Reagents.* Prepare the following solutions: 20 mg of NADH in 2 ml of 0.1 M potassium phosphate buffer, pH 7.4; 16 mg sodium oxaloacetate (OAA) in 5 ml of the same buffer.

(ii) *Assay.* Set up a recording spectrophotometer at 340 nm, and set the chart speed for 5 cm/min (or 2 inches/min). Add to a glass or quartz cuvette 960 μl of 0.1 M potassium phosphate buffer, pH 7.4, 10 μl of NADH solution and 10 μl of mitochondrial extracts containing 0.01–0.05 mg of protein. Mix and record the change in absorbance for about 1 min. Add 20 μl of OAA solution and record the absorbance (it should decrease) for another 2 min or until the absorbance falls below 0.2. For total activity, add 50 μl of 10% Triton X-100 (v/v in water, freshly prepared) to the assay before addition of mitochondria.

For the calculation, determine the change in absorbance/min before the addition of OAA (due to NADH dehydrogenase) and subtract this value from the change in absorbance/min after addition of OAA and divide the difference by 6.22 to find the μmol NADH utilized/min/portion of protein added. There may be a large decrease in absorbance with no subsequent apparent change in absorbance after addition of the final substrate. This is ordinarily due to the fact that the enzyme is in such abundance in mitochondria that the reaction is over (i.e. the NADH is completely oxidized) before insertion of the cuvette into the spectrophotometer. Should this occur, do not assume that there is no enzyme in the sample until you have diluted the sample 10-fold and re-assayed its activity.

3.3.9 *Pyruvate dehydrogenase complex (14)*

(i) *Assay method.* The pyruvate dehydrogenase complex catalyses the oxidative decarboxylation of pyruvate in the presence of NAD^+ and CoA-SH to acetyl-S-CoA, CO_2 and NADH, and so the enzyme activity can be determined from the NADH formation by measuring the increase in absorbance at 340 nm.

(ii) *Reagents.* The following reagents are used: 0.5 M Tris-HCl, pH 7.8; 20 mM $MgCl_2$; 4 mM thiamine pyrophosphate; 40 mM pyruvate (sodium or potassium salt); 20 mM cysteine-HCl; 2 mM CoA-SH; and 10 mM NAD^+.

(iii) *Procedure.* The enzyme activity is measured in a 1 ml quartz cuvette (1 cm light path) at 340 nm which is thermostatted in the cuvette holder of the spectrophotometer at 30°C. Add to a cuvette 600 μl of distilled water, 100 μl of Tris-HCl buffer, 50 μl of $MgCl_2$, 50 μl of thiamine pyrophosphate, 50 μl of pyruvate, 50 μl of cysteine, 50 μl of CoA-SH and 50 μl of NAD^+. Mix and keep the cuvette in the thermostatted cuvette holder until it reaches thermal equilibrium. Start the reaction with the addition of enzyme and determine the enzyme activity from the initial linear part of the curve. If the reaction has a lag phase, then the enzyme activity has to be determined from the linear part of the curve. If the enzyme is assayed from a crude tissue extract, then a blank cuvette has to be used which contains everything (substrate, enzyme) except pyruvate. If the tissue extract contains lactate dehydrogenase, then this assay system cannot be used because the NADH formed in this reaction is used up by lactate dehydrogenase.

4. ISOLATION OF MITOCHONDRIAL ENZYMES

Since mitochondria represent on average about 15% of the protein in a mammalian cell, preparation of relatively pure mitochondria should remove 85% of the cellular proteins which otherwise would contaminate the enzyme preparation. Unfortunately, no mitochondrial preparation yields 100% of the mitochondria originally present in the tissue, the average yield being in the range of 30–50% depending on the method of purification. In general, a decision must be made as to whether it is better to start with about half the enzyme originally present in tissue to obtain a 5-fold increase in purity or take all the enzyme present in a more contaminated form. This must be decided on a case-by-case basis, but it is usually better to isolate the mitochondria and accept the lower quantity of enzyme. It is usually quite easy to scale up a mitochondrial preparation starting with larger amounts of tissue to overcome the loss inherent in isolating mitochondria. Should you be limited in the availability of starting material, this option may not be open to you. Where cytosolic isozymes exist, it may be necessary to start with mitochondria.

The second question specific to the purification of mitochondrial proteins is the choice of disruption protocol. The individual techniques have been discussed earlier (Section 2.1), but it should not be assumed that the technique of choice for determination of enzyme activity in mitochondria is the best one for its purification. Those techniques which depend on physical methods are generally best since the presence of detergents is always a problem in the purification of proteins. Additionally, other than in extreme cases, the question of enzyme stability is not always of first importance when determining enzyme levels in mitochondria, whereas for purification, stability is of primary importance.

Enzyme stability in purification of mitochondrial proteins is a function of many factors which include buffer composition and concentration, pH, presence or absence of substrates (including metal ions, especially if used as co-factors in the reaction of interest) and protein concentration in the extract and during each step of the purification. An additional factor is the presence of proteases and susceptibility of the protein of interest to such proteases which are present in all biological extracts. Also, a change in any of these factors can potentially change the response to one or all of the others.

4.1 An overview of protein purification (15)

Purification of soluble enzymes can usually be accomplished using the following four procedures: solubility separations; ion-exchange chromatography; gel filtration (also known as molecular sieving); and affinity chromatography. Classically these are applied to purification in the order given whether or not such a sequence is particularly appropriate.

Ammonium sulphate or PEG are the reagents of choice for the initial step since they accomplish purification and concentration of enzymes. The concentration of the enzyme (obtaining the activity in a small volume) determines the priority of using these reagents since one rarely gets a high degree of purification with them.

Ammonium sulphate fractionation consists of adding a measured amount of the salt to your enzyme solution allowing time for the salt to dissolve and centrifuging the resultant mixture. Ammonium sulphate concentration is expressed as percent of saturation of ammonium sulphate, that is at room temperature 76.7 g salt/100 ml is 100%, not 100 g/100 ml. This is temperature dependent so that tables of percent saturation versus grams of ammonium sulphate to be added are useful only at the designated temperature, usually either room temperature or 4°C. The rate at which the salt is added is seldom specified but should be between 1 and 5 min. Simply adding the salt all at once leads to large concentration gradients of salt which can cause undesired precipitation. Addition of moderate portions with gentle stirring over a few minutes is usually sufficient, making sure the salt is dissolved before centrifugation. The smallest ammonium sulphate 'cut' feasible in terms of efficient fractionation is 10%, that is 25–35% (of saturation) and large cuts (25%) seldom achieve any purification.

PEG precipitation is a useful substitute for conventional salt purification. It normally gives good yields and has the advantage that since the polymer is uncharged the redissolved precipitate can be directly applied to ion-exchange or affinity columns. Disadvantages of this technique are that the polymer is fairly slow to dissolve and that solutions above 30% are viscous and require ultracentrifugation to obtain quantitative collection of precipitate although most proteins are precipitated at concentrations less than 20% (g per 100 ml).

To circumvent the problem of the slowly dissolving polymer, PEG purifications are usually done by the addition of stock solutions. This is done by the simple ratio-proportionality:

$$V2 = \frac{V1\ (C2 - C1)}{C0 - C1}$$

where V2 is the amount of stock PEG solution to be added, V1 is the volume of the enzyme solution, C0 is the decimal equivalent of the concentration of the stock PEG solution (50 g/100 ml should be 0.5), C1 is the decimal equivalent of the concentration of PEG in the enzyme solution (if any), and C2 is the decimal equivalent of the desired PEG concentration to be obtained. This same equation can be used with a saturated solution of ammonium sulphate by expressing percent of saturation as the decimal equivalents, but only if the saturated salt solution and enzyme solution are at the same temperature.

Most proteins are in the molecular weight range of 50 000–150 000. If your protein is outside this range, gel filtration, with its separation based on size, will be most effective in separating the protein from the rest of the species present. If it is in this range, ion-exchange will probably give better results since there is a broader spread of differences in charge among proteins. This distinction is important because the effectiveness of subsequent steps is enhanced by the removal of the most contaminants by the first column you run.

If you do not know the molecular weight of your protein, do ion-exchange first. Separation of proteins by use of ion-exchange columns is based on the charge of your protein. If the enzyme has an isoelectric point of 7.5 or less, use a cationic exchanger, such as DEAE (diethylaminoethyl) cellulose. If the pI is above 7.5, use an anionic exchanger such as CM (carboxymethyl) cellulose. If you do not know the charge on your protein, use the cationic exchanger first and see if the protein is retained by the column. An ion-exchange column should be short and fat with a length–diameter ratio of no more than 5. This gives best flow-rates and most efficient use of the column material.

Proteins are removed from ion-exchange columns by increasing the ionic strength of the eluting solution. Other than the individual preference of your protein, it does not matter which salt is used.

(i) Gradients of increasing salt concentration are done by trickling a high salt concentration solution into a lower salt concentration solution with stirring, and putting the resultant solution continuously onto the column. This is done using a gradient maker; there are many good commercial ones available.
(ii) Batch elution; where you add solutions of increasing ionic strength in bulk to the top of the column. These should be about 1.5 times your total column volume. If you do not know the concentration of salt which is needed to remove your protein from the column, use batch elution.

Gel filtration columns should have a height to radius ratio greater than 10, and 20 is better for most materials used for protein purification. The volume loaded onto a gel filtration column is important and should really never exceed 5% of column volume, and less is better.

Affinity chromatography is very useful for purifying specific enzymes. Since the ligand used on the solid support in affinity chromatography is a substrate or effector of the enzyme, there are always various hydrolytic enzymes or other enzyme activities which may react in some way with the bound ligand. This technique therefore is usually used at the end of most preparations when the interfering organelles that could destroy the ligand have already been removed. However, it is usually more efficient to insert this technique at the start of a preparation since the specificity of the technique is so great and then throw away the column after one use.

There are many good commercial preparations of most affinity columns you will need. The type of attachment of the ligand to the column is important since, for instance, ATP bound to a column with a hexyl group spacer will have different binding properties than ATP bound by a succinylimido-group spacer. There is no way to predict which will be most effective for purifying any given enzyme.

Enzymes bound to affinity columns are removed either by passing the free ligand through the column (e.g. ATP for an ATP column), or by increasing ionic strength. Elution with the substrate is not necessarily better than increasing ionic strength, especially if co-factors are shared among large numbers of enzymes. On ATP, NAD^+ and other co-factor based columns, elution of the bound enzyme by increasing salt concentration will generally give better results than elution with the substrate.

An alternative to conventional affinity chromatography is artificial ligand chromatography. These include the various organic dyes which interact with enzymes and have the advantage of being non-biodegradable. The most common ligand in use is the blue dye column (reactive blue 2, also known as Cibacron blue). These columns bind almost all nucleotide-binding enzymes, but are not necessarily restricted to these classes. No matter what the enzyme, they are almost always worth a try (see ref. 16 for lists of enzymes known to bind here, and refs 17 and 18 for some specialized techniques).

Two other techniques used routinely in enzyme purification are dialysis and enzyme concentration. In the dialysis procedure, the enzyme solution is placed into a bag of semi-permeable material and the bag is placed into a buffer solution that you want the enzyme to be in. Tubing should be soaked for about 30 min in a dilute (1 mM) solution of EDTA (to remove metal ion impurities) and glycerol to make the tubing supple. The tubing is then filled with water while pinching the ends with your fingers and checked closely for leaks.

Should there be a high salt concentration in your enzyme solution (re-dissolved ammonium sulphate pellets fall into this category), leave space in the dialysis bag for volume expansion. It is possible to explode a dialysis bag when the concentration inside is much higher in ionic strength than that outside the sac.

Concentration of enzyme solution can be done by covering the dialysis bag containing the enzyme with solid PEG or any dried gel filtration material. Alternatively, addition of PEG or glycerol to the dialysis buffer will also lead to concentration of the enzyme. There are also commercial concentrators such as the Amicon Diaflo system which work effectively.

Various methods of protein determination yield different results. However, any method which yields consistent results can be used, with absorbance of the enzyme solution at 280 nm probably being the most convenient. To determine whether a purification procedure is effective, one needs to know the increase in specific enzyme activity (rate of reaction/mg protein) and the yield of activity. As long as the enzyme assay and protein estimation procedure are done consistently before and after the purification step, this calculation can be done. The choice of enzyme assay, if available, should be made on the basis of speed when purifying the protein.

Usually enzymes are purified in the cold and with the addition of the heavy metal chelator EDTA in all buffers to inhibit proteolysis and increase enzyme stability. Proteolytic inhibitors have been used to protect enzymes from endogeneous proteases. These include: benzamidine hydrochloride (stable in water); phenylmethanesulphonyl chloride (unstable in water — make up fresh in ethanol and add directly to the enzyme solution in small volumes); tosyllysine chloroketone and tosylphenyl chloroketone (stable in water). For each of these first four inhibitors try using an initial concentration of 0.5 mM. The arginal peptides pepstatin and leupeptin (stable in water) can also be used at around 10 μg/ml. These compounds are all toxic, so use normal precautions for handling them.

There are several other methods used in protein purification such as extraction with organic solvents, thiol reagent and lectin columns and various permutations around preparative isoelectric focusing.

4.2 Examples of enzyme isolation: isolation of citrate synthase, malate dehydrogenase and fumarase from yeast mitochondria

As an example of the isolation of mitochondrial enzymes, we describe the isolation of citrate synthase, fumarase and malate dehydrogenase from yeast. Mitochondria were isolated according to the method of Daum *et al.* (19).

(i) Dilute the mitochondria suspended in solution (150 mM KCl, 2 mM Hepes, 0.5 M 2-mercaptoethanol, pH 7.0) with 5 vol of 10 mM Tris-HCl, pH 7.4, to a final mannitol concentration of 0.1 M. Stir gently at 0°C for 20 min. Centrifuge at 32 000 *g* for 20 min. The supernatant is the intramembrane space, and the pellet consists of 'shocked' mitochondria.

(ii) Resuspend the 'shocked' mitochondria in 10 mM Tris-HCl, pH 7.4 to a protein concentration of about 2 mg/ml using five strokes of the pestle of a loose-fitting glass–Teflon homogenizer and incubate for 5 min on ice.

(iii) After the 5 min, add 1/3 the suspension volume of 1.8 M sucrose, 8 mM ATP,

8 mM $MgCl_2$, pH 7.4. Homogenize with three strokes and incubate for 5 min on ice. Sonicate the suspension at the maximal output one can without causing foaming and splattering for three periods of 5 sec with incubation on ice between sonications and during sonication (we found that a plastic gradient tube cut about 2/3 of the way down serves as a nice sonication vessel).

(iv) Centrifuge the above sonicated solution for 60 min at 100 000 *g* at 5°C.

(v) The pellet resulting from step (iv) consists of the mitochondrial membranes, and the supernatant is the matrix fraction.

(vi) The purification of the enzymes is usually done the day after the isolation of the mitochondria so the matrix fraction is stored overnight in an ice bucket in a 4°C refrigerator.

(vii) Pool the mitochondrial fractions of the enzymes (intermembrane space plus the matrix) and concentrate by ultrafiltration using a PM30 membrane (Amicon). It is possible to go from about 50 ml of solution down to about 5 ml, and then dilute the solution back up to 50 ml with 10 mM Tris-acetate, 0.5 mM 2-mercaptoethanol, pH 7.4 (buffer A) and concentrate again to about 5–10 ml.

(viii) Dilute the final concentrate about 2-fold so that its conductivity measurement is equal to that of buffer A.

(ix) Load the concentrate onto a blue dye/agarose column (~4 ml of gel) pre-equilibrated with buffer A at 15°C (at each step remove aliquots for enzyme assays in order to calculate yields and specific activities at each step).

(x) After all the concentrate is loaded and has entered the column, wash the column with at least two column volumes of buffer A.

(xi) Elute the citrate synthase with 0.5 mM CoA and 0.5 mM OAA in buffer A containing 0.25 M KCl. Collect 1 ml fractions. Pool all tubes containing citrate synthase activity for further purification.

(xii) Wash the blue/agarose column well with buffer A (until the conductivity is the same as the buffer) and elute fumarase with buffer A containing 10 mM L-malate and 0.25 M KCl in the same manner as in step (xi).

(xiii) Wash the column with buffer A and then with 10 mM Tris-HCl, 0.5 mM 2-mercaptoethanol, pH 8.6 (buffer B). Elute the mitochondrial malate dehydrogenase with 5.5 mM NADH and 5.5 mM D-malate in buffer B containing 0.25 M KCl. Collect 1 ml fractions and pool only those tubes containing the highest activity.

(xiv) For the further purification of citrate synthase the breakthrough solution of the blue dye/agarose column from step (x) (it contains ~50% of the original citrate synthase activity) can be loaded onto an agarose–NAD^+ Type I (2 ml) column at room temperature pre-equilibrated with buffer A.

(xv) Wash the column with at least two column volumes of buffer A and elute the citrate synthase with 0.5 mM CoA and 0.5 mM OAA in buffer A. Collect 1 ml fractions.

(xvi) Remove the citrate synthase eluted from the blue dye/agarose column in step (xi) by ultrafiltration. It can then be further purified on the agarose–NAD^+ Type I column as in step (xiv).

5. ACKNOWLEDGEMENTS

The new methods used in our laboratory have been developed with the assistance of many of our colleagues. We are indebted to Ms Penny Perkins for secretarial assistance. Support of the research was by the Veterans Administration, USPHS, NSF and the Welch Foundation.

6. REFERENCES

1. Srere,P.A. (1985) *Organized Multienzyme Systems.* Welch,G.R. (ed.), p. 1.
2. Robinson,J.B. and Srere,P.A. (1985) *J. Biol. Chem.,* **260**, 10800.
3. Matlib,M.A., Shannon,W.A., Jr. and Srere,P.A. (1977) *Arch. Biochem. Biophys.,* **178**, 396.
4. Quistorff,B., Grunnet,N. and Cornell,N.W. (1985) *Biochem. J.,* **226**, 289.
5. Janski,A.M. and Cornell,N.W. (1980) *Biochem. J.,* **186**, 423.
6. Srere,P.A. (1969) In *Methods in Enzymology,* Lowenstein,J.M. (eds), Academic Press, NY, Vol. 13, p. 7.
7. Fansler,B. and Lowenstein,J.M. (1969) In *Methods in Enzymology,* Lowenstein,J.M. (eds), Academic Press, NY, Vol. 13, p. 26.
8. Plaut,W.E. (1969) In *Methods in Enzymology,* Lowenstein,J.M. (eds), Academic Press, NY, Vol. 13, p. 34.
9. Cleland,W.W., Thompson,V.W. and Barden,R.E. (1969) In *Methods in Enzymology,* Lowenstein,J.M. (eds), Academic Press, NY, Vol. 13, p. 30.
10. Porpaczy,Z., Sumegi,B. and Alkonyi,I. (1983) *Biochim. Biophys. Acta,* **749**, 172.
11. Cha,S. and Parks,R.E., Jr. (1969) *J. Biol. Chem.,* **239**, 1961.
12. Hill,R.L. and Bradshaw,R.A. (1969) In *Methods in Enzymology,* Lowenstein,J.M. (eds), Academic Press, NY, Vol. 13, p. 91.
13. Kitto,G.B. (1969) In *Methods in Enzymology,* Lowenstein,J.M. (eds), Academic Press, NY, Vol. 13, p. 106.
14. Sumegi,B. and Alkonyi,I. (1983) *Eur. J. Biochem.,* **136**, 347.
15. Scopes,R. (1982) *Protein Purification.* Springer-Verlag.
16. Dean,P.D.G. and Watson,D.H. (1979) *J. Chromatogr.,* **165**, 301.
17. Robinson,J.B., Strottmann,J.M. and Stellwagen,E. (1981) *Proc. Natl. Acad. Sci. USA,* **78**, 2287.
18. Robinson,J.B., Strottmann,J.M., Wick,P.G. and Stellwagen,E. (1980) *Proc. Natl. Acad. Sci. USA,* **77**, 5847.
19. Daum,G., Bohni,P.C. and Schatz,G. (1982) *J. Biol. Chem.,* **257**, 13028.

CHAPTER 7

Methods for studying the genetics of mitochondria

W.W.HAUSWIRTH, L.O.LIM, B.DUJON and G.TURNER

1. INTRODUCTION TO THE GENETICS OF MITOCHONDRIA FROM ANIMAL AND PLANT CELLS

Efficient techniques for isolating and analysing the DNA genomes of mammalian and higher plant mitochondria have led to rapid advances in our understanding as to how these organelles function in cellular metabolism. As with many other genetic systems, the ability to clone either the entire mitochondrial genome intact (in the case of most mammals) or significant portions of the genome (in the case of plants) has provided a plentiful and convenient supply of mitochondrial DNA (mtDNA) for many purposes, including genomic sequencing and, as described in Chapter 8, *in vitro* analyses of mtDNA replication and transcription. In the first part of this chapter the current protocols used for the isolation and purification of mtDNA from mammalian tissue and from the tissues of higher plants, particularly the monocotyledons are described. The potential of tissue culture for broader genetic manipulation demands an inclusion of similar protocols applicable to tissue culture systems in both cases. Because an analysis of mitochondrial genetic function is not complete without a thorough understanding of the transcriptional process, a selection of the procedures used for isolating mitochondrial RNA (mtRNA) are included. Finally, in view of the fact that mitochondrial genomes from mammals and higher plants represent the two extremes of mtDNA size and gene organization, separate protocols for analysis of each system are presented.

2. ISOLATION OF mtDNA FROM MAMMALIAN TISSUE AND TISSUE CULTURE CELLS

The mitochondrial genomes of animals in general, and mammals in particular, represent a very conservative use of DNA sequences to encode mitochondrial proteins and RNA. Within the approximately 16 000 bp of the mammalian genome almost all of the genes present in most larger mitochondrial genomes, including those of yeast, the fungi and plants are found. Initial studies of animal mtDNA size and shape coincided with the discovery of supercoiled, circular molecules. Vinograd and his co-workers (1) utilized density gradients to band supercoiled mtDNA molecules according to their buoyant density noting that such molecules would separate from corresponding relaxed circular and linear molecules of the same size and base content. This property remains the primary technique for the initial purification of mtDNA from the bulk of nuclear DNA.

The minimal size and supercoiled structure of mammalian mtDNAs allow rapid and efficient purification of the mitochondrial genome in many species. However, there remain two major problems in the genetic analysis of animal mtDNA. Firstly, genetics both in the organism and in tissue culture are very limited because few phenotypes have been noted that can be clearly attributed to a specific mitochondrial gene. Secondly, it has not yet been possible to devise a mtDNA-mediated transformation system for introducing *in vitro* altered mitochondrial genes into cells in order to dissect their functions. Therefore most of what is understood about animal mitochondrial genetics has been derived from naturally occurring mutants in animal populations. Breeding and tissue culture studies, particularly in plants, have added to the range of cytoplasmic genetics. However, relative to the yeasts and fungi where many thousands of mutants have been isolated and partially characterized (see Section 8), there remains a paucity of distinct mitochondrial genotypes in higher organisms. Nevertheless, for both plants and animals, it has been possible to demonstrate significant mtDNA heterogeneity between individuals of the same species (reviewed in ref. 2) and, in the case of some animal species, between maternally related individuals (3). This phenomenon of intraspecific mtDNA variation appears to be a general feature which has made possible a number of population genetic studies (e.g. ref. 2). Therefore, efficient and rapid isolation of mtDNA from tissue remains an important and viable tool for understanding cytoplasmic genetics in animals.

Depending on the amount of tissue to be analysed, one should use slightly different protocols for purifying mitochondria. For kilogram amounts of tissue (up to ~5 kg) a method adapted from Denslow and O'Brien (4) should be used. For samples less than about 25–50 g, a method which gives somewhat better organelle yield is used.

2.1 Large-scale preparation of animal mitochondria

(i) Slice the tissue into about 1 cm cubes and reduce it in size further in a meat grinder.
(ii) For each 1 kg of tissue, mix with 1.6 l of homogenizing medium (0.34 M sucrose, 1 mM EDTA, 5 mM Tris-HCl, pH 7.5) and pass the slurry through a medium mesh nylon screen.
(iii) Add 800 ml of homogenizing medium to the solid tissue and liquify it in a food blender at high speed for 10 sec. Re-screen this as before and combine with the liquid from step (ii). Discard all remaining solid tissue.
(iv) Add 500 ml of homogenizing medium to the combined slurries and then homogenize with a Tekmar SDT blender for 2 min at full speed.
(v) Centrifuge the solution at 700–900 *g* for 10 min at 4°C.
(vi) Collect the supernatant and pellet at 12 000–15 000 *g* for 10 min at 4°C.
(vii) Resuspend the mitochondrial pellet in 1 litre of homogenizing medium, re-homogenize in the Tekmar for 30 sec at 1/2 speed and then re-pellet as in step (vi).
(viii) Resuspend the pellet in 750 ml of homogenizing medium and repeat step (vii).
(ix) Resuspend the pellet in 750 ml of homogenizing medium containing 0.4 g of digitonin (Sigma) with stirring for 15 min and then re-pellet as in step (vi).

(x) Divide the washed and pelleted mitochondria into lots for storage at −70°C. Typically, 2 g of mitochondria can be used per 10 ml of solution to be ultra-centrifuged for mtDNA isolation (see Section 2.3). Therefore lots are divided into 2 g units or multiples thereof for storage.

All steps should be carried out in the cold room at 4°C. For tissues such as liver, brain or kidney, about 20−50 g of mitochondria can be isolated per kilogram of tissue. The yield is reduced about 50% when dealing with muscle tissue, probably due to incomplete tissue disruption during the early steps. It is found that immediate purification of mitochondria from fresh tissue is the best technique for long-term storage of samples for subsequent mtDNA analysis. Reasonable yields of mtDNA can be obtained after storage for more than 5 years, whereas tissue stored at the same temperature without mitochondrial purification takes up considerably more freezer space usually and gives only 10% or less of the mtDNA obtainable from fresh tissue or from purified mitochondria.

2.2 **Small-scale preparation of animal mitochondria**

Smaller mammals and biopsy material often provide only 5−50 g of tissue. In these cases, a procedure is used which is quicker and requires less expensive equipment. The procedure can be done on the bench-top if all steps are carried out in ice.

(i) Mince the tissue using single-edged razor blades fitted to an electric knife.
(ii) Suspend the tissue in 1 ml of homogenizing buffer (0.34 M sucrose, 1 mM EDTA, 10 mM Tris-HCl, pH 7.5) per gram of tissue and homogenize in a Dounce homogenizer (loose-fitting pestle) 10 strokes.
(iii) Add EDTA to a final concentration of 10 mM and centrifuge the solution in a swing-out rotor at 700 *g* for 5 min at 4°C. Carefully remove the supernatant and re-centrifuge at 700 *g*.
(iv) Pellet the mitochondria from the second low-speed supernatant by centrifugation at 15 000 *g* for 30 min at 4°C.
(v) Resuspend the mitochondrial pellet in 1 ml of homogenizing buffer per gram of original tissue, re-homogenize in a Dounce as above, and re-pellet as in step (iv).
(vi) Store the pellet at −70°C.

2.3 **Isolation of mtDNA from purified mitochondria**

Two variations for isolating mtDNA from purified mitochondria are described: procedure 1 maximizes the mtDNA yield but takes longer and requires significant volumes of aqueous phenol whereas procedure 2 is less laborious, but gives only about 50% as much mtDNA.

2.3.1 *Procedure 1*

(i) Suspend 2 g of mitochondria in 100 ml of SSC (0.15 M NaCl, 15 mM sodium citrate, pH 7.6) containing 0.1 mM EDTA by Dounce homogenization.
(ii) Add 20% SDS (w/v) to a final concentration of 1% and extract the solution twice with an equal volume of phenol. Combine the two phenol phases and then back-extract with 100 ml of SSC containing 0.1 mM EDTA and 1% SDS; combine

this extract with the original aqueous phases.

(iii) Precipitate nucleic acids from the aqueous phase by adding 3 M sodium acetate, pH 6 to a final concentration of 0.3 M and then 2.5 vols of 95% ethanol at −20°C. Shake well and store at −20°C overnight or at −70°C for 2 h.

(iv) Pellet the nucleic acids by centrifugation at 15 000 *g* for 45 min at 0°C and dry the pellet under vacuum.

(v) Resuspend the dried pellet in 20 ml of 0.1 M Tris-HCl, pH 8.0, 1 mM EDTA, 200 mg/ml ethidium bromide and add 0.9 g of CsCl/ml. The refractive index is adjusted to 1.3900 with solid CsCl or Tris-HCl buffer as necessary.

(vi) Centrifuge the DNA at 100 000 *g* at 20°C for 48 h in a fixed-angle rotor (50 Ti or 60 Ti Beckman or equivalent rotors).

(vii) Visualize the lower band (form I) under u.v. illumination and collect it by side-puncture with a 3 ml syringe. (Often the upper DNA band contains a significant fraction of form II mtDNA from the purified mitochondria, therefore this band can also be collected if Southern blot analysis is planned.)

(viii) Remove the ethidium bromide by three extractions with equal volumes of butanol saturated with a 5M solution of CsCl or NaCl.

(ix) To the aqueous phase, add 2.5 vols of water/ml. (This dilution is important so that CsCl is not precipitated upon ethanol addition.) Add two volumes of 95% ethanol, store the solution at −20°C and pellet the mtDNA as in step (iv).

(x) Dissolve the DNA pellet in 200 μl of distilled water, extract twice with equal volumes of phenol−chloroform (1:1, v/v), re-precipitate by adding 3.0 vols of 95% ethanol, cool and centrifuge as before. Desalt the pellet by washing it with 70% ethanol. Re-centrifuge the washed pellet and dry as above. (This is conveniently done in 1.5 ml microcentrifuge tubes.)

(xi) Dissolve the pellet in 1 mM EDTA, 10 mM Tris-HCl, pH 7.5 and store at −20°C. (For long-term storage, the precipitated DNA may be left in 70% ethanol at −20°C.)

2.3.2 *Procedure 2*

(i) Suspend 2 g of mitochondria in 6 ml of SSC containing 0.1 mM EDTA and lyse by the addition of 0.33 ml of 20% SDS.

(ii) Add 7.0 g of CsCl and 0.4 mg of ethidium bromide and adjust the refractive index to 1.3900.

(iii) Follow steps (vi)−(xi) of procedure 1 (preceding section).

Many workers recommend a second CsCl banding step or a velocity sedimentation following the initial equilibrium sedimentation. The authors find, however, that for routine restriction enzyme analysis further purification is rarely necessary. For cloning or other uses where very pure mtDNA is required a second CsCl banding efficiently removes most of any remaining contaminants.

2.4 **Very small-scale preparation of mtDNA**

Occasions may arise where it is desirable to analyse mtDNA from only a few cells. Normal procedures for isolating mtDNA from partially purified organelles are not possible in these instances. However, by adapting protocols originally intended for isolating nuclear DNA and utilizing Southern blot analysis, mtDNA can be routinely isolated and limited analysis carried out on as few as 50 somatic cells (e.g. a small tissue culture colony) or an individual, mature mammalian oocyte with its mtDNA typically amplified 100–200 times (5,6). In such instances the following procedure should be used.

(i) Pellet the cells from growth medium in a 1.5 ml microcentrifuge tube, resuspend in 20 μl of 0.1 M Tris-HCl, pH 8.0, 50 mM EDTA. Freeze and thaw the suspension.
(ii) To disrupt any remaining cells, homogenize the cells using a small (1 ml) glass homogenizer (10 strokes).
(iii) Place the homogenate and 100 μl of 5 mM EDTA, 0.1 M Tris-HCl, pH 8.0 used to rinse the homogenizer into a 1.5 ml microcentrifuge tube, lyse by the addition of 20% SDS to a final concentration of 1%, and digest with 0.1 mg of proteinase K (Boehringer-Mannheim) for 2 h at 37°C. (The proteinase K is self-digested by incubation at 37°C for 30 min prior to addition to remove any contaminating nucleases.)
(iv) Incubate further at 65°C for 2–4 h, then extract twice with phenol–chloroform (1:1, v/v) and precipitate the nucleic acids by adding an equal volume of 2 M ammonium acetate, pH 7 and 3 vols of 95% ethanol.
(v) After chilling in a dry ice–ethanol bath for at least 30 min, pellet the nucleic acids by centrifugation for 10 min in a microcentrifuge, resuspend in 25 μl of 1 M NaCl, 10 mM EDTA, 0.1 M Tris-HCl, pH 8.0, ethanol precipitate as before and centrifuge again. The pellet is stored at –70°C in 70% ethanol or dissolved in 1 mM EDTA, 10 mM Tris-HCl, pH 7.5 prior to further analysis.

2.5 Direct mtDNA analysis from animal tissue

It may be necessary to analyse mtDNA from animal tissues quickly without purifying the mtDNA from other nucleic acids. Again, blot analysis works well if a general nucleic acid extraction is performed on the tissue. The authors use a scaled-up version of the procedure described in Section 2.4, typically beginning with about 10 g of tissue and 10 ml of 0.34 M sucrose, 1 mM EDTA, 10 mM Tris-HCl, pH 7.5.

(i) Mince the tissue and homogenize as described in Section 2.2, steps (i) and (ii).
(ii) Proteinase K digest, extract and purify nucleic acids as described in Section 2.4, steps (iii)–(v).
(iii) mtDNA can then be analysed on blots using mitochondrial gene-specific probes (e.g. see Section 5.2.2).

2.6 Isolation of displacement loops of mtDNA

Both the H- and L-strands of mtDNA are replicated unidirectionally from separate origins in animal cells (reviewed in ref. 7). After initiation, most newly initiated H-strands terminate less than 700 nucleotides downstream. This leads to a partial relaxation of parental supercoiled molecules because the short, newly synthesized H-strand (variously termed D-loop strand, 7S DNA or DH-DNA) remains associated with the template, thus creating a triple-stranded structure known as a displacement loop (D-loop) molecule. The major form of mammalian mtDNA isolated from cells is this triple-stranded, partially replicated D-loop molecule.

Isolation of D-loop DNA from the mitochondrial genome requires direct purification of mitochondria and mtDNA from fresh tissue, because D-loop DNA is rapidly lost from the genome during storage. Therefore, initial purification of D-loop DNA is based on its association with the unit genome and its co-purification.

(i) Mince and homogenize 50–100 g of fresh tissue and pellet mitochondria as described in Section 2.2, steps (i)–(v).
(ii) Purify mtDNA according to Section 2.3.1 steps (i)–(iii).
(iii) A variable fraction (10–50%) of this mtDNA will contain associated D-loop DNA. Release the D-loop DNA from the genome by heating at 90°C for 5 min in 10 mM EDTA, 10 mM Tris-HCl, pH 8.0 and analyse by labelling either the 5′ end (8) or 3′ end of the DNA (9) followed by gel analysis (10).
(iv) Further purification of D-loop DNA requires its separation from the unit circular genome by gel electrophoresis. The lengths of the D-loop DNA vary from one species to another (7,10), but all mammals thus far analysed contain D-loop DNA of approximately 450–700 nucleotides. D-loop DNA from step (iii) can be separated by electrophoresis on a vertical 6% polyacrylamide gel (1 mm thick, 45 cm long) for 16 h at 3–4 V/cm.
(v) Visualize labelled D-loop DNA by autoradiography and excise it from the gel by aligning the autoradiographic film with the gel.
(vi) Crush the gel fragment with a glass rod and then incubate it for 12 h at 37°C in 3–5 gel volumes of 0.5 M ammonium acetate, 10 mM EDTA, 0.1% SDS and 5 mg/ml carrier tRNA.
(vii) Pass the solution and gel over a 1 ml Sephadex G-50 column (11) and precipitate the DNA in the gel-free solution by the addition of 2.5 vols of 95% ethanol as in Section 2.3.1, steps (x) and (xi).

RNA will interfere with 5′ end-labelling using polynucleotide kinase and must be eliminated as a carrier during ethanol precipitation if 5′ end-labelling is planned. In this case, care must be taken so that the very small pellet of D-loop DNA is not lost during the final wash.

2.7 Isolation of mtDNA from tissue culture cells

Although the range of species for which efficiently growing tissue culture systems are available is limited, some aims, such as studying mitochondrial genetic properties in virus-infected or neoplastically transformed cells, make tissue culture cells the preferred source in some instances. This method is based on that of Tapper *et al.* (12). Spinner culture cells are the most convenient source of material because their culture can be scaled up as required. However, adherent cells can also be used, particularly if they are adapted to growth in roller-bottle cultures. In either case, cells in late log-phase growth (or 60–80% confluent) are pelleted from the medium. All steps are performed on ice or in the cold room at 4°C.

(i) Suspend the initial cell pellet in 10 vols of 0.133 M NaCl, 5 mM KCl, 0.7 mM Na_2HPO_4, 25 mM Tris-HCl, pH 7.5 and re-pellet the cells by centrifugation at 250–300 *g* for 10 min.

(ii) Resuspend the cells in 10 vols of 10 mM NaCl, 1.5 mM $CaCl_2$, 10 mM Tris-HCl, pH 7.5 and allow the cells to swell for 10–15 min on ice.
(iii) Homogenize the cells by 6–10 strokes in a Dounce homogenizer with a tight-fitting pestle. It is important at this stage to monitor the cells under the microscope to ensure that all the cells have been disrupted. If not, continue to homogenize until disruption is complete.
(iv) Add an equal volume of 0.68 M sucrose, 2 mM EDTA, 20 mM Tris–HCl, pH 7.5.
(v) Pellet the nuclei and cell debris from the homogenate by two sequential centrifugations at 1200 *g* for 10 min at 4°C in a swing-out rotor.
(vi) Pellet the mitochondria from the supernatant by centrifugation at 20 000 *g* for 20–30 min at 4°C.

At this point mtDNA may be isolated by CsCl-equilibrium centrifugation as described in Section 2.3.1, steps (v)–(xi). Alternatively, a significant additional purification of mitochondria and its associated nucleic acids can be achieved by banding the mitochondria on an isopycnic sucrose gradient as described next.

(vii) Resuspend the mitochondrial pellet in 20% sucrose, 10 mM EDTA, 50 mM Tris–HCl, pH 7.5 at 1 ml/l equivalent of cells ($\sim 10^9$ cells).
(viii) Carefully layer up to 5 ml onto a 1.0 M/1.5 M sucrose step gradient (15 ml/15 ml) in a tube for a Beckman SW 27 rotor or equivalent. After centrifugation at 80 000 *g* for 30 min at 4°C, collect the mitochondria from the 1.0 M/1.5 M sucrose interface. Dilute the mitochondria with an equal volume of 10 mM EDTA, 50 mM Tris–HCl, pH 7.5, and pellet them as in step (vi).
(ix) Isolate the mtDNA as in Section 2.3.1, steps (v)–(xi).

3. ISOLATION OF MITOCHONDRIAL RNA FROM MAMMALIAN TISSUE AND TISSUE CULTURE CELLS

In general, the isolation of mitochondrial RNA (mtRNA) requires more care in the preparation of solutions and manipulations than for the isolation of mtDNA due to the ubiquitous presence and stability of RNases. Accordingly, all solutions are pre-treated with diethylpyrocarbonate (DEPC) prior to use. To each solution add 1% by volume of a 10% (v/v) DEPC in 95% ethanol solution. After incubation at 37°C for 2 h, heat-stable solutions are autoclaved. Thermally unstable solutions are kept at room temperature overnight before use. In addition, RNA degradation can be minimized by always working on ice, wearing disposable plastic gloves and working with reasonable dispatch until the RNA is in ethanol solution.

The procedures for isolating mtRNA described here are independent of the cellular source. The most important point is that, regardless of whether animal tissue or tissue culture cells are being used, the most intact RNA is isolated from fresh cells, immediately processed. If absolutely necessary, small amounts of animal tissue (<25 g) can be quick-frozen and stored in liquid nitrogen or on dry ice for several days.

Mitochondria suitable for mtRNA isolation may be obtained by any of the above procedures (Section 2). However, it has been noted (12) that the mtRNA isolated from post-nuclear supernatants contains a significant amount of nuclear RNA contamination (~50%). If a cleaner preparation is desired, step-gradient purification of the mitochondria [Section 2.7, steps (vii)−(ix)] offers a substantial improvement in purity of the mtRNA. Two protocols for isolating mtRNA from mitochondria are given here.

3.1 **Phenol purification of mtRNA (12)**

(i) Suspend 2 g of freshly purified mitochondria in 5 ml of 0.1 M NaCl, 1 mM EDTA, 10 mM Tris−HCl, pH 7.5, add to 5 ml of re-distilled phenol and add SDS to a final concentration of 1% at 70°C in a water bath and mix gently.

(ii) Maintain the solution at 70°C until a single phase is obtained, cool to room temperature, add 5 ml of chloroform with gentle mixing and then centrifuge at 20 000 *g* for 20 min at 4°C.

(iii) Re-extract the aqueous phase twice more at 70°C with phenol followed by chloroform extraction as above, but without SDS.

(iv) Adjust the aqueous phase to 0.3 M sodium acetate, pH 7.0 and precipitate the nucleic acids with 2.5 vols of 95% ethanol. Rinse the pellet with 70% ethanol and store the mtRNA pellet at −20°C in 70% ethanol.

3.2 **CsCl purification of mtRNA**

(i) Prepare 2 g of freshly purified mitochondria and fractionate the nucleic acids by SDS lysis and CsCl equilibrium centrifugation as described in Section 2.3.1, steps (v) and (vi).

(ii) Gently withdraw all of the CsCl solution from the centrifuge tube using a pipette. The mtRNA will be in the pellet due to its higher density (>1.9 g/ml) in relation to mtDNA. (If required, mtDNA may be isolated at the same time as described in Section 2.3.)

(iii) Resuspend the pellet in 2 ml of 1 mM EDTA, 10 mM Tris−HCl, pH 7.5, remove the ethidium bromide by three extractions with equal volumes of butanol saturated with CsCl or NaCl solution. Precipitate the mtRNA with 2 vols of ethanol, centrifuge, rinse with 70% ethanol and store as above.

Both procedures yield intact mtRNA as judged by the presence of primary transcript 5′ ends. Careful analysis of the form I mtDNA and mtDNA between form I and II in CsCl gradients has revealed some mtRNA covalently attached to the 5′ end of D-loop DNA (13). This material can be analysed directly by pooling these fractions from a CsCl gradient, extracting, and precipitating as described in Section 2.3.

4. ISOLATION OF MITOCHONDRIAL DNA AND RNA FROM HIGHER PLANT TISSUES

The sizes of plant mitochondrial genomes greatly exceed those of animals or fungi. Various studies indicate that the mtDNA from different plant species may vary in size from about 160 to over 2000 kb (reviewed in ref. 14). Physical mapping using restriction enzymes suggests that the mitochondrial genomes of corn (*Zea mays*) (15) and Chinese cabbage (*Brassica compestris*) (16) exist as multiple circular molecules created from a single master circular genome through intramolecular recombination at any of several pairs of repeated sequences. In general, only 5−10% of these mtDNA genomes exist as repeated sequences. As a consequence of this complexity, rapid progress in plant mitochondrial genetics has awaited complete physical maps for some species (15,16). An additional genetic complexity exhibited by some plants is the presence of multiple, small, linear and/or circular mtDNA molecules which appear to replicate independently of the main mtDNA genome. These components of the mitochondrial genome have been termed mitochondrial plasmids. In several cases the physical status of these plasmid DNAs, that is whether or not copies are integrated into the main genome, has been correlated with but not proven to be the cause of the phenotype of male sterility (17,18). Therefore, although the size of the main mtDNA genome isolated from plant tissue suggests that it is best analysed after cloning subgenomic fragments, analysis of plasmids can frequently be carried out by gel electrophoresis on directly isolated mtDNA (see Section 5.2.2). Methods for isolating both genomic and plasmid mtDNA in *Z. mays* are given. These protocols have been shown to be adaptable to many species.

4.1 Preparation of plant mitochondria (19)

(i) Soak approximately 200 seeds in flowing tap water overnight and grow in the dark for 5–6 days.
(ii) Excise the etiolated shoots from the remainder of the plant, grind at 4°C in 250 ml of 0.5 M mannitol, 1 mM EGTA, 0.1 M Tes-NaOH, pH 7.5 containing 0.5% cysteine and 0.2% bovine serum albumin and then filter the homogenate through six layers of gauze.
(iii) Centrifuge the filtrate at 3000 *g* for 10 min at 4°C and discard the pelleted material.
(iv) Centrifuge the supernatant at 15 000 *g* for 10 min at 4°C, resuspend the pellet into two 10 ml aliquots of grinding buffer containing 10 mM $MgCl_2$, 200 mg/ml DNase I and incubate for 60 min at 4°C.
(v) Adjust the suspension to 10 mM EDTA, gently layer over a 25/45% (10 ml/10 ml) sucrose step gradient and centrifuge in a Beckman SW 27 rotor or equivalent for 45 min at 80 000 *g* at 4°C.
(vi) Pool the mitochondria which band at the 25/45% sucrose interface, dilute 2-fold with distilled water and pellet at 20 000 *g* for 10 min at 4°C.
(vii) Store the mitochondrial pellet at −70°C or process immediately for mtDNA and/or mtRNA (see next sections).

If plant tissue culture cells are used as the source of mitochondria a modification of the method of Sparks and Dale (20) as adapted from Quetier and Vedel (21) and Levings and Pring (22) should be used. For every gram of log-phase cells grown in the dark and harvested by filtration through Miracloth, add 1 ml of grinding buffer at 4°C. The cells in suspension are disrupted by passage through a French pressure cell at 3000 p.s.i. (200 bar) at 4°C. At this point, mitochondria can be isolated as in steps (iii)–(vii) above.

4.2 Preparation of plant mtDNA

(i) Suspend each gram of mitochondrial pellet in 1 ml of 5 mM EDTA, 50 mM Tris–HCl, pH 8.0 and add 55 μl of 20% SDS.
(ii) Adjust the volume to 5.7 ml with 1 mM EDTA, 10 mM Tris–HCl, pH 7.5, transfer to a tube for a Beckman 50 Ti or equivalent rotor, add 6.0 g of CsCl

and 0.3 ml of 10 mg/ml ethidium bromide (refractive index 1.3900) and band the DNA by centrifugation in a fixed-angle rotor at 100 000 *g* at 20°C for 36 h.
(iii) Collect the DNA band as in Section 2.3.1, steps (vii)−(xi).

Although these procedures are designed for etiolated shoot tissue, they should suffice for other tissues such as roots as well, although usually the organelle yield per weight of tissue is reduced. Isolation of mitochondria from green tissue is a considerably larger problem because of the approximately 50-fold excess of chloroplast contamination which will dominate subsequent preparations of mitochondrial nucleic acids. Extra cycles of centrifugation at 3000 *g* to pellet chloroplasts, step (iii) of Section 4.1, markedly improves the purity of mitochondria relative to chloroplasts. However, chloroplast contamination will remain and great care must be taken in the analysis of resultant nucleic acids to ensure its organellar identity. A good technique for assessing chloroplast nucleic acid contamination is to hybridize Northern or Southern blots of such preparations with cloned gene probes specific for each type of organelle.

4.3 Preparation of plant mtRNA

All of the general precautions for avoiding and eliminating RNase contamination outlined in Section 3 apply to methods for isolating plant mtRNA as well. In addition, mtRNA from plant tissues tends to be somewhat more susceptible to degradation as compared with that from animal tissues. Accordingly, all mtRNA preparations should begin with fresh tissue. Several alternative procedures are presented and their relative merits discussed below.

4.3.1 *Procedure 1 (23)*

(i) Harvest tissue from 5-day-old etiolated corn plants and immediately grind in a cold (4°C) mortar and pestle with cold grinding buffer (0.5 M mannitol, 1 mM EGTA, 0.5% cysteine, 0.2% bovine serum albumin, 0.1 M Tes-NaOH, pH 7.5). Use 1 ml of buffer/g of tissue.
(ii) Filter the slurry of buffer and ground tissue through Miracloth overlayered with two layers of gauze.
(iii) Centrifuge the filtrate at 1500 *g* for 15 min at 4°C and discard the pellet.
(iv) Centrifuge the supernatant at 10 000 *g* for 15 min at 4°C.
(v) Rinse the pellet of mitochondria gently with cold grinding buffer then resuspend in 0.2% SDS, 0.1 M Tris-HCl, pH 9.0 and incubate on ice for 15 min.
(vi) Extract the lysate twice with equal volumes of phenol-chloroform-isoamyl alcohol (50:50:1, by vol.).
(vii) To the aqueous phase add 0.1 vol of 3 M sodium acetate, pH 7.0 and 2.5 vols of 95% ethanol at −20°C and keep this solution at −20°C for at least 2 h.
(viii) Pellet total nucleic acids by centrifugation at 15 000 *g* for 10 min at 4°C, wash the pellet with 70% ethanol, re-pellet and dry under vacuum.
(ix) Pelleted samples may be stored for many months in 70% ethanol at −70°C.

4.3.2 *Procedure 2*

(i) Mitochondria are gradient purified as in Section 4.1 with the modification that

an RNase-free grade of DNase is used at step (iv).
(ii) Then follow steps (v)–(ix) of procedure 1 (Section 4.3.1).

4.3.3 *Procedure 3*

(i) Rather than immediately precipitating total nucleic acids from the lysed mitochondrial pellets as in Sections 4.3.1 and 4.3.2, organelle lysis and hot phenol extraction is carried out as for animal mtRNA (Section 3.1).
(ii) Follow steps (v)–(ix) as in Section 4.3.1.

The three procedures described yield different patterns of mtRNA species and purity. The choice of method depends on the use intended for the mtRNA. Procedure 1 gives the highest total yield of intact mtRNA and is preferred for blot-hybridization analysis where gene-specific probes are to be used. However, in the 1–10 kb range these preparations contain both plasmid mtDNA and nuclear nucleic acid contamination. Additionally, there appears to be a selective relative under-representation of rRNAs. Procedure 2 eliminates nuclear contamination but retains the other disadvantages of procedure 1. If procedure 3 is coupled with procedures 1 or 2, good yields of rRNAs are achieved but the overall mtRNA yield is reduced by 50% and the mtRNA obtained is usually somewhat degraded. As noted in Section 4.2, all such preparations will contain some chloroplast RNA contamination. If the levels are unacceptably high, most of the contaminating chloroplasts can be removed by several pelletings of the homogenate at 3000 *g* between steps (iii) and (iv) of the procedure given in Section 4.1.

5. ANALYSIS OF mtDNA AND mtRNA FROM ANIMAL AND PLANT TISSUES

Mitochondrial DNA and RNA isolated as described above are amenable to all modern molecular techniques and manipulations. The one major exception is that mitochondrial genes cannot in general be expressed individually by cloning them into bacterial expression vectors because of their modified genetic code. It is not the purpose of this section to provide protocols for all conceivable circumstances; however, several protocols particularly well suited for the genetic analysis of animal and plant mtDNA are given here.

5.1 Rapid detection of restriction site polymorphisms in animal tissue

The high rate of mtDNA evolution has suggested its use in detecting and relating interbreeding populations within an animal species (2). A rapid and sensitive assay for determining mtDNA sequence similarity between individuals of a species has been developed by Brown (24) and is based on a survey of restriction enzyme sites. It remains the method of choice for an initial genetic study of mtDNA within a population. A modified version used by the authors is as follows.

(i) Isolate mitochondria from tissue using either the procedures in Section 2.1 or 2.2, depending on the amount of starting material.
(ii) Purify the mtDNA according to the procedure given in Section 2.3.
(iii) For each restriction enzyme, digest 20–50 ng of mtDNA with 2–5 units of

enzyme for 2 h using the buffer and temperature conditions recommended by the supplier in a total volume of 20 μl.

(iv) For radiolabelling the resultant mtDNA restriction fragments, different protocols are used depending upon whether the DNA ends produced by the restriction endonuclease contain a 5′ single-stranded extension, a blunt end, or a 3′ single-stranded extension.

5.1.1 *5′ single-stranded extension*

(i) Add 10 μCi (370 kBq) of [α-^{32}P]dNTP (>300 mCi/mmol, >11 GBq/mmol) to the digested DNA. The added nucleotide triphosphate is determined by the first nucleotide that would be added to the 3′-OH end of the restriction site in a DNA polymerase reaction (e.g. for *Bam*HI [α-^{32}P]dGTP is used).

(ii) Add one unit of the large fragment of *Escherichia coli* DNA polymerase I (Klenow fragment) and incubate at 20°C for 20 min. (All restriction enzyme buffer conditions, except those carried out in the absence of NaCl or KCl, are suitable for DNA polymerase. For these low salt buffers, add NaCl to a final concentration of 50 mM before beginning the incubation.)

(iii) Stop the reaction by adding 80 μl of 0.3 M sodium acetate, pH 6.0, 10 mM EDTA and 20 μg of *E. coli* tRNA (Sigma).

(iv) Remove unreacted triphosphates by passage over a 1 ml Sephadex G-50 column prepared in a 1 ml disposable plastic syringe which has been pre-washed twice with 0.3 M sodium acetate, pH 6.0, 10 mM EDTA, 200 μg/ml *E. coli* tRNA (11).

(v) Adjust the mini-column eluate to 400 μl with 0.3 M sodium acetate, pH 6.0, precipitate and rinse the DNA as in Section 2.4, step (v).

(vi) Resuspend the pellet in 3 μl of 10% sucrose, 5 mM EDTA, 0.2% bromophenol blue (w/v), 0.2% xylene cyanol (w/v) for electrophoretic analysis.

(vii) Load each sample into one well of a 6% polyacrylamide (0.15% bis-acrylamide) gel in 10 mM EDTA, 90 mM Tris–borate, pH 8.3. The gel dimensions are typically 15 cm wide, 40 cm long and 0.2 cm thick. Each well should be 5 mm wide. The gel is run at 250 V for 16 h using Tris–borate/EDTA buffer.

(viii) Visualize the individual fragments by autoradiography on Cronex XAR-5 X-ray film for a few hours at room temperature or at −20°C overnight. Intensifying screens are not needed. An example of this sort of analysis for mtDNA from maternally related Holstein cows is shown in *Figure 1*.

5.1.2 *Blunt ends*

(i) Add 10 μCi (370 kBq) of [α-^{32}P]dNTP to the digested DNA as in step (i) above. The added nucleotide triphosphate is determined by the 3′-terminal nucleotide at the restriction endonuclease-generated blunt end.

(ii) Add one unit of Klenow fragment DNA polymerase and incubate the reaction for 20 min at 37°C.

(iii) Follow steps (ii)–(viii) as in Section 5.1.1.

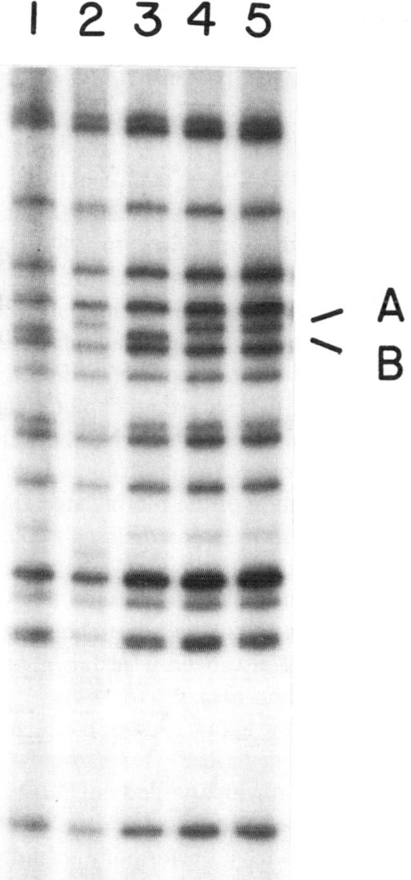

Figure 1. *Dde*I restriction fragment comparison of mtDNA from maternally related Holstein cows **(lanes 1–5)**. The fragments were prepared and labelled according to Section 5.1, using [α-^{32}P]dTTP. Note the polymorphic restriction fragments designated 'A' and 'B'.

5.1.3 *3′ single-stranded extension*

(i) Add one unit of Klenow fragment polymerase and incubate at 20°C for 10 min.
(ii) Add 10 μCi (370 kBq) of [α-^{32}P]dNTP. The added nucleotide triphosphate is determined by the nucleotide complementary to the terminal 5′ nucleotide at the restriction endonuclease-generated end (e.g. for *Pst*I the nucleotide is [α-^{32}P]dCTP).
(iii) Incubate the reaction for an additional 20 min at 37°C.
(iv) Follow steps (iii)–(viii) as in Section 5.1.1.

The protocols detailed above will lead to some minor length heterogeneity at each labelled end due to the 3′-5′ exonuclease activity associated with the DNA polymerase. This is only a problem when high-resolution sequencing gels are used to analyse very short restriction fragments (<200 bp) and can be minimized by adding 1 mM of the unlabelled dNTP which is immediately 5′ to the incorporated radiolabelled dNTP (25).

5.2 **Detection of mtDNA by the biotin–avidin method**

Identification and detection of mtDNA have traditionally been studied by the use of nucleotides radiolabelled with ^{32}P or ^{35}S. DNA bands are visualized by autoradiography and quantified by densitometry of the X-ray film. In 1981, Langer *et al.* (26) reported the synthesis of biotin-labelled polynucleotides which are efficient substrates for several types of DNA and RNA polymerases. Hybridization characteristics of the biotinylated nucleotides are similar to those for non-biotinylated nucleotides. This discovery represents an important step in the development of non-radioactive methods for detecting nucleic acids. In 1983, Leary *et al.* (27) showed that biotinylated nucleotides, or DNA hybridized to biotinylated probes after Southern, Northern or dot-blot hybridization, could be visualized by a colorimetric technique using immunochemical reagents. The method requires the initial conjugation of the biotinylated DNA with avidin, which is subsequently detected by an enzymatic reaction yielding a coloured precipitate. In these studies streptavidin and alkaline phosphatase showed a higher sensitivity than avidin and peroxidase. The sensitivity of detection for plasmid DNA is in the picogram range and it is therefore as sensitive as a radiolabelled probe. In addition to blot hybridization, biotinylated nucleotides have also been used for *in situ* hybridization (28–30) and complementary strand separation (31).

The following section describes the application of the biotin–avidin method for the detection of liver mtDNA from rat or mouse. Our particular interest was in detection of the conversion of covalently closed, supercoiled mtDNA to relaxed and linear forms as a result of drug regimens in isolated mitochondria. The following protocols for mitochondria and mtDNA isolation have been optimized for preparation from mouse or rat liver and subsequent detection by Southern blotting and biotinylated probe hybridization. More general protocols are presented in Section 2.

5.2.1 *Preparation of mitochondria and mtDNA*

(i) Remove and homogenize the liver with a Potter–Elvejhem tissue homogenizer at 4°C in 9 vols/g tissue of homogenizing medium containing 0.25 M sucrose, 10 mM Hepes-NaOH, pH 7.4 and 1 mM EDTA.

(ii) Centrifuge the homogenate at 950 *g* for 10 min at 4°C to pellet nuclei and unbroken cells.

(iii) Re-centrifuge the supernatant fraction at 8500 *g* for 10 min at 4°C to obtain the mitochondrial pellet.

(iv) Wash the pellet with homogenizing medium, resuspend the mitochondria in 0.15 ml of homogenizing medium to yield approximately 100 mg protein/ml as determined by the Biuret method.

(v) Solubilize the mitochondria (1 mg) in 250 μl of homogenizing medium containing 0.5% SDS, add 100 mg/ml proteinase K and incubate at 37°C for 30 min and then at 65°C for 10 min.

(vi) Extract the mtDNA solution with a 5-fold volume of phenol–chloroform–

isoamyl alcohol (25:25:1, by vol.) saturated with 0.2% 2-mercaptoethanol, 0.1 M Tris–HCl, pH 8.0.

(vii) Re-extract the aqueous phase with an equal volume of chloroform–isoamyl alcohol (24:1, v/v) to remove residual phenol.

(viii) Add 2.4 vol of 95% ethanol at −20°C to the aqueous phase and chill in a dry ice–ethanol bath for 30 min.

(ix) Collect the nucleic acid precipitate by centrifugation in a microcentrifuge at 13 000 *g* for 10 min at 4°C.

(x) Dry the pellet under vacuum and resuspend in 1 mM EDTA, 10 mM Tris–HCl, pH 7.5. This preparation does not require further purification for analysis by Southern blot hybridization. MtDNA (250 ng estimated on the basis of 0.5 μg/mg protein) closed circular, open circular and linear forms are separated by gel electrophoresis. MtDNA on the blot is detected by a probe containing the entire mouse mtDNA genome cloned into the plasmid vector pSP64 (Promega).

5.2.2 *Electrophoresis*

(i) Run samples in a 0.7% agarose gel with 10 mM EDTA, 90 mM Tris-borate, pH 8.3 buffer at 100 V for 3 h. DNA bands are usually not visible by ethidium bromide staining.

(ii) Denature the mtDNA by soaking the gel in 100 ml of 1.5 M NaCl and 0.5 M NaOH for 60 min, and then neutralize the gel in 100 ml of 1.0 M Tris–HCl, pH 8 and 1.5 M NaCl for 60 min.

(iii) Transfer the mtDNA onto nitrocellulose paper by the method of Southern (32) with 10 × SSC (SSC is 0.15 M NaCl, 15 mM sodium citrate, pH 7.6). The side of the blot containing the DNA must be marked because the reagents and dyes cannot penetrate through the blot.

(iv) Dry the nitrocellulose blot at 80°C in a vacuum oven for 2 h.

5.2.3 *Nick translation of the probe with biotinylated dUTP*

(i) Nick translate the probe (1 μg) with 0.2 mM each of dCTP, dGTP, dATP and 0.4 mM biotinylated dUTP in the presence of 200 pg DNase I and 2 units of *E. coli* DNA polymerase I at 16°C for 90 min.

(ii) Stop the reaction with 5 μl of 0.3 M EDTA (pH 8).

(iii) Remove the unincorporated nucleotides with a Sephadex G-50 spin column (11) previously equilibrated with 1 mM EDTA, 10 mM Tris–HCl, pH 7.5 and yeast or *E. coli* tRNA.

5.2.4 *Hybridization*

(i) Re-hydrate the blot in 6 × SSC for 30 min, and then pre-hybridize with 30 ml of 6 × SSC containing 0.6 ml of 50 × Denhardt's solution (1% Ficoll 400, 1% polyvinylpyrrolidone and 1% bovine serum albumin) and 900 μg of heat-denatured yeast tRNA in an oven at 68°C (or 42°C for rat samples) for at least 6 h.

(ii) Hybridize the blot for at least 16 h in 10 ml of 6 × SSC containing 0.2 ml of 50 × Denhardt's solution, 1 μg of heat-denatured probe and 100 μg per ml of heat-denatured yeast tRNA at 68°C (or 42°C for rat samples).

5.2.5 *Detection of biotinylated mtDNA (adapted ref. 27)*

(i) After hybridization, rinse the blot twice in 80 ml of 2 × SSC containing 0.1% SDS and then twice in 80 ml of 0.2 × SSC containing 0.1% SDS. Each wash should be for at least 3 min at room temperature.
(ii) Wash the blot twice in 80 ml of 0.16 × SSC containing 0.1% SDS for 15 min at 55°C. For blots with rat mtDNA samples, wash at 42°C.
(iii) Rinse the blot in 2 × SSC and dry briefly between filter papers.
(iv) Re-hydrate the blot in 70 ml of Buffer 1 containing 0.1 M Tris–HCl, pH 7.5, 0.1 M NaCl, 2 mM $MgCl_2$ and 0.05% Triton X-100 for 1 min.
(v) Block non-specific sites on the blot with 80 ml of Buffer 1 containing 3% bovine serum albumin for 30 min at 42°C and then dry in a vacuum oven at 80°C for 10 min.
(vi) Place the blot with the DNA side up in a box of similar size as the blot. Alternatively the blot may be placed inside a 250 ml centrifuge bottle with the blot adhering to the wall of the bottle so that during rotation, complete coating of the blot occurs. Re-hydrate the blot in 80 ml of Buffer 1 containing 3% bovine serum albumin for at least 30 min and decant.
(vii) Add streptavidin (12 μg in 3 ml of Buffer 1) and shake gently to cover the entire blot. Incubate the blot for 10–20 min at room temperature and rinse in 80 ml of Buffer 1 three times to remove excess streptavidin.
(viii) Add biotinylated calf intestinal alkaline phosphatase (3 μg/3 ml) and incubate for 10 min (or 20 min for blots in bottles).
(ix) Rinse the blot twice each in 80 ml of Buffer 1, and then twice each with 80 ml of Buffer 3 containing 0.1 M Tris–HCl, pH 9.5, 0.1 M NaCl and 50 mM $MgCl_2$.
(x) Add nitroblue tetrazolium (NBT, 2.5 mg) and 5-bromo-4-chloro-3-indolyl phosphate (BCIP, 1.25 mg) in 7.5 ml of Buffer 3 to the blot. This results in the formation of a purple precipitate. If precipitate forms in the solution before the bands are visible, remove the dye solution, rinse the blot with 80 ml of Buffer 3 and repeat this step.
(xi) After colour development is complete, stop the reaction with a buffer solution containing 20 mM Tris–HCl, pH 7.5 and 5 mM EDTA.

All the reagents (streptavidin, alkaline phosphatase, BCIP and NBT) are available from Bethesda Research Laboratories (BRL, Gaithersburg, MD). Avidin conjugated with alkaline phosphatase is available from Cooper Biochemical, Inc. (Malvern, PA) and streptavidin conjugated to alkaline phosphatase from BRL may be used to change the two-step addition of streptavidin and alkaline phosphatase to one step. The authors find that the use of avidin instead of streptavidin yields a higher background. NBT (75 mg/ml in 70% dimethylformamide) and BCIP (50 mg/ml in 100% dimethylformamide) from Sigma Chemical Co. (St. Louis, MO) are also effective.

The sensitivity of detection is dependent on the amount of DNA and time of colour development. At 100 ng of mtDNA, the bands are visible within 15 min. MtDNA of less than 100 ng requires development for more than 3 h. However, with the longer development time, the background is also increased.

An example of the described method is shown in *Figure 2* with mtDNA isolated from

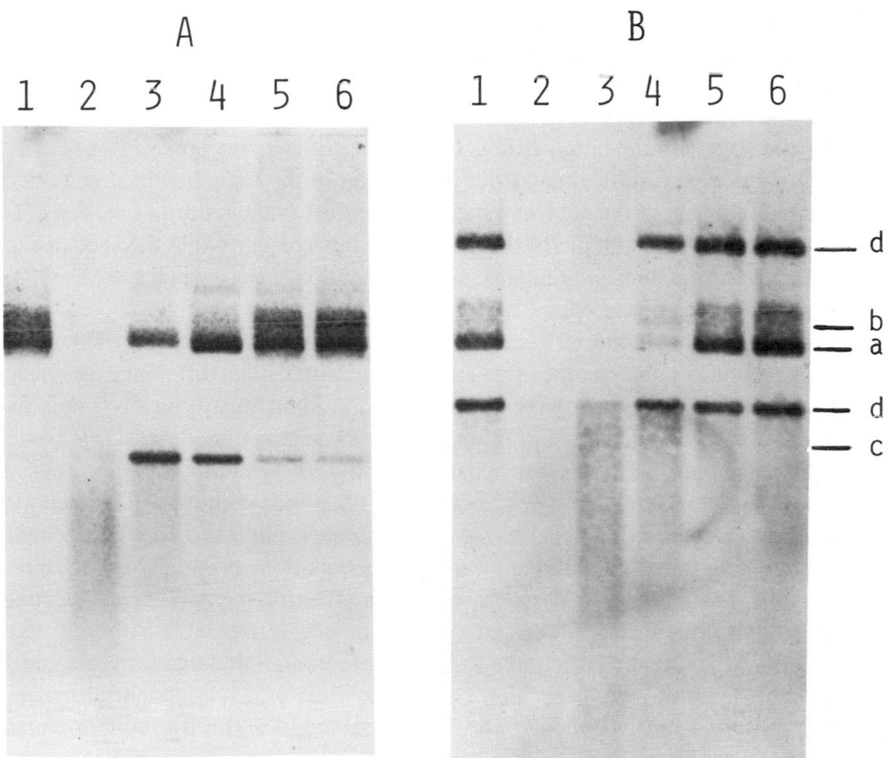

Figure 2. Analysis of mtDNA treated with bleomycin and visualized according to Section 5.2. **Lane 1** is untreated control. **Lanes 2–6** are mtDNA treated with bleomycin at 10 μM, 1 μM, 100 nM, 10 nM and 1 nM, respectively. Bands a, b, c and d are closed circles, nicked circles, double-stranded linear and single-stranded forms, respectively. In *Figure 2B* the samples were treated with 0.1 M NaOH for 2 min before loading onto the gel.

mitochondria treated with bleomycin. There was a dose-dependent increase in linear mtDNA fragments and a loss of closed circular forms (*Figure 2A*) with alkali treatment (final concentration 0.1 M NaOH). Nicked circles can be distinguished from closed circles and linear forms because they migrate at different mobilities compared with double-stranded forms (compare lane 1 in *Figure 2A* and lane 1 in *Figure 2B*).

This sort of analysis appears suitable for any animal system in which a loss of the *in vivo* supercoiled mtDNA is to be monitored. The method works equally well for visualization of restriction fragments at least as small as 500 nucleotides. Finally, analysis of plant mtDNA plasmids can be accomplished using biotinylated probes and running undigested mtDNA as isolated in Section 4.2 on 1–2% agarose–Tris/borate/EDTA gels.

6. INTRODUCTION: NATURE OF THE MITOCHONDRIAL MUTATIONS OF SACCHAROMYCES CEREVISIAE

The baker's yeast *Saccharomyces cerevisiae* is a facultative aerobe that can tolerate all types of alterations to the mtDNA, including its complete loss, which would be lethal for the majority of other organisms. Most genetic techniques to study mitochondrial mutations are, therefore, based on the primary distinction between respiratory competent cells (RC) and respiratory deficient cells (RD). This distinction is easily made by testing the ability of cells to grow on media containing a non-fermentable carbon source as the sole carbon source (RC cells grow, RD cells do not).

The RD phenotype results either from a mutation in the mitochondrial genome or from a mutation in any of the nuclear genes which affect mitochondrial functions. The first type of mutation is the topic of this section of this chapter while the second type, which will be referred to as *pet* mutants, will be mentioned only when a clear distinction from the first type is needed.

Mutations in the mitochondrial genome of *S. cerevisiae* (reviewed in ref. 33) belong to three general classes, the specific properties of which imply that different experimental procedures be used to manipulate them. Firstly, point mutations (typically single base substitutions) or short deletions within one of the structural mitochondrial genes encoding components of the respiratory chain, of the ATP synthase complex or of the mitochondrial protein synthesis machinery. Such mutations confer an RD phenotype and will be referred to in this chapter as *mit*$^-$ mutants. Note that this class encompasses mutations originally termed *mit*$^-$ as well as mutations originally termed *syn*$^-$, *pho*$^-$ or *gin*$^-$. The *pho*$^-$ mutants, with mutations affecting the ATP synthase genes, are not physiologically RD, yet they cannot grow on non-fermentable carbon sources. For this reason, they can be considered to be like other *mit*$^-$ mutants for most genetic tests. Secondly, point mutations within one of the structural mitochondrial genes as above but conferring an RC phenotype with resistance to a specific inhibitor of the mitochondrial function concerned (typically an antibiotic resistance or a drug resistance). They will be referred to in this chapter as *ant*^{R} mutants even though this class encompasses resistance to drugs other than antibiotics. Thirdly, very large deletions in the mitochondrial genetic map accompanied by reiteration of the conserved DNA segment or, in the extreme case, complete loss of all of the mtDNA. They obviously all confer an RD phenotype and are non-revertible. They will be referred to in this chapter as either *rho*$^-$ mutants if a segment of mtDNA is still present or *rho*^{0} mutants for those devoid of mtDNA.

A normal cell of *S. cerevisiae* contains 20–35 molecules of mtDNA (34). The actual figure may be slightly variable from strain to strain and varies according to growth conditions, but in all cases a few dozen mtDNA molecules are present in a cell and are all active at the same time. The distribution of mtDNA molecules between individual mitochondria is certainly important for the physiology of the cell but can be largely disregarded in the genetic methods described here because the molecules of mtDNA are apparently redistributed in an efficient manner during the time of a genetic cross. An explanation for this fact is probably that, in rapid growth conditions, mitochondria tend to fuse with one another, forming a very small number of large and branched mitochondria or even only one giant mitochondrion (35). The multiplicity of the mtDNA molecules results in a cell being in either one of two states, namely either *homoplasmic*, in which case all of the mtDNA molecules are identical in terms of the genetic markers they carry, or *heteroplasmic*, in which different mtDNA molecules carrying different genetic markers are present simultaneously. This definition is theoretical but, in practice, successive subcloning experiments as described in Section 8.2.1 provide an excellent operational definition of homoplasmic and heteroplasmic cells that is most useful in all mitochondrial genetics of *S. cerevisiae*.

7. MEDIA FOR WORKING WITH YEAST MITOCHONDRIAL MUTANTS

7.1 **Composition and preparation of media**

Only media directly relevant to mitochondrial genetics are described. Other yeast media can be found elsewhere if necessary. Quotations of manufacturers or commercial suppliers of chemicals do not imply that they are superior to others but only that they have been successfully used in the procedures described.

7.1.1 *Complete media without antibiotics*

All complete media listed in this section contain 10 g/l of yeast extract (Difco Lab.) and 10 g/l of bactopeptone (Difco Lab.). They only differ in the carbon source and the presence or absence of sodium potassium phosphate buffer.

- YPglu 20 g/l of D-glucose
- N0 Same as YPglu but buffered at pH 6.24 using 50 mM final concentration of sodium potassium phosphate buffer
- YP10 100 g/l of D-glucose
- YPdif 1 g/l of D-glucose plus 20 ml/l of glycerol
- YPgal 20 g/l of D-galactose
- YPgly 20 ml/l of glycerol
- N3 Same as YPgly but buffered at pH 6.24 using 50 mM final concentration of sodium potassium phosphate buffer
- N1 20 ml/l of ethanol (add after autoclaving). Buffer at pH 6.24 using 50 mM final concentration of sodium potassium phosphate buffer

All ingredients are added to their final concentrations prior to autoclaving (except for N1 medium). For solid media add 22 g/l of bacto agar (Difco Lab.) prior to autoclaving. Note that some agar brands may contain trace amounts of fermentable carbon sources and should not be used for YPgly, YPdif, N1 or N3 plates. Allow agar-containing media to cool down to about 60°C before pouring plates. When the agar is set, allow all the plates to dry for about 2–3 days before use. The use of a well aerated 37°C incubator reduces drying time to about 36 h. Liquid or solid media can be stored for months in the refrigerator to prevent desiccation. Note that rich solid media containing glucose in Petri dishes are highly susceptible to bacterial contamination. Adding an anti-bacterial antibiotic such as sodium benzyl penicillinate (100 000 U/l) after autoclaving and before pouring plates greatly diminishes the risk of contamination without affecting yeasts. If necessary, penicillinate can be added to other media as well.

7.1.2 *Complete media containing antibiotics or inhibitory drugs*

The following antibiotics or drugs have been used for the isolation of mitochondrial mutants:

specific inhibitors of the mitochondrial ribosome: chloramphenicol, erythromycin, spiromycin, paromomycin;

specific inhibitors of the ATP synthase complex: oligomycin (or rutamycin), venturicidin, ossamycin;
specific inhibitors of the cytochrome bc_1 complex: antimycin A, funiculosin, diuron, mucidin, myxothiazol.

Always use buffered media (N3 or N1) since many antibiotics or drugs are only effective on yeast mitochondria within a narrow pH range. Final concentrations are given in *Table 1*.

Table 1. Antibiotics and drugs containing media for *Saccharomyces cerevisiae*.

This table gives the final concentrations in media (N3 or N1) of the antibiotics or drugs and of their carriers.

(i) Antibiotics added as dry powders

N3C	4 g/l of chloramphenicol (Roussel UCLAF, France) in N3 (ref. 36)
N3E	5 g/l of erythromycin base (Roussel UCLAF, France) in N3 (ref. 36)
N3S	5 g/l of spiramycin base (Rhône-Poulenc, France) in N3 (ref. 37)
N3P	2 g/l of paromomycin sulphate (Parke Davies, California) in N3 all buffered at pH 6.5 using 50 mM sodium potassium phosphate buffer (ref. 38)

(ii) Antibiotics or drugs added from stock solutions

N30	3	mg/l of oligomycin + 0.5% (v/v) of methanol in N3 (ref. 39)
N3V	1	mg/l of venturicidin + 0.5% (v/v) of methanol in N3 (ref. 40)
N3oss	2	mg/l of ossamycin + 0.5% (v/v) of methanol in N3 (ref. 41)
N1A	0.1	mg/l of antimycin A + 0.1% (v/v) of ethanol in N1 (ref. 42)
N1F	5	mg/l of funiculosin + 1% (v/v) of ethanol in N1 (ref. 42)
N3D	35	mg/l of diuron + 2% (v/v) of acetone in N3 (ref. 43)
N3M	0.3	mg/l of mucidin + 0.1% (v/v) of ethanol in N3 (ref. 44)
N3myx	2	mg/l of myxothiazol + 2% (v/v) of ethanol in N3 (ref. 45)

These media correspond to the standard concentrations of antibiotics or drugs used for growth of resistant strains and for distinguishing between resistant and sensitive colonies by replica plating. If other concentrations are used for particular purposes keep the final concentration of carrier constant.

(iii) Preparation of stock solutions

(a) *Oligomycin*: dissolve oligomycin (mix of oligomycins A, B and C; Sigma, Missouri) in methanol at a final concentration of 600 μg/ml. Add 5 ml of stock solution per litre of N3 medium.
(b) *Venturicidin*: dissolve venturicidin A (B.D.H. Biochemicals, UK) in methanol at a final concentration of 0.2 mg/ml. Add 5 ml of solution per litre of N3 medium.
(c) *Ossamycin*: dissolve ossamycin (Bristol-Myers Co.) in methanol to a final concentration of 0.4 mg/ml. Add 5 ml of solution per litre of N3 medium.
(d) *Antimycin*: dissolve antimycin A (Serva, FRG or Boehringer, FRG) in ethanol at a final concentration of 100 μg/ml. Add 1 ml of stock solution per litre of N1 medium.
(e) *Funiculosin*: dissolve funiculosin (Sandoz A.G., Switzerland) in ethanol at a final concentration of 0.5 mg/ml. Add 10 ml per litre of N1 medium.
(f) *Diuron*: dissolve 3-(3,4-dichlorophenyl)-1,1-dimethylurea (E.I.duPont de Nemours and Co) in acetone at a final concentration of 7 mg/ml. Take 5 ml of stock solution per litre of N3 medium. Add 15 ml of acetone, mix and pour in medium.
(g) *Mucidin*: dissolve the 'Mucidermin' spray solution (Spofa, Prague, Czechoslovakia) in ethanol to a final concentration of 0.3 mg/ml of mucidin. Add 1 ml of solution for one litre of N3 medium.
(h) *Myxothiazol*: dissolve myxothiazol (Gesellschaft für Biotechnologische Forschung, Braunschweig, FRG) in ethanol to a final concentration of 0.1 mg/ml. Add 20 ml of solution per litre of N3 medium.

(i) *Media in which antibiotics are added as dry powders.* This is the case for N3C, N3E, N3S and N3P media. Prepare N3 medium, autoclave and let the medium cool down to about 60°C prior to adding antibiotics. Weigh the dry antibiotic powder and add to the medium directly to its final concentration. Note that these antibiotics are used at final concentrations close to their limit of solubility. Stir the medium with a magnetic stirrer until the powder is completely dissolved (it is convenient to place a magnetic stirring bar in the medium prior to autoclaving). Dissolution is quick and clumping is prevented if the powder is absolutely dry. It is therefore advisable to store the antibiotic powders in a dry container at room temperature; they are stable for years under such conditions.

(ii) *Media in which antibiotics or drugs are added from stock solutions.* This is the case for all other media listed in *Table 1* which contain antibiotics or drugs not soluble in media without any carrier. Prepare stock solutions as indicated (stock solutions can be stored at −20°C if properly sealed). Prepare N3 or N1 medium, autoclave and allow medium to cool down to below 60°C prior to adding antibiotics or drugs. Add the stock solution as indicated in *Table 1* and stir immediately.

7.1.3 *Minimal media*

In principle, minimal media can be prepared with a variety of carbon sources as for the complete media. In practice, however, yeast strains grow very slowly on minimal media containing a non-fermentable carbon source as the sole carbon source. For this reason, only the glucose medium has been of general use for mitochondrial genetics. For solid media, add 22 g/l of bacto agar (Difco Lab.). Note that some agar brands may contain trace amounts of metabolites and should not be used for minimal media.

(i) *Standard minimal media*

- W0 6.7 g/l of yeast nitrogen base without amino acids (Difco Lab.) and 20 g/l of D-glucose
- W10 Same as W0 but containing 100 g/l of D-glucose

Both ingredients are added to their final concentrations prior to autoclaving.

(ii) *Minimal media for specific applications (46)*

- G0 See *Table 2* for composition
- Ggal Same as G0 but containing 20 g/l of D-galactose

All minimal media can be supplemented as described in *Table 3* for the most commonly used auxotrophic markers.

7.2 Use of media and precautions in growing mitochondrial mutants

7.2.1. *YPglu*

This is the most common medium providing maximum growth rate and highest final yield of cells. Wild-type cells as well as all types of mitochondrial mutants can grow in it. However, growth rates always favour RC cells over RD ones, hence leading to an under-representation of the latter type; this may become a problem if quantitative

Table 2. Preparation of G0 minimal medium with or without sulphate.

A. Prepare separately the following stock solutions

10 × Salts: weigh the following, dissolve in 1 litre final volume with water and autoclave.

	With sulphate	*Without sulphate*
$(NH_4)H_2PO_4$ (mol. wt 115.03)	60 g	60 g
$MgSO_4 . 7H_2O$ (mol. wt 246.48)	5 g	–
$MgCl_2 . 6H_2O$ (mol. wt 203.31)	–	4.1 g
$(NH_4)_2SO_4$ (mol. wt 132.14)	20 g	–
NH_4Cl (mol. wt 53.50)	–	8.1 g
KH_2PO_4 (mol. wt 136.09)	10 g	10 g
NaCl (mol. wt 58.45)	1 g	1 g
$CaCl_2$ (mol. wt 110.99)	1 g	1 g

Trace elements: weigh the following, dissolve in 1 litre final volume with water and autoclave.

	With sulphate	*Without sulphate*
H_3BO_3 (mol. wt 61.83)	500 mg	500 mg
$CuSO_4 . 5H_2O$ (mol. wt 249.68)	60 mg	–
$CuCl_2 . 2H_2O$ (mol. wt 170.48)	–	40 mg
KI (mol. wt 166.01)	100 mg	100 mg
$MnSO_4 . H_2O$ (mol. wt 169.00)	400 mg	–
$MnCl_2 . 4H_2O$ (mol. wt 197.91)	–	470 mg
$Na_2MoO_4 . 2H_2O$ (mol. wt 241.95)	200 mg	200 mg
$ZnSO_4 . 7H_2O$ (mol. wt 287.54)	400 mg	–
$ZnCl_2$ (mol. wt 136.29)	–	190 mg

Ferric chloride: prepare a stock solution at 0.2 mg/ml of $FeCl_3$ (mol. wt 162.21) in water and filter sterilize

Vitamins and co-factors:

1. Weigh the following and dissolve in 99 ml final volume with water:
   calcium pantothenate (mol. wt 476.53) 40 mg
   thiamine hydrochloride (vit.B1) (mol. wt 337.28) 40 mg
   pyridoxine hydrochloride (vit.B6) (mol. wt 205.64) 40 mg
   nicotinic acid (niacin) (mol. wt 123.11) 10 mg
2. Add 1 ml of a 0.4 mg/ml stock solution of biotin (mol. wt 244.31) in 50 mM $NaHCO_3$.
3. Filter sterilize.

Inositol: dissolve 400 mg of inositol (mol. wt 180.16) into 100 ml of water and filter sterilize

B. Preparation of final medium

1. Measure 100 ml of '10 × Salts' stock solution with or without sulphate as appropriate.
2. Add 20 g of D-glucose.
3. Dissolve and complete to 950 ml with water.
4. Autoclave.
5. Allow the medium to cool down to 60°C and then add:
   5 ml of vitamins and co-factors stock solution
   5 ml of inositol stock solution
   1 ml of trace elements stock solution with or without sulphate as appropriate.
   1 ml of ferric chloride stock solution
   10 ml of each stock solution of required supplements (*Table 3*) and/or sterile water to complete to 1 litre final volume.
6. Mix.

Table 3. Supplements to minimal media corresponding to the most commonly used auxotrophic markers in yeast mitochondrial genetics.

1. Prepare stock solutions as below.
2. Autoclave.
3. Store in the refrigerator or freeze.
4. Add 10 ml of stock solution per litre of minimal medium after autoclaving when temperature is below 60°C.

	Stock solutions	*Auxotrophic mutations commonly used in mitochondrial genetics*
adenine:	2 mg/ml in 50 mM HCl	*ade1, ade2*[a]
uracil:	2 mg/ml in 0.5% (w/v) $NaHCO_3$	*ura1, ura3*
L-arginine	1 mg/ml in water	*arg4*
L-histidine:	1 mg/ml in water	*his1, his4C*
L-isoleucine:	6 mg/ml in water	*ilv5*
L-leucine:	6 mg/ml in water	*leu1, leu2*
L-lysine HCl:	1 mg/ml in water	*lys2*
L-methionine:	1 mg/ml in water	*met13*
L-phenylalanine:	2 mg/ml in water	*phe*
L-tryptophan:	2 mg/ml in water	*trp1, trp2*
L-tyrosine:	1 mg/ml in 50 mM HCl	*tyr6*
L-valine:	2 mg/ml in water	*ilv5*

[a]*ade1* and *ade2* can be particularly useful for mitochondrial genetics as *rho*$^+$ cells turn red while *rho*$^-$ cells remain white (*mit*$^-$ mutants are generally pink).

measurements are needed. The relative proportion of RD cells is even more significantly reduced when culture is extended until stationary phase since growth on this medium shows a biphasic pattern.

During the first phase of growth, when metabolizing glucose, yeasts produce ethanol but do not use it as long as the glucose concentration is high enough. During this first phase of growth both RC and RD cells can grow. When glucose comes near to exhaustion, a second phase of growth (usually 2–3 additional cell doublings) is possible only for the RC cells, using the previously produced ethanol as the carbon source.

The major concerns in growing yeast cells in this medium are therefore as follows. Firstly, for *rho*$^+$ strains: spontaneous *rho*$^-$ mutants (which are frequent, see below) are not eliminated and tend to accumulate in the culture. Secondly, for *mit*$^-$ mutants: spontaneous *mit*$^+$ revertants (when they can arise) will be favoured (because they are RC) and will rapidly outgrow the *mit*$^-$ population which grows more slowly. This medium is useful both as a liquid medium and for plates.

7.2.2 *NO*

This is similar to YPglu. It is useful only when pH becomes important, such as in ethidium bromide mutagenesis or inhibition of mitochondria by antibiotics or drugs.

7.2.3 *YP10*

Since the glucose concentration of this medium is extremely high, the stationary phase is reached before glucose exhaustion. It follows that a biphasic growth pattern is not observed and that RC cells are not so strongly favoured as in YPglu medium. For this reason YP10 is to be preferred to YPglu when revertible *mit*$^{-}$ mutants are grown or when a quantitative estimate of the frequency of *rho*$^{-}$ mutants is needed. It is useful both as a liquid medium and for plates. It is recommended that liquid medium cultures are not shaken to reduce further the selective advantage of RC cells.

7.2.4 *YPgly, N3 and N1*

These media allow the growth of RC cells only. Growth is slower than on YPglu or YP10 media but all spontaneous *rho*$^{-}$ mutants are immediately stopped. These media are therefore recommended if pure *rho*$^{+}$ cultures are needed. They are useful both as liquid media and for plates. Ensure good aeration of liquid cultures by vigorous agitation.

7.2.5 *Antibiotic-containing media*

These media allow the growth of RC antibiotic-resistant mutants only. These media are therefore recommended when a pure culture of such mutants is needed (e.g. as a pre-culture to start an experiment). They are useful both as liquid media and for plates.

7.2.6 *YPdif*

This is a differential medium which increases differences of growth between RC and RD cells. The limited amount of glucose is rapidly exhausted while glycerol permits the growth of RC cells only. It is useful only for plates to distinguish between *rho*$^{+}$ and *rho*$^{-}$ colonies. Incubation should be increased to about 5–6 days at 28°C to maximize differential growth. *Rho*$^{+}$ colonies are large, thick and yellowish, while *rho*$^{-}$ colonies are small, flat and white (hence the frequent designation of *rho*$^{-}$ mutants as *petite* mutants). Extended incubation of such plates for about 10 days may result, under specific genetic conditions, in the formation of a cluster of RC cells in an RD colony. These are easily visible as abscessed colonies.

7.2.7 *YPgal*

This medium is used to eliminate the effect of glucose repression. Both RC and RD cells can grow in it. This is useful as a liquid medium and for plates. Some strains may not grow on this medium if they are from glucose-grown cultures. In this case grow the cells first in YPgal containing 0.1% glucose prior to transferring them to YPgal.

7.2.8 *W0 and W10*

These are the minimal media used to select prototrophic diploids formed by crosses of auxotrophic parents. They are useful as liquid media and for plates. Both RC and RD cells grow in these media. However, the growth of RD cells is significantly slower than that of RC cells, resulting in a very strong counter-selection of RD cells. If quantitative estimates of the relative frequencies of RC and RD cells are needed (e.g. in crosses involving *rho*$^{-}$ or *mit*$^{-}$ mutants), use only W10 liquid medium without shaking and avoid prolonged incubation.

7.2.9 *G0*

This has the same properties as W0 but is much more laborious to prepare! Use this medium only for specific requirements such as minimal medium without sulphate (see *Table 2*).

8. BASIC GENETIC TECHNIQUES FOR YEAST

8.1 **Isolation of mitochondrial mutants**

8.1.1 *Induction and isolation of rho$^-$ and rho^0 mutants*

Rho$^-$ and *rho^0* mutants are frequent. Cultures of *rho$^+$* strains routinely contain about 1% of such mutants under normal conditions. The frequency of mutants can reach up to 100% after mutagenesis.

Isolation of rho^0 mutants. Rho^0 mutants form a specific subclass of RD mutants which completely lack all mtDNA. They are particularly useful in two aspects. Firstly, they provide a null control against which other mitochondrial mutations can be tested by crosses and secondly they can be used in coordination with the *kar1* mutation to introduce a specific mitochondrial genotype within a known nuclear background (see Section 8.2.3). A specific induction of *rho^0* mutants can be obtained by growing cells in the presence of ethidium bromide as follows.

(i) Grow the cells of the desired haploid strain (with appropriate auxotrophic markers) in N3 medium (if RC) or in YPglu medium (if RD).
(ii) Pellet the cells by centrifugation at 1000 *g* for 5 min, wash and resuspend in sterile water at a density of 10^7 cells/ml (use a haemacytometer or calibrated spectrophotometer).
(iii) Inoculate 0.05 ml of the cell suspension into 5 ml of N0 medium. Add 0.01 ml of an ethidium bromide stock solution at 10 mg/ml in sterile water (to obtain a final concentration of 20 μg/ml of ethidium bromide). Wrap the tube or flask containing the culture in aluminium foil and incubate at 28°C for 24 h with shaking to prevent cell sedimentation.
(iv) Transfer 0.05 ml of the previous culture into 5 ml of fresh N0 medium containing 20 μg/ml of ethidium bromide and incubate as in step (3).
(v) Repeat the transfer once again as in step (4). Note that one culture in the presence of ethidium bromide is usually sufficient to convert all of the cells into *rho^0* cells. The purpose of three successive cultures in the presence of ethidium bromide is to ensure that all cells are indeed *rho^0*.
(vi) Pellet the cells, wash and resuspend in sterile water. Count the cell density and dilute to 10^3 cells/ml. Plate 0.1 ml aliquots on YPglu and incubate the plates at 28°C for 2–3 days. Note that only one dilution is necessary since the ethidium bromide treatment does not reduce cell viability.
(vii) Replicate the YPglu plate onto YPgly or N3 medium to ensure that all the clones are RD (this is always the case).
(viii) Pick up a single colony on the YPglu plate, grow in 5 ml of YPglu and store mutant (*Table 4*).

Table 4. Long-term storage of yeast mitochondrial mutants.

Storage medium

Mitochondrial mutants can be stored indefinitely at −70°C in the following medium
1% (w/v) yeast extract (Difco)
1% (w/v) Bactopeptone (Difco)
2% (w/v) D-glucose
25% (v/v) glycerol
Prepare this medium, distribute 0.5−1 ml aliquots in small screw-cap tubes and autoclave.

Preparation of strains for storage

This is most important since cells should be cultivated in the appropriate medium to eliminate undesirable mutants or revertants. For *ant*^{R} mutants use the corresponding antibiotic- or drug-containing medium. If several *ant*^{R} mutations are present simultaneously, use either the N3 medium or the antibiotic or drug corresponding to the less stable mutations (i.e. the mutation conferring the strongest selective disadvantage on N3 medium). Check purity of culture as in *Figure 3*. For *mit*$^-$ mutants, use YP10. If the strain is pet9, verify the absence of *mit*$^+$ revertants by the qualitative replica cross to a PET 9 *rho*^{0} strain as in *Figure 4*. If the strain is PET, verify the absence of RC revertants by plating on N3 and check the purity of the culture by crossing subclones to a *rho*$^-$ mutant containing the wild-type *mit*$^+$ allele, using the qualitative replica cross method. Clones with original *mit*$^-$ genotype will give rise to RC diploids while spontaneous *rho*$^-$ mutants (which tend to accumulate in PET9 *mit*$^-$ cultures) will not. For *rho*$^-$ mutants, use YPglu or YP10 and test the purity of the culture as in Section 8.2.2.

Freezing and thawing of cells

If culture is on solid medium, dissociate a loopful of cells into storage medium. If the culture is liquid, add 0.1−0.2 ml of culture directly into storage medium. Do not grow cells in this medium. Freeze immediately at −70°C.

To use the stocks, thaw and inoculate into YPglu (*rho*$^-$), YP10 (*mit*$^-$), N3 or antibiotic-containing medium (*ant*^{R}). Incubate for 1−2 days at 28°C.

If *rho*^{0} mutants are needed from a large number of strains (such as when testing the mitochondrial inheritance of *ant*^{R} mutants, see *Table 5*) the following method may provide a quicker alternative.

(i) Place the strains to be mutagenized in grids on N3 medium (e.g. a 5 × 5 grid) and incubate at 28°C for 2−3 days.
(ii) Replicate on N0 medium containing 40 μg/ml of ethidium bromide. Incubate at 28°C in darkness for 2 days.
(iii) Transfer by replica plating onto fresh N0 medium containing 40 μg/ml of ethidium bromide. Incubate as in step (ii).
(iv) Repeat the transfer three times in succession.
(v) After the fourth passage on ethidium bromide, replica plate on YPglu and N3 media. Incubate at 28°C for 3 days.
(vi) Check for the complete absence of growth on N3 and use the YPglu plate for further tests, as necessary.

Caution: ethidium bromide is mutagenic. Wear gloves and dispose of wastes properly. Only add ethidium bromide to the medium after autoclaving; the ethidium bromide stock solution can be easily filter sterilized.

Table 5. Criteria for determining the mitochondrial inheritance of yeast mutations.

Verification of a *rho*^{0} genotype

Apply the following criteria to verify that an RD clone is a *rho*^{0} mutant and not an undesirable *rho*$^-$, *mit*$^-$ or *pet* mutant.

1. Check for the total absence of mtDNA by running total yeast DNA in a CsCl analytical gradient (mtDNA is lighter than nuclear DNA). This is the original definition of *rho*^{0} mutants (51). The procedure is, however, time consuming and not every laboratory is equipped with an analytical ultracentrifuge.
2. Spot approximately 1 μg of total yeast DNA from a minilysate (see Section 9.2) on a nitrocellulose sheet, bake at 80°C under vacuum and hybridize using CsCl-purified *rho*$^+$ mtDNA (Section 9.1) as a probe. Include spots of 1 μg total DNA from a *rho*$^+$ and from a known *rho*^{0} mutant as controls.
3. Cross the presumed *rho*^{0} clone with a series of *mit*$^-$ tester strains representative of the various mitochondrial genes and loci (same procedure as for testing *rho*$^-$ mutants, see *Figure 4*). A *rho*^{0} clone will *only* produce RD diploids whatever the *mit*$^-$ tester used.
4. Cross the presumed *rho*^{0} mutant to a *rho*$^+$ tester and verify that zygotic suppressiveness is zero to within the limits of statistical significance given by the total number of colonies scored (see Section 8.5).

ant^R mutants

Apply the following criteria to distinguish between mitochondrial mutations and undesirable nuclear mutations.

1. Cross the *ant*^{R} mutant to a *rho*$^+$ tester strain with the corresponding *ant*^{S} allele and possessing another (non-allelic) *ant*^{R} mitochondrial mutation, used as a control. Use the quantitative random cross method only (Section 8.3.2). Verify that the mitotic segregation of the *ant*^{R}/*ant*^{S} phenotype occurs among diploids and that the percentage transmission of the *ant*^{R} allele is coordinated with that of the control allele (see *Table 8*).
2. Cross the *ant*^{R} mutant to a *rho*^{0} tester strain, using the quantitative random cross method. Verify that all diploid clones are *ant*^{R}.
3. Isolate a *rho*^{0} derivative of the *ant*^{R} mutant and cross it to a *rho*$^+$ tester strain, using the quantitative random cross method. Check that all of the diploid clones are *ant*^{S}.
4. Sporulate one homoplasmic *ant*^{R} diploid clone from the cross (step 1) and check that all spores of tetrads are *ant*^{R}.

mit$^-$ mutants

Apply the following criteria to distinguish between mitochondrial mutations and undesirable *rho*$^-$ (or *rho*^{0}) or *pet* mutations.

1. Cross the presumed *mit*$^-$ mutant to a *rho*$^+$ *mit*$^+$ tester strain, using the quantitative random cross method of Section 8.3.2. Check the mitotic segregation of the RC/RD phenotype among the diploids derived from the cross.
2. Cross the presumed *mit*$^-$ mutant to a *rho*^{0} tester strain and check that the entire diploid population is RD.
3. Cross the presumed *mit*$^-$ mutant to a series of well-characterized *rho*$^-$ mutants representing, together, every part of the mitochondrial genome. Check that RC diploids are formed with at least one *rho*$^-$ mutant.

rho$^-$ mutants

Apply the following criteria to distinguish between a *rho*$^-$ mutant and an undesirable *pet* or *rho*^{0} mutant.

1. Cross the presumed *rho*$^-$ mutant to a *rho*^{0} PET tester strain and check for the complete absence of RC cells in the diploid progeny.
2. Cross the presumed *rho*$^-$ mutant to a series of *mit*$^-$ tester strains representative of the various mitochondrial genes of loci. A *rho*$^-$ mutant will either give rise to RC diploids in the progeny *of some but not all* crosses or will only give rise to RD diploids in all crosses. In the latter case, use criteria 3 and 4 to distinguish from a *rho*^{0} mutant.
3. Cross the presumed *rho*$^-$ mutant to a *rho*$^+$ tester strain and measure the zygotic suppressiveness as described in Section 8.5.
4. Prepare a minilysate from the presumed *rho*$^-$ mutant, digest with restriction endonuclease, electrophorese and hybridize using purified mtDNA from a *rho*$^+$ strain as probe. Check for the presence of mtDNA and the existence of large deletions in the mitochondrial genome as compared with the *rho*$^+$ strain.

Isolation of independent spontaneous rho⁻ mutants (primary clones) from a rho⁺ strain. Because *rho⁻* mutants are frequent, it is not necessary to induce them, so long as the number of mutants needed is not too large. The major difficulty with spontaneous mutants, however, is to ensure that they are of independent origin. The following method can be used for *rho⁻* mutants.

(i) Grow the *rho⁺* strain in N3 until stationary phase in order to eliminate all pre-existing *rho⁻* mutants.
(ii) Dilute to 10^3 cells/ml in sterile water.
(iii) Plate 0.1 ml aliquots on YPdif and incubate at 28°C for at least 5−6 days (prepare enough plates for the number of *rho⁻* mutants desired, the frequency of spontaneous mutants is usually between 0.5 and 5%).
(vi) Score the plates and pick up RD colonies (petite) individually. These constitute primary clones and need subcloning (see the following note on subcloning).

Induction of a range of rho⁻ mutants (primary clones) from a rho⁺ strain. Rho⁻ mutants can be induced efficiently by a wide variety of chemicals or treatments (see ref. 33 for a review). Most studies on the induction of *rho⁻* mutations (or *rho⁰* mutations since the two were rarely distinguished) have focused on u.v. irradiation, ethidium bromide mutagenesis or berenil mutagenesis. Below is the method used to determine the kinetics of induction of *rho⁻* mutants by ethidium bromide and to obtain a collection of independent mutants (47).

(i) Grow the desired haploid strain (with appropriate auxotrophic markers) in N3 medium until stationary phase (remember that *rho⁻* diploid cells never sporulate so that it is useless to induce a complete collection of *rho⁻* mutants from a diploid strain).
(ii) Pellet the cells by centrifugation at 1000 *g* for 5 min, wash and resuspend in sterile water at a density of 10^7 cells/ml.
(iii) Mix the following in a sterile tube or conical flask wrapped in aluminium foil: 3.85 ml of sterile water; 0.5 ml of 0.5 M sodium potassium phosphate buffer, pH 6.5; 0.5 ml of cell suspension; 0.1 ml of 500 μg/ml cycloheximide stock solution in sterile water and 0.05 ml of 500 μg/ml ethidium bromide stock solution in sterile water.
(iv) Incubate at 28°C with gentle shaking to prevent sedimentation.
(v) Take 0.1 ml aliquots at various time intervals [e.g. immediately after mixing (zero time) and every hour thereafter until 6 h]. Dilute in sterile water to 10^3 cells/ml and plate 0.1 ml aliquots on YPdif medium. Note that the kinetics of induction may vary from strain to strain. When using a given strain for the first time it may be advisable to take aliquots of the mutagenesis for longer time periods.
(vi) Incubate YPdif plates at 28°C for at least 5−6 days.
(vii) Score the plates and draw the kinetics of induction as the logarithm of the frequencies of RC colonies remaining as a function of time of treatment. A straight line should be found for a large time interval.
(viii) Pick up the RD (petite) colonies individually at a time corresponding to a moderate

induction (this reduces the proportion of undesirable *rho*^{0} mutants). These constitute primary clones and need to be subcloned (see the following note on subcloning).

The following method provides a quicker alternative if it is not necessary to study the kinetics of induction.

(i) Grow the desired haploid strain (with appropriate auxotrophic markers) in N3 medium until stationary phase.
(ii) Pellet the cells by centrifugation at 1000 *g* for 5 min, wash and resuspend in 5 ml of N0 medium at a density of 10^7 cells/ml. Add 0.01 ml of an ethidium bromide stock solution at 10 mg/ml (this gives a final concentration of 20 μg/ml ethidium bromide). Wrap the tube in aluminium foil and incubate at 28°C for 60 min (this corresponds to a moderate induction for most strains. Do not extend the incubation time to ensure that mutants remain independent).
(iii) Pellet the cells as in step (2), wash and resuspend in sterile water at a cell density of 10^7 cells/ml.
(iv) Plate 0.1 ml aliquots on YPdif and incubate at 28°C for at least 5–6 days.
(v) Score the plates and pick up RD colonies individually. These constitute primary clones and need to be subcloned (see the following note on subcloning).

Subcloning of rho$^-$ *cells.* RD colonies picked out by any of the above methods represent primary clones (i.e. clones arising from the mutagenized cells themselves). Primary clones are very often mixtures of *rho*^{0} and *rho*$^-$ cells, the latter themselves being mixtures of different *rho*$^-$ mutations. It is therefore necessary to subclone each primary clone at least once prior to using it as a pure *rho*$^-$ mutant of one particular genotype. Primary clones can be either subcloned individually or mixed and subcloned altogether (see below). Test all clones (primary or secondary) as described in Section 8.3.1.

Direct Isolation of large collections of secondary clones of rho$^-$ *mutants directly.* Because subcloning of individual *rho*$^-$ primary clones is time consuming, the following procedure is recommended to obtain a large collection of secondary clones directly.

(i) Pick out RD colonies from one of the above procedures, taking care that equivalent numbers of cells are taken from each colony to limit any bias. Dissociate in a separate tube containing 10 ml of sterile water.
(ii) When a sufficient number of primary clones has been picked up (use moderate induction to limit the frequency of *rho*^{0} clones), mix thoroughly. This suspension constitutes a random mixture of primary clones. Do not grow such a mixture to avoid bias. Count the cell density, dilute to 10^3 cells/ml and plate 0.1 ml aliquots on YPglu.
(iii) Incubate for 2 or 3 days at 28°C. Each colony represents a secondary clone. Secondary clones usually arise from homoplasmic cells and are pure within the limit of stability of the particular *rho*$^-$ mutant considered (see Section 8.2.2).
(iv) Test for secondary clones by the qualitative replica cross procedure (Section 8.3.1).

Deletion of additional mtDNA from a characterized rho⁻ mutant. Rho⁻ mutants are susceptible to further deletion and can be treated for this purpose essentially like *rho*⁺ strains. Spontaneous deletions may occur with a significant frequency in some *rho*⁻ mutants of low stability. In other cases, or if you need a large collection of different deletions, use ethidium bromide mutagenesis (as described previously) or manganese mutagenesis (see Section 8.1.3). Note that firstly, for most *rho*⁻ mutants, the kinetics of loss of mtDNA fragments is slower than for *rho*⁺ strains, and secondly mutant clones are indistinguishable by their phenotype from non-mutant clones. Plate the mutagenized cultures on YPglu or YP10 only (YPdif is useless for this!). The mutant clones picked up are primary clones and need subcloning.

8.1.2 *Isolation of ant*ᴿ *mitochondrial mutants*

Mitochondrial point mutations conferring an *ant*ᴿ phenotype are rare (of the order of $10^{-7}-10^{-8}$ per cell per division). However, because they remain RC and, in addition, become resistant to a specific inhibitor, *ant*ᴿ mutants can easily be selected for. Thus, there is no major need for a mutagenesis to increase their frequency (although manganese mutagenesis as described in Section 8.1.3 can be used).

The following method can be used to select a collection of independent mitochondrial *ant*ᴿ mutants.

(i) Grow the desired haploid strain with at least one auxotrophic requirement in N3 medium (remember that it is necessary that the parental strain be RC to select *ant*ᴿ mutants: use only a *rho*⁺ *mit*⁺ *PET* strain).
(ii) Dilute in sterile water to 10^3 cells/ml and plate 0.1 ml aliquots on N3 or streak the culture on N3 plates for single colonies. Incubate at 28°C for 3 days.
(iii) Pick up single colonies (subclones) and inoculate individually in 5 ml of N3 medium. Incubate at 28°C for 2 days. Prepare as many subclones as you eventually want mutants.
(iv) Pellet the cells by centrifugation at 1000 *g* for 5 min and resuspend in 1 ml of sterile water.
(v) Plate 0.1 ml of each suspension on N3 or N1 medium containing the appropriate antibiotic or drug.
(vi) Incubate the selective plate at 28°C for at least a week to allow appearance of resistant colonies over the lawn of inhibited cells. Some antibiotics require longer incubation times (e.g. N3C or N3M plates require up to 3 weeks of incubation). Note the gradual appearance of colonies by examining the plates at regular intervals (e.g. every other day). Prevent dessication during prolonged incubation by placing the plates in a humidity-controlled incubator or put the plates in a container with wet pads in a Petri dish.
(vii) Pick up a single mutant colony per original subclone to ensure that mutants are of independent origin.
(viii) Inoculate in 5 ml of N3 medium and incubate at 28°C for 2 days.
(ix) Dilute in sterile water to 10^3 cells/ml and plate 0.1 ml on the N3 medium containing the antibiotic or drug (or streak out for single colonies). This subcloning step ensures that the colony picked up on the selective plate was actually due to a resistant mutation and not a physiological adaptation of sensitive cells to

a prolonged incubation on the selective medium at high cell density.

(x) Pick up a single subclone from each mutant, inoculate in N3 or N1 medium containing the antibiotic or drug, incubate at 28°C for 2–3 days and store the mutants (*Table 4*).

Although this procedure is extremely simple, the following remarks will help to ensure a successful isolation of *ant*^{R} mitochondrial mutants.

Firstly, for several antibiotics (e.g. chloramphenicol, spiromycin, paromomycin) the concentrations of antibiotics necessary to inhibit the wild-type sensitive strains can be relatively close to the concentrations used to select resistant mutants. Hence, a residual growth of the sensitive cells on the selective plates may occur. The level of resistance of wild-type cells varies considerably from strain to strain. It is advisable to select strains with the lowest level of resistance before isolating mitochondrial mutants or before using them as tester strains in crosses.

Secondly, although most *ant*^{R} mitochondrial mutants are healthy RC cells, some show growth rates slower than the wild-type on N3 or YPgly media (e.g. chloramphenicol-resistant or mucidin-resistant cells). It is recommended to grow *ant*^{R} mutants always in the appropriate antibiotic-containing medium prior to using them for genetics or molecular tests in order to eliminate possible spontaneous revertants.

Thirdly, not all resistant mutants isolated by the above procedure are due to mutations in the mitochondrial genome and it is *absolutely necessary* that the mutants are tested genetically to determine if the mutation is mitochondrial or nuclear prior to further consideration (see *Table 5*). Note that mitochondrial mutants are generally resistant to much higher concentrations of antibiotics or drugs than their nuclear counterparts; hence the use of high concentrations of antibiotics in media. Resistance to high concentrations is *not*, however, sufficient to prove that the mutation is mitochondrial and not nuclear.

8.1.3 *Isolation of mit*$^-$ *mitochondrial mutants*

Point mutations in the mitochondrial genome conferring an RD phenotype are very rare as compared with the production of *rho*$^-$ or *rho*^{0} mutants. It is therefore necessary to increase their frequency by mutagenesis and to eliminate the *rho*$^-$ or *rho*^{0} mutants. This is best achieved by combining manganese mutagenesis (48) with the use of the *op1* (*pet9*) nuclear mutation which is lethal if combined with a *rho*$^-$ or *rho*^{0} mutation (49) but not if combined with a *mit*$^-$ mutation. The following method has been developed (50).

(i) Grow a *rho*$^+$*mit*$^+$*pet9* haploid strain with at least one auxotrophic marker in YP10 medium.

(ii) Pellet the cells by centrifugation at 1000 *g* for 5 min, wash and resuspend in sterile water, count the cells and dilute in water to 10^7 cells/ml.

(iii) Inoculate 0.1 ml of the cell suspension in 5 ml of YP10 medium. Add 0.1 ml of a 350 mM $MnCl_2$ sterile solution (to a final concentration of 7 mM $MnCl_2$ in the culture medium). Note that Mn^{2+} is toxic, so do not exceed a final concentration of 10 mM.

(iv) Incubate at 28°C for 24 h with shaking to prevent cell sedimentation.

(v) Count the cells and dilute in sterile water to 2×10^3 cells/ml. Plate 0.1 ml aliquots on YP10 plates. Remember that *mit*$^-$ mutants are rare even after induction, so make sure that you plate enough Petri dishes to obtain several thousands of colonies in total.

(vi) Incubate at 28°C for 3 days. The population of colonies appearing is composed of non-mutant and of the desired *mit*$^-$ mutants (*rho*$^-$ and *rho*^{0} mutants, which are also induced by Mn^{2+}, are eliminated). However, all colonies are RD, due to the *pet9* mutation (in fact *pet9* does not confer an actual RD phenotype, but cells grow so extremely slowly on N3 or YPgly that they do not form visible colonies).

(vii) Pick up colonies individually and place in 8 × 8 grids on YP10. Incubate at 28°C for 3 days.

(viii) Replicate the grids on YP10, N3, W0 media and on W0 plates covered with a lawn of PET9 *rho*^{0} tester strain of opposite mating type and auxotrophic requirements (see Section 8.3.1). Incubate at 28°C for 3 days.

(ix) Check for the absence of growth of haploids on the N3 and W0 replicas and the confluent growth of diploids on W0. Replica plate the diploids on N3. Incubate the plates at 28°C for 3 days.

(x) Score diploid colonies that fail to grow on the N3 replica and pick up the corresponding haploid clone on the YP10 grid.

(xi) Dissociate and dilute to 10^3 cells/ml in sterile water. Plate 0.1 ml aliquots on YP10. Incubate and repeat steps (vii)–(x) on subclones.

(xii) Pick out a single subclone from each mutant. Inoculate in YP10 and incubate at 28°C for 2 days. Store the mutant subclone (*Table 4*).

8.1.4. *Isolation of conditional mit*$^-$ *mitochondrial mutants*

A number of temperature-dependent mitochondrial mutants have been described. Those conferring a conditional RD phenotype occur in many, if not all, mitochondrial genes. Temperature-dependent *mit*$^-$ mutants can be treated as other *mit*$^-$ mutants at their restrictive temperature but as RC strains at their permissive temperature (although their growth rate on N3 medium may be affected). For the isolation of such mutants one must first define the restrictive and permissive temperatures according to ones requirements (e.g. 35°C and 25°C, respectively if heat-sensitive mutants are desired or 18°C and 28°C, respectively if cold-sensitive mutants are required) and then apply either of the following methods.

If the pet9 mutation is used. Steps (i)–(viii) are the same as in Section 8.1.3.

(ix) Replica plate diploids on N3 medium in duplicate. Incubate one N3 plate at the restrictive temperature and the other at the permissive temperature.

(x) Score the diploid colonies that fail to grow on the N3 replica at the restrictive temperature but do grow at the permissive temperature. Pick out the corresponding haploid clone on the YP10 grid.

(xi) and (xii) Same as in Section 8.1.3.

If a direct phenotypic screening is used.

(i) Grow a *rho*$^+$*mit*$^+$PET9 haploid strain with at least one auxotrophic mutation in N3 medium.

(ii)−(vii) Same as in Section 8.1.3 except that the population of colonies is RC.

(viii) Replicate the grids on N3 medium in duplicate. Incubate one plate at the restrictive temperature and the other at the permissive temperature.

(ix) Score haploid colonies that grow on N3 at the permissive temperature but not at the restrictive temperature. Pick out the corresponding clone on YP10 (do not worry about *rho*$^-$ or *rho*^{0} mutants as they will never be conditional mutants!).

(x) Cross with a *rho*^{0} tester strain and verify that the diploids grow on N3 at the permissive temperature but not at the restrictive temperature (this step eliminates undesirable nuclear *pet*$^-$ mutations).

(xi) and (xii) Same as in Section 8.1.3.

8.2 **Segregation tests**

8.2.1 *Determination of the homoplasmic/heteroplasmic state of cells and the purity of cultures*

Heteroplasmic cells may originate from crosses, mutagenesis or spontaneous mutations (mainly *rho*$^-$ mutations which are frequent and unstable, see below). It is of primary importance in mitochondrial genetics to know the homoplasmic/heteroplasmic state of a cell. This can only be determined *a posteriori* from the pure/mixed character of the clone derived from that cell, after mitotic segregation of the mtDNA molecules has taken place. The mechanism by which segregation takes place is not entirely clear (see ref. 33 for review) but it does appear that, under normal conditions, the process is rapid, that is heteroplasmic cells always give rise to mixtures of different homoplasmic cells during formation of a visible colony. A single subcloning is therefore usually sufficient to isolate homoplasmic cell lines from heteroplasmic situations. Rare cases of persistent heteroplasmons resulting from local peculiarities of mtDNA (e.g. a duplication) have been reported. However, even under such exceptional circumstances the formation of a visible colony is sufficient to result in mixed clones composed of numerous homoplasmic cells in addition to the persistent heteroplasmic type.

The purity of cultures can be determined by plating out diluted aliquots on a non-selective medium (e.g. YPglu or YP10 for haploids or W0 or W10 for diploids) and by replicating the subclones on selective media (e.g. N3 for distinguishing between RC and RD cells or antibiotic-containing media for distinguishing between *ant*^{R} and *ant*^{S} cells). The pure/mixed character of the culture is demonstrated by the homogeneity/heterogeneity of the subclones (it is necessary to examine a sufficient number of subclones for statistical significance), the purity of each subclone itself being determined by subsequent subcloning and/or by the appearance of replica colony on the selective media (i.e. confluent growth, sectored growth, papillate growth or complete absence of growth). *Figure 3* shows an example of this.

For correct interpretation of the appearance of the replica colonies, it is necessary to note that, from an average size colony, $2-5 \times 10^5$ cells are deposited by the velvet on the replica. This figure sets the lower limit of detection of the replica method (i.e. absence of growth means a clone in which the frequency of cells able to grow on the selective medium is below $2-5 \times 10^{-6}$). This figure is particularly convenient for mitochondrial genetics because firstly it is well below the frequency of the rarer homoplasmic cell type segregated out of a heteroplasmic cell, and secondly it is at least

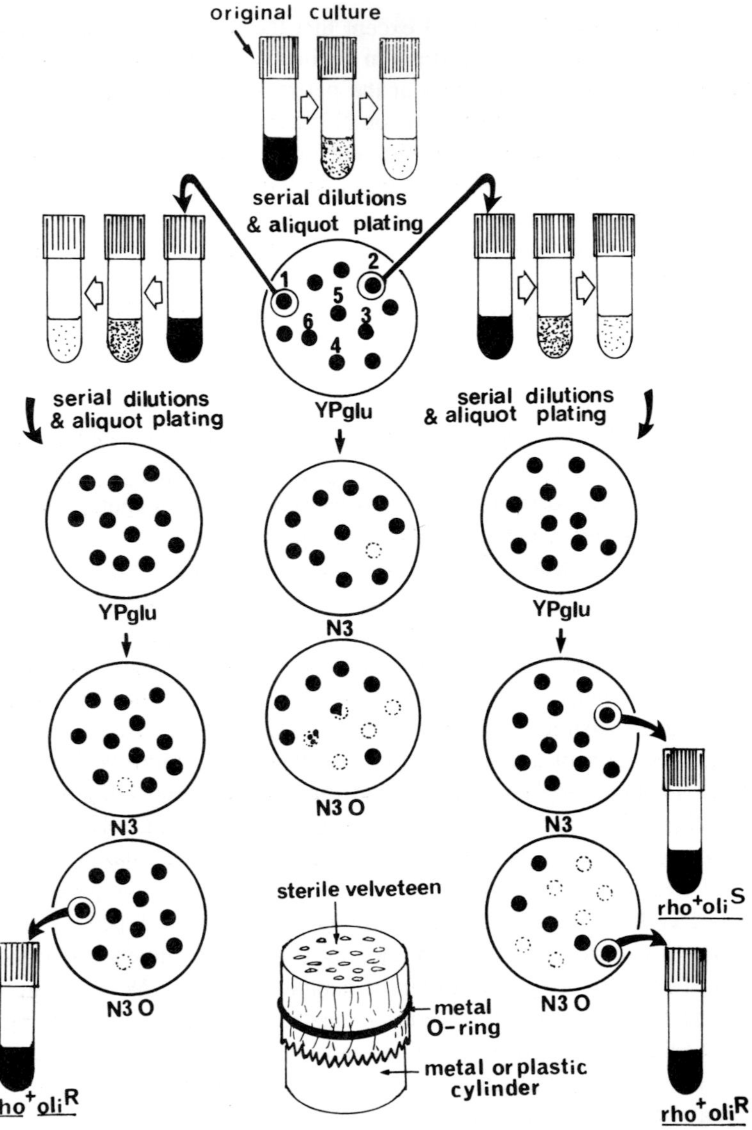

Figure 3. Subclonings and replica tests determine the homoplasmic/heteroplasmic state of cells and the purity of cultures. The example shows the test by replica and the isolation of homoplasmic subclones from a hypothetical *rho*$^+$ population containing homoplasmic and heteroplasmic cells for the *oli*^{R} mutation. **Clone 1** is a pure *rho*$^+$ *oli*^{R} clone originated from a homoplasmic cell. Subcloning reveals 100% of clones of the same type (except for the spontaneous *rho*$^-$ mutants). **Clone 2** is a mixed clone originated from a heteroplasmic cell containing simultaneously *rho*$^+$ *oli*^{R} and *rho*$^+$ *oli*^{S} mtDNA molecules. Subcloning permits the isolation of pure *rho*$^+$ *oli*^{R} subclones and pure *rho*$^+$ *oli*^{S} subclones. **Clone 3** is either a pure *rho*$^-$ (or *rho*^{0}) clone or a mixed clone containing various *rho*$^-$ mutants. Subcloning and qualitative replica crosses are necessary to distinguish between these possibilities. **Clone 4** is a homoplasmic *rho*$^+$ *oli*^{S}. **Clone 5** is a mixed clone as clone 2 but in which the proportion of *rho*$^+$ *oli*^{R} cells is lower. **Clone 6** is a mixed clone as clone 5 but in which the proportion of *rho*$^+$ *oli*^{R} cells is even lower. Note that the conclusions drawn about the homoplasmic/heteroplasmic state of the parental cell of each clone are valid only if each colony originates from a single cell. Clumpy strains make the analysis more difficult and should be avoided whenever possible.

one order of magnitude above the spontaneous mutation rate for point mutations or revertants. Complete absence of growth of the replica trace on a selective medium is therefore demonstrative of a pure clone derived from a homoplasmic cell. Confluent growth of the replica, on the contrary, may indicate either a pure clone (i.e. entirely composed of cells able to grow on the selective medium) or a mixed clone in which the frequency of such cells is sufficiently high. Sectors or papillate growth are obviously indicative of mixed clones derived from heteroplasmic cells.

Now, the classical distinction between RC cells whose mitochondrial genotype can be determined directly from their phenotype and RD cells whose mitochondrial genotype can only be determined after marker rescue obviously holds true for the above analysis of subclones.

8.2.2 *Methods to check the purity of and/or purify a rho$^-$ culture*

All *rho*$^-$ mutants are intrinsically unstable and spontaneously undergo secondary mutation, deletions, rearrangements or complete loss of their mtDNA. Because all *rho*$^-$ (and *rho*^{0}) mutants have the same RD phenotype there is no direct method to eliminate such undesirable mutations and the purity of each culture needs to be determined carefully. The probability that the original genotype remains unaltered is generally constant per cell division (this is the intrinsic stability of the *rho*$^-$). Consequently, the overall frequency of cells of the original genotype in a culture (the purity of the culture) shows an exponential decrease relative to the total number of cell generations of the culture. The intrinsic stability varies to a large extent from one *rho*$^-$ mutant to the next and is also influenced by the nuclear genotype of the strain used. In practice, it is impossible to work with *rho*$^-$ mutants with a stability lower than 0.95/cell/generation (in this case only 30% of the cells have the original genotype after the growth of a visible colony!). But even with *rho*$^-$ mutants of high stability the proportion of cells with the original genotype rapidly decreases in the population during cultivation (e.g. *rho*$^-$ mutants with stability values equal to 0.99 still give rise to more than 20% of cells with altered genotypes after the growth of a visible colony). For this reason it is strongly recommended not to grow a *rho*$^-$ mutant from a previous culture without checking the purity of that culture and the number of generations that it has undergone; in addition it is advisable to store aliquots of the tested cultures at −70°C (see *Table 4*) for subsequent use.

The purity of a *rho*$^-$ culture can be determined by a qualitative replica cross and/or by minilysate analysis.

Replica cross test.

(i) Dilute the culture in sterile water to 10^3 cells/ml, store the undiluted culture in the refrigerator at 5°C.
(ii) Plate 0.1 ml of diluted culture on YPglu, incubate at 28°C for 3 days.
(iii) Place colonies on grids on YPglu (~200−300 colonies in total are needed if statistical significance of the composition of the population is desired) and incubate for 3 days.
(iv) Test colonies by the qualitative replica cross using appropriate *rho*$^+$ tester strains as shown in *Figure 4*.

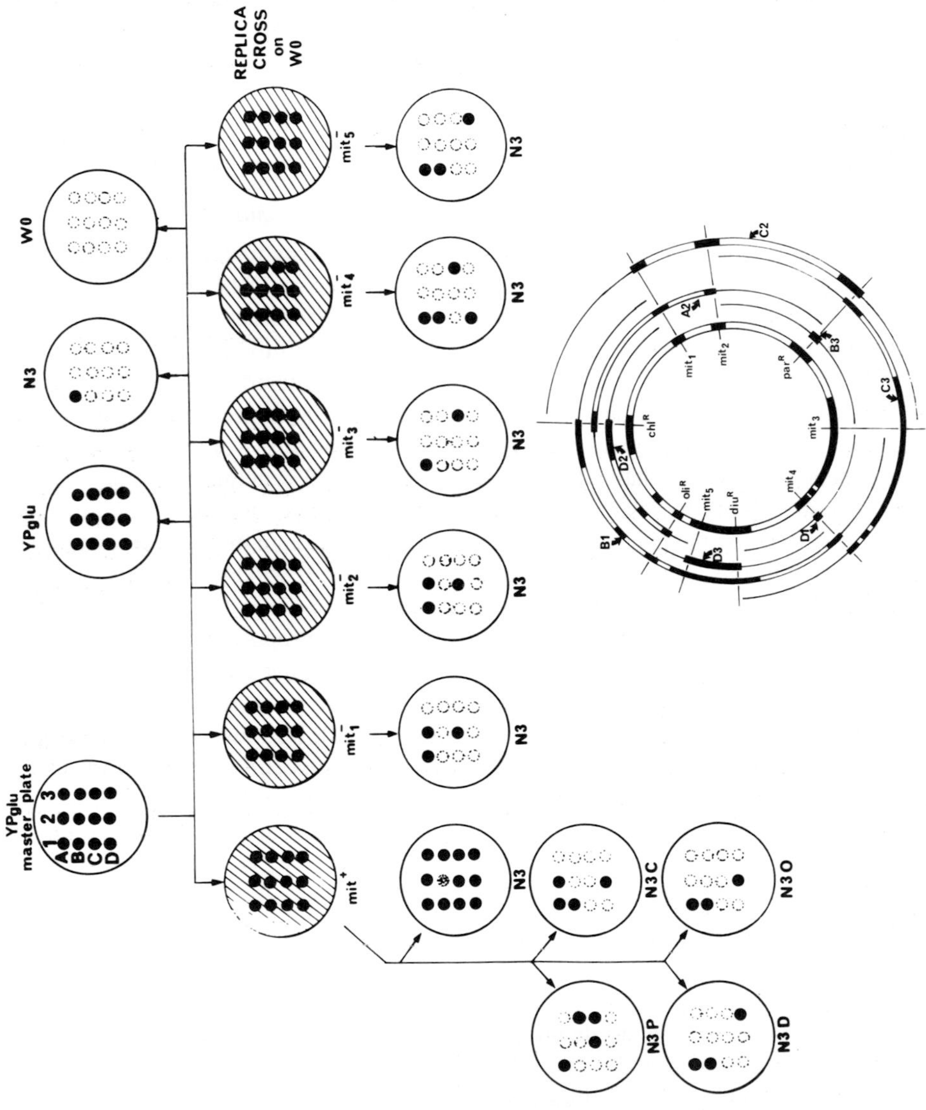
YPglu master plate
A
B
C
D
1
2
3
YPglu
N3
W0
REPLICA CROSS on W0
mit+
mit1-
mit2-
mit3-
mit4-
mit5-
N3
N3C
N3O
N3P
N3D
chlR
oliR
mit5
diuR
mit4
mit3
parR
mit2
mit1
A2
B1
B3
C2
C3
D1
D2
D3

Minilysate analysis.

(i) and (ii) As for the replica cross.
(iii) Pick up colonies one by one and inoculate in 5 ml of YPglu.
(iv) Prepare minilysates as in Section 9.2 and test the mtDNA by restriction digest and hybridization using CsCl-purified mtDNA as a probe.

If purity is satisfactory, aliquot the original culture and store aliquots at −70°C (*Table 4*). Always prepare enough aliquots of your clone for the foreseeable future. It is less work to prepare more aliquots than to repeat the testing of cultures. Use one aliquot to inoculate the culture needed for each experiment.

If purity is not satisfactory, pick up colonies with the required genotype on original YPglu plates and inoculate in 10 ml of YPglu or YP10 medium. Incubate at 28°C for 2 days. Test the purity of the subclones as before.

8.2.3 *How to transfer a known mitochondrial genotype into a new nuclear genetic background*

This can be performed by either of the following methods but the results are not always equivalent.

Method A: sporulation. This method can only be applied to *rho*$^+$*mit*$^+$ genotypes. Cross your strain with the mitochondrial genotype of interest to a *rho*^{0} strain with opposite mating type and different auxotrophic requirements. Do a quantitative random cross using method A steps (i)−(iv) of Section 8.3.2. Pick out the diploids and sporulate as described in *Table 9*. Note that all haploid segregants have the same mitochondrial genotype but different nuclear genotypes (unless the *rho*^{0} tester was already isogenic to your strain of interest except for mating type).

Figure 4. The qualitative replica cross. The example shows the genetic analysis of individual *rho*$^-$ clones by crosses to a series of well-defined tester strains (all are alpha HIS1 MET1 ade 2 pet9 *rho*$^+$ and either *mit*$^+$ *or mit*$^-$) and qualitative analysis of the resulting diploids by replica. **Clone A1** is the *rho*$^+$ control (use the parental strain from which the collection of *rho*$^-$ mutants is derived). In the hypothetical example shown the genotype of the parental strain is *a his1 met1 ADE2 PET rho*$^+$ *chl*^{R} *oli*^{R} *par*^{R} *diu*^{R} *mit*$^+$. **Clone A3** is a *rho*^{0} control (derived from the same parental strain as the collection of *rho*$^-$ mutants tested). Complete genotypes of other clones can be deduced as follows:
clone A2, *rho*$^-$ *chl*^{R} *oli*^{0} *par*^{0} *diu*^{0} *mit1*$^+$ *mit2*$^+$ *mit3*^{0} *mit4*^{0} *mit5*^{0}
clone B1, *rho*$^-$ *chl*^{R} *oli*^{R} *par*^{0} *diu*^{R} *mit1*^{0} *mit2*^{0} *mit3*^{0} *mit4*$^+$ *mit5*$^+$
clone B2, *rho*$^-$ HS
clone B3, *rho*$^-$ chl^0 *oli*^{0} *par*^{R} *diu*^{0} *mit1*^{0} *mit2*^{0} *mit3*^{0} *mit4*^{0} *mit5*^{0}
clone C1, either *rho*$^-$ chl^0 *oli*^{0} *par*^{0} *diu*^{0} *mit1*^{0} *mit2*^{0} *mit3*^{0} *mit4*^{0} *mit5*^{0} or *rho*^{0}
clone C2, *rho*$^-$ *chl*^{0} *oli*^{0} *par*^{R} *diu*^{0} *mit1*$^+$ *mit2*$^+$ *mit3*^{0} *mit4*^{0} *mit5*^{0}
clone C3, *rho*$^-$ *chl*^{0} *oli*^{0} *par*^{R} *diu*^{0} *mit1*^{0} *mit2*^{0} *mit3*$^+$ *mit4*$^+$ *mit5*^{0}
clone D1, *rho*$^-$ *chl*^{0} *oli*^{0} *par*^{0} *diu*^{0} *mit1*^{0} *mit2*^{0} *mit3*^{0} *mit4*$^+$ *mit5*^{0}
clone D2, *rho*$^-$ *chl*^{R} *oli*^{R} *par*^{0} *diu*^{0} *mit1*^{0} *mit2*^{0} *mit3*^{0} *mit4*^{0} *mit5*^{0}
clone D3, *rho*$^-$ *chl*^{0} *oli*^{0} *par*^{0} *diu*^{R} *mit1*^{0} *mit2*^{0} *mit3*^{0} *mit4*^{0} *mit5*$^+$.
The insert gives a schematic representation of the mitochondrial map of *S. cerevisiae* (major genes in full) and places the segment retained by each *rho*$^-$ clone with its maximum limits (thin lines). Such a map can only be drawn if *rho*$^-$ clones tested are secondary clones or clones of subsequent orders. For primary clones, there is no indication if the different genetic markers are retained on the same mtDNA molecules of the same cells. Only pick out the interesting *rho*$^-$ clones on original YPglu plates (other plates have diploid cells issued from the cross, not the desired *rho*$^-$ mutant cells).

Table 6. Standard yeast strains for mitochondrial genetics.

KL14-4A: a *his1 trp2* $rho^+ omega^+ chl^R_{321} oli^R_1 par^R_{454}$
KL14-4A/I21: *leu2* derivative of KL14-4A
Both strains are used as standard rho^+ strains for crosses and mapping purposes. Large collections of rho^- mutants are available from both strains. Parental strains of several original ant^R mutations. (38, 76)

D273-10B/A: *alpha* met $rho^+ omega^+$
Standard rho^+ strain for crosses and mapping purposes. Large collections of rho^- mutants and of mit^- mutants available from this strain. (77)

777-3A: *alpha* ade1 pet9 rho^+
Standard pet9 strain (rho^- lethal). Large collection of mit^- mutants available from this strain. (50)

AB1-4A/8: a his4C ADE1 PET9 derivative of 777-3A rho^+
AK51-17A: *alpha* met ADE1 PET9 derivative of 777-3A rho^+
Both strains are used for crosses and/or transfer of mit^- mutations from 777-3A to isogenic PET9 nuclear backgrounds. (78)

MH32-12D: a ade2 his1 rho^+ chl^R_{321} ery^R_{221} oli^R_{144} par^R_{454}
MH41-7B: a ade2 his1 rho^+ chl^R_{321} ery^R_{514} oli^R_{145} par^R_{454}
Large collections of rho^- mutants available from these strains. (79)

J69-1B: *alpha* ade1 his rho^+
Parental strain of numerous mit^- mutants (especially in the ATP synthase genes). (80)

ID41-6/161: a ade lys $rho^+ omega^-$ chl^R_{321} oli^R_{1-4} par^R_1
Parental strain of numerous mit^- mutants. (81)

55R5-3C: a ura1 $rho^+ omega^-$
Parental strains of numerous original ant^R mutants. Standard $omega^-$ strain. (36)

DP1-1B: *alpha* his1 trp1 $rho^+ omega^+$
Parental strain of several original ant^R mutants. Standard $omega^+$ strain. (36)

IL828-3C. a his1 ura1 $omega^+$ chl^R_{321} ery^R_{221} oli^R_1
IL828-5D: *alpha* his1 $omega^+$ chl^R_{321} ery^R_{221} oli^R_1
IL871-1D: a ura1 trp1 $omega^-$ chl^R_{321} ery^R_{221} oli^R_1
IL871-1A: *alpha* his1 trp1 $omega^-$ chl^R_{321} ery^R_{221} oli^R_1
Standard *omega* tester strains. (38,39,82)

JC8/55: a leu1 can^R kar1 rho^0 (83)
CK50A/50: *alpha* leu2 trp1 tyr6 phe can^R kar1 rho^0. (Perrodin and Slonimski, unpublished)
Both strains are used to transfer mitochondrial mutations into new nuclear backgrounds by cytoduction.

Method B: cytoduction. This method can be applied to all types of mitochondrial mutations. It is based on the property of the *kar1* mutant to delay nuclear fusion after cytoplasmic fusion during mating. As a result, a fraction of zygotic buds are haploids with the nucleus of one parent but the cytoplasm of both. Combination of the *kar1* mutation with the rho^0 mutation offers, therefore, a simple method to transfer a known mitochondrial genome into a new genetic background.

(a) For $rho^+ mit^+$ strains.

(i) Cross your strain with the mitochondrial genotype of interest to a *kar1* rho^0 strain with opposite mating type and different auxotrophic requirements (*Table 6*). Use synchronous mating as in Method *B* of Section 8.3.2, steps (i)–(vi).

(ii) When zygotes are formed take aliquots at various time intervals (e.g. every 2 h), dilute to 10^3 cells/ml and plate 0.1 ml aliquots on N3. Incubate at 28°C for 2–3 days.

(iii) Replicate on W0 and W0 plates supplemented for the auxotrophic requirements of the *kar1 rho*^{0} parent. Incubate at 28°C for 3 days.

(iv) Score the RC clones with auxotrophic requirements identical to the *kar1 rho*^{0} parent. They are cytoductants with the nucleus of the *kar1* parent and the mtDNA of your strain.

(v) If you want to transfer your mitochondrial genotype of interest into a specific non-*kar1* nucleus, cross this cytoductant to a KAR1*rho*^{0} strain with the desired nuclear genetic background as in steps (i) and (ii) above, then replicate on W0 and W0 plates supplemented for the auxotrophic requirements of the KAR1*rho*^{0} parent. Pick out the RC clones with auxotrophic requirements identical to those of your KAR1 parent. These are the final cytoductants with both mitochondrial and nuclear genotypes of interest.

(b) For *rho*$^-$ mutants.

(i)–(iii) Same as in Method (a) except that aliquots are plated on YPglu instead of N3 medium.

(iv) Pick up RD clones with auxotrophic requirements identical to the *kar1* parent. Distinguish the desired *kar1 rho*$^-$ cytoductant from non-mated *kar1 rho*^{0} parent by either the qualitative replica cross using a known *mit*$^-$ tester strain (Section 8.3.1) or the minilysate procedure (Section 9.2). In all cases verify the identity of mtDNA of the cytoductant selected with the original *rho*$^-$ mutant, using the minilysate procedure, to eliminate the possibility of secondary mutations of the *rho*$^-$ mtDNA.

(v) Same as in Method (a) except that the RD clones are picked out and screened as in step (iv).

(c) For *mit*$^-$ mutants.

(i)–(iii) Same as Method (*a*).

(iv) Pick up RD clones with auxotrophic requirements identical to the *kar1* parent. Distinguish the desired *kar1 mit*$^-$ cytoductant from the non-mated *kar1 rho*^{0} parent by the qualitative replica cross method using a known *rho*$^-$ tester strain (Section 8.3.1).

(v) Same as in step (iv).

Note that the frequency of cytoductants is variable from cross to cross. Plate 200–300 colonies from the mating mixture at each time interval until 6–8 h after zygote formation.

8.3 Tests for recombination

8.3.1 *Qualitative replica crosses*

Because all *rho*$^-$ (or *rho*^{0}) mutants have the same RD phenotype, determination of their mitochondrial genotype requires rescuing their alleles, by recombination, into an RC cell (52). The rationale is as follows.

mit$^+$ *alleles*: When crossed to a given *mit*$^-$ mutant, each *rho*$^-$ mutant which retains the fragment of mtDNA containing the corresponding *mit*$^+$ allele will give rise to at least some RC diploids. On the contrary, each *rho*$^-$ mutant that has lost the fragment of mtDNA containing the corresponding *mit*$^+$ allele will give rise to RD diploids only. In this case the genotype of the *rho*$^-$ mutant for that particular allele is written *mit*^{0}. Because the test of diploids is qualitative, the frequencies of RC recombinants generated are not taken into account so long as they are above the limit of detection of the replica method (see Section 8.2.1).

ant^R *alleles*: The same principle applies to rescuing *ant*^{R} mutations carried by the *rho*$^-$ mutants except that crosses are performed to a *rho*$^+$ tester strain and that the diploids are tested for the *ant*^{R}/*ant*^{S} phenotype instead of the RC/RD phenotype. *Rho*$^-$ mutants that have lost the fragment containing the *ant*^{R} allele are indicated *ant*^{0}.

Determination of the markers retained or lost by rho$^-$ *mutants (Figure 4).*

(i) Place the *rho*$^-$ mutants in grids on YPglu medium (grids 8 × 8 are convenient for this purpose) and incubate at 28°C for 2–3 days. Remember to incorporate the *rho*$^+$ parental strain and its *rho*^{0} derivative on each grid as controls.

(ii) Inoculate *pet9 rho*$^+$ *mit*$^-$ tester strains (with opposite mating-type and auxotrophic requirements) in 10 ml of YP10 medium. Prepare a set of tester strains with *mit*$^-$ mutations representative of the various genes or loci for which you want to determine the loss/retention in the *rho*$^-$ tested (see *Tables 6* and *7*). If the *rho*$^-$ mutants tested originate from a *rho*$^+$ strain which has *ant*^{R} mutations, also inoculate a *pet9 rho*$^+$ *mit*$^+$ tester strain in 10 ml of YP10 medium. Incubate at 28°C for 2–3 days without agitation.

(iii) Plate 0.1 ml of the YP10 cultures on W0 medium to form a lawn (plating should be very regular for good mating efficiency).

(iv) Replica plate the YPglu grids onto W0, N3 and YPglu media first, then onto the lawns of tester strains (remember to use a new velvet for each lawn to avoid cross-contamination; a unique master plate can tolerate 5–10 successive impressions if good quality velvets are used). Incubate at 28°C for 3 days. Keep the master plates in the refrigerator at 5°C.

(v) Sort the W0 replica cross plates. Check for the absence of growth of replicas on N2 and W0 plates. Check for the confluent growth of diploids over the replica colonies and for the absence of growth of the lawn elsewhere (a slight residual growth of the auxotrophic lawn may be visible; it is due to the very high density of the inoculum and is not a problem so long as a *pet9* strain is used).

(vi) Replica plate the diploids on N3 medium alone (for the *mit*$^-$ testers) and/or on N3 containing antibiotic (for the *mit*$^+$ tester). Incubate the new replicas at 28°C for 3 days. Store all other plates in the refrigerator at 5°C.

(vii) Sort all the plates and score the result of each *rho*$^-$ mutant individually (see example in *Figure 4*).

This method indicates which alleles are present in each *rho*$^-$ clone but does not demonstrate that the segments between retained alleles are present. Always apply the minilysate procedure or purify the mtDNA to check continuity of the fragments retained as compared with the *rho*$^+$ mtDNA.

Table 7. A set of yeast mitochondrial mutations useful for mapping. Mutations are listed clockwise on map (see ref. 33).

Gene		*Mutations*	*Phenotype*	*Approximate map location (units on KL14-4A map)*
21S rRNA[a]		cs901	conditional syn⁻	95–97%
		ery^{R}_{221}	erythromycin resistance	97–98%
		chl^{R}_{321}	chloramphenicol resistance	0%
Asp tRNA		ts170	conditional syn⁻	5–6%
Tyr tRNA		tsm8	conditional syn⁻	6–7%
Sub II cyt. ox. (*oxi1* gene)		M13-249	respiratory deficient	15–17%
Sub III cyt. ox. (*oxi2* gene)		G199	respiratory deficient	20–22%
16S rRNA		par^{R}_{454}	paromomycin resistance	35–38%
Sub I cyt. ox. (*oxi3* gene)		G20	respiratory deficient	45–58%
		G922	respiratory deficient	
Sub 6 ATPase (*oli2* gene)		pho^{-}_{1}	phosphorylation deficient	60–65%
		oli^{R}_{144}	oligomycin resistance	
Cyt.b (cob-box gene)	exon 1	G706	respiratory deficient	72–80%
		$ana^{R}_{1\text{-}32}$	antimycin resistance	
		$diu^{R}_{2\text{-}732}$	diuron resistance	
		$muc^{R}_{1\text{-}771}$	mucidin resistance	
	intron 2	G1370	respiratory deficient	
	exon 4	G1988	respiratory deficient	
		$ana^{R}_{2\text{-}25}$	antimycin reistance	
		$diu^{R}_{1\text{-}731}$	diuron resistance	
	intron 4	G1659	respiratory deficient	
		M7832	respiratory deficient	
	exon 6	M6-200	respiratory deficient	
		$muc^{R}_{2\text{-}772}$	mucidin resistance	
Sub 9 ATPase (*oli1* gene)		oli^{R}_{1}	oligomycin resistance	82–83%

[a]Under the control of the *omega* locus.

Mapping of mit⁻ mutants using petite deletion mapping. This is the reciprocal of the previous method.

(i) Place the *mit⁻* to be mapped in grids on YP10 plates. Incubate at 28°C for 2–3 days.

(ii) Inoculate *rho⁻* testers (with opposite mating type and auxotrophic requirements) into 10 ml of YP10 medium. Prepare a set of *rho⁻* mutants with overlapping fragments of mtDNA that cover either the entire genome or the segment in which the *mit⁻* mutations are expected. Incubate at 28°C for 2–3 days.

(iii)–(vii) Same as for the previous method.

Each *mit⁻* mutant that gives rise to RC diploids with a given *rho⁻* tester is a mutation that maps within the fragment retained by that *rho⁻*. Each *mit⁻* mutant that gives rise to RD diploids only with a given *rho*$^-$ tester is a mutation that maps outside of the fragment retained by that *rho*$^-$. Use an appropriate set of *rho*$^-$ mutants with overlapping fragments for a quick and precise mapping of the *mit*$^-$ mutation. For significant results it is critical that your set of *rho*$^-$ mutants has been submitted to precise restriction mapping (or, better, to complete DNA sequencing). Any *rho*$^-$ mutant whose retained fragment might be rearranged as compared with the *rho*$^+$ map should be avoided. Remember that, since the petite deletion mapping is a recombination test and not a complementation test, it is not necessary for functional segments of the mtDNA to be retained in the *rho*$^-$ mutants. However, any limitation or inhibition of recombination will result in a lack of RC recombinants in diploids which may be misleading. Check the consistency of the results, using different *rho*$^-$ testers, to deduce the final mapping assignments.

8.3.2 *Quantitative random cross*

This is the standard type of cross for the quantitative determination of the percentage transmission of mitochondrial alleles and of the recombination frequencies between pairs of alleles. It is based on the fact that both recombination and segregation of mtDNA molecules are rapid and efficient. The cross can be made either by placing the two strains in contact directly on the minimal medium selective for diploids or by synchronous mating in complete medium, followed by selection of diploids in minimal medium. The first method is quicker but results in a severe counter-selection of RD cells due to their slow growth on W0 medium. This method is therefore reserved for crosses between *ant*^{R} and *ant*^{S} cells. The second method must be used for all crosses involving *mit*$^-$ or *rho*$^-$ mutations in which a quantitative measurement of the frequency of RD cells in the progeny is needed.

Method A: crosses involving RC cells only.

(i) Grow the parental strains in N3 medium (if *ant*^{S}) or in the antibiotic-containing medium (if *ant*^{R}). Remember that the two parents should have opposite mating type and auxotrophic requirements.

(ii) Dilute each culture 10 times in sterile water.

(iii) Mark a W0 plate with three points (**a, x, b**) about 2 cm from each other. Place one drop of culture of one parent strain over points **a** and **x**, wait for complete absorption, then place one drop of the other parent strain over points *b* and *x*. Wait for complete absorption.

(iv) Incubate the W0 plate at 28°C for 3 days.

(v) Check for the confluent growth of prototrophic diploids over the **x** mark and the complete absence of growth of both haploid parents over the positions of **a** and **b**.

(vi) Pick up the entire diploid patch, using a sterile loop, and dissociate it in 10 ml of sterile water (picking up the entire patch ensures that the number of zygotic clones is as high as possible such that the interclonal variance, which is high, is properly averaged).

(vii) Dilute in sterile water to 10^3 cells/ml and plate 0.1 ml aliquots on W0 plates.

Prepare 5–10 plates to obtain enough total colonies for statistical significance.
(viii) Incubate the plates at 28°C for 3 days.
(ix) Replica plate on YPglu, N3 and antibiotic-containing media as appropriate for the *ant*^{R} mutations involved in your cross. Incubate at 28°C for 3 days (or longer for some antibiotic-containing media, check the different levels of growth of resistant and sensitive colonies at regular intervals). Keep all the plates in the refrigerator until growth of resistant colonies is visible on the slowest growing plates incubated.
(x) Sort the plates. Score the mitochondrial genotype of each colony individually.
(xi) Calculate and interpret as described in Section 8.3.3.

Method B: crosses involving one (or two) RD parents.

(i) Grow the parental strains in N3 or antibiotic-containing medium, for the RC parent, and in YP10, for the RD parent(s).
(ii) Plate 0.1 ml of each culture on W0 plates to check for the absence of prototrophic revertants. Incubate at 28°C for 3 days.
(iii) Pellet the cells in the remainder of the cultures, wash and resuspend in YP10 at a density of 10^7 cells/ml.
(iv) Mix 5 ml of the suspension from each parent and incubate at 28°C for 1.5 h with shaking.
(v) Pellet the cells by centrifugation at 1000 *g* for 5 min. Let the tube stand for 15 min.
(vi) Gently resuspend the pellet in the same medium taking care not to disrupt cell aggregates. Incubate further at 28°C for 2–3 h. Monitor the appearance of zygotes under the microscope.
(vii) When the zygotes have formed, centrifuge as before, wash and resuspend the pellet in 10 ml of sterile water.
(viii) Inoculate 0.5 ml of the suspension into 10 ml of W10 and incubate at 28°C for 2 days *without* shaking.
(ix) Inoculate 0.1 ml of the previous culture into 10 ml of W10 and incubate again at 28°C for 2 days *without* shaking.
(x) Dilute to 10^3 cells/ml in sterile water and plate 0.1 ml aliquots on W0 medium. Incubate at 28°C for 3 days.
(xi) Replica plate onto YPglu, N3 and antibiotic-containing media as appropriate for the mutations studied. Incubate at 28°C for 3 days (or longer for some antibiotic-containing media; check the different levels of growth of the resistant and sensitive colonies at regular intervals). Keep all of the plates in the refrigerator until growth of resistant colonies is visible on the slowest growing incubated plates.
(xii) Sort the plates. Score the total number of RD colonies, then score the genotype of each RC colony individually.
(xiii) Calculate and interpret the results as described in the next section.

8.3.3 *Calculation of linkage between mitochondrial genetic markers*

Carry out a quantitative random cross (Methods *A* or *B* as appropriate). Never use the

Table 8. How to compute transmissions, recombination and allelic distributions in three-factor mitochondrial crosses of *Saccharomyces cerevisiae*.

First case

Crosses involving RC parents only and in which the different ant^R loci determine distinguishable phenotypes: (e.g. $rho^+mit^+ant^R_1ant^S_2ant^S_3 \times rho^+mit^+ant^S_1ant^R_2ant^R_3$)

Types found among progeny:

	ant1	*ant2*	*ant3*	*designation*
RC	R	S	S	P1 (parental)
	R	S	R	R1 (recombinants)
	R	R	S	R2 (recombinants)
	R	R	R	R3 (recombinants)
	S	S	S	R4 (recombinants)
	S	S	R	R5 (recombinants)
	S	R	S	R6 (recombinants)
	S	R	R	P2 (parental)

RD (corresponding to spontaneous rho^- mutants, if any)

Calculations

1. % transmission of alleles from parent 1:
   %tra(ant1) = (P1+R1+R2+R3)/Total RC × 100
   %tra(ant2) = (P1+R1+R4+R5)/Total RC × 100
   %tra(ant3) = (P1+R2+R4+R6)/Total RC × 100
   Transmissions of all alleles of the same parent should be equal within limits of statistical significance independently of their genetic location (except for alleles linked to *omega*). This parameter measures the parental contribution and is characteristic of a given cross: this is the *coordinated transmission*.
2. % recombinants between allelic pairs = first type + second type
   %rec(ant1-ant2) = (R2+R3)/Total RC × 100 + (R4+R5)/Total RC × 100
   %rec(ant1-ant3) = (R1+R3)/Total RC × 100 + (R4+R6)/Total RC × 100
   %rec(ant2-ant3) = (R1+R5)/Total RC × 100 + (R2+R6)/Total RC × 100
   The two reciprocal recombinant types between two allelic pairs should be equal within limits of statistical significance (except for alleles linked to *omega*).
3. Distribution of third allelic pair among recombinants between the two others and conclusions (valid only for alleles unlinked to *omega*):

	if dis > 1	if dis < 1
dis (ant3) = R2/R3 + R5/R4	ant3-ant1 linked	ant3-ant2 linked
dis (ant2) = R3/R1 + R4/R6	ant2-ant3 linked	ant2-ant1 linked
dis (ant1) = R1/R2 + R6/R5	ant1-ant2 linked	ant1-ant3 linked

   If *dis* not significantly different from 1: absence of linkage.

Second case

Crosses involving a mit^- parent (e.g. $rho^+mit_1^-ant^S_2ant^S_3 \times rho^+mit_1^+ant^R_2ant^R_3$)

Types found among progeny:

	ant2	*ant3*	*designation*
RC	S	S	R4 (recombinants)
	S	R	R5 (recombinants)
	R	S	R6 (recombinants)
	R	R	P2 (parental)

RD corresponding to all other types

Calculations

1. % transmission of alleles from parent 1:
   %tra(mit) = RD/Total RC + RD (valid only if Method *B* of Section 8.3.2 properly used)
2. % recombinants between allelic pairs (absolute figures)
   %rec(mit1-ant2) = 2 × (R4+R5)/Total RC+RD × 100
   %rec(mit1-ant3) = 2 × (R4+R6)/Total RC+RD × 100
3. Linkage calculations:
   if (R4+R5)/(R4+R6) >1 then mit1-ant3 linked
   if (R4+R5)/(R4+R6) <1 then mit1-ant2 linked
   if (R4+R5)/(R4+R6) not significantly different from 1: absence of linkage

---

results from individual zygotic clones because the interclonal variance is high for reasons unrelated to genetic distances. Interpret the results according to the type of cross as follows.

(i) *Crosses involving only RC cells and in which the different ant^R mutations determine distinguishable phenotypes.* This is the simplest case since all genotypes can be directly determined from the phenotypes of the homoplasmic diploid clones issued from the cross. Crosses can be multifactorial but, in practice, three point crosses are more convenient to score and provide adequate information. An example is given in *Table 8*.

In order to deduce properly the genetic linkage from the number of recombinant clones counted, it is necessary to take into account the parental contribution. This figure is a complex parameter that depends simultaneously upon the number of mtDNA molecules contributed by each parent to the zygote, the kinetics with which such parental mtDNA molecules segregate out into homoplasmic buds, the efficiency with which mtDNA molecules of the two parents find each other to make recombination possible, the bud positions and, perhaps, other cellular factors as well. Such phenomena are not quantitatively equivalent in strains having the same mitochondrial mutations but different nuclear genetic backgrounds. As a result the parental contribution varies from cross to cross. In most instances, the parental contribution is such that the final frequency of recombinants between two unlinked allelic pairs does not exceed 10−15% for each reciprocal recombinant type. Genetic linkage is demonstrated for frequencies of recombinants significantly below the 10−15% limit so long as the parental contribution in the particular cross studied is not extremely biased in favour of one or the other parent.

Examination of the distribution of alleles of a third allelic pair among recombinants between the two others eliminates the problem of the parental contribution, hence the interest of three point tests. In this case an equal distribution of the two alleles of the third pair is expected when the third allelic pair is unlinked to either of the other two. An unequal distribution of the third allelic pair reflects linkage to one or the other (or even both) of the first two. The distribution favours the combination which remains parental for the two alleles brought in *cis* in the cross (see *Table 8*). The greater the bias, the stronger the linkage and the shorter the genetic distance.

(ii) *Crosses involving RC cells in which two different ant^R mutations determine the same phenotype.* In this case, only the wild-type recombinant between the two allelic pairs determining the same phenotype can be distinguished from the two parental types. Distribution of alleles of the third pair among such recombinants can be used to determine genetic linkage between this pair and the other two, as explained previously.

(iii) *Crosses involving both mit^+/mit^- and ant^R/ant^S allelic pairs.* In this case, a fraction of the progeny will be composed of RD clones, the mitochondrial genotypes of which cannot be determined. Distribution of the alleles of each ant^R/ant^S allelic pair among the remaining RC subpopulation simultaneously depends upon the genetic linkage between the mit^+/mit^- and ant^R/ant^S allelic pairs and the parental contribution in this particular cross.

If the quantitative random cross has been performed using the best non-selective conditions of Method B, then the parental contribution can be estimated by calculating the percentage of RD clones out of the total number of clones. However, even in this case, it is better to use an additional ant^R/ant^S allelic pair, unlinked to the first one, as a control. The distributions of both ant^R/ant^S allelic pairs among the mit^+ progeny will be equal in the absence of linkage. Genetic linkage between the mit^+/mit^- pair and either one of the ant^R/ant^S pairs is demonstrated if the two distributions differ significantly (see *Table 8*).

(iv) *Crosses involving two mit^+/mit^- allelic pairs.* In this case, only the wild-type mit^+mit^+ recombinant will be RC. For the same reasons as above, the percentage of this recombinant is limited to 10–15%. Genetic linkage is established for the percentage of RC recombinants significantly below this limit so long as the parental contribution in the particular cross is not extremely biased in favour of one or the other parent. In this case, however, the parental contribution cannot be estimated independently since both parents are indistinguishable RD.

(v) *Crosses involving a rho^- mutant.* In this case, a fraction of the progeny will be RD and if the quantitative random cross has been performed using the best non-selective conditions of Method *B* then the percentage of RD clones among the total represents the parental contribution of the rho^- mutant. Sometimes this percentage is referred to as a delayed suppressiveness and should be clearly distinguished from the zygotic suppressiveness described in Section 8.5 even though both are obviously not independent. Distribution of ant^R/ant^S alleles among the rho^+ subpopulation can be (and have been) examined but this results in complex data beyond the scope of this chapter (see ref. 33 for review).

(vi) *Important aspects of the analysis of genetic crosses.* Firstly, conclusions about genetic linkage drawn above are valid only if precautions are taken in the cross not to select against any particular genotype during growth of the diploid cells issued from that cross. This is generally the case for most ant^R mutations using Method *A*. If there is any doubt as to a possible selective advantage under such conditions, then use Method *B*.

Secondly, remember that the observable percentage of recombinants between any two allelic pairs is low (upper limit 10–15%) and becomes even lower in the case of genetic linkage or when the parental contribution is excessively biased. Always score a total number of colonies large enough to obtain statistically significant data for each recombinant type counted. (Finding five colonies of one type and three of the other does not demonstrate that the ratio is 5/3. In fact, the ratio is almost indeterminate from such data. Precision increases if you count 50 colonies of one type and 30 of the other and improves further if you count more colonies.)

Thirdly, mitochondrial recombination is very active and genetic linkage is only found for relatively short physical distances (e.g. shorter than 2000–3000 bp or even less). For longer distances use petite deletion mapping (Section 8.3.1).

8.3.4 *Determination of the omega allele of a rho$^+$ strain*

The optional intron of the 21S rRNA gene encodes a double-strand specific endonuclease that cleaves the intron-less forms of that gene at a specific site as a prerequisite for the insertion of a copy of that intron. This molecular phenomenon exerts a strong effect on the recombination of flanking genetic markers in crosses between intron-plus strains (*omega*$^+$) and intron-minus strains (*omega*$^-$). Consequently, the quantitative random cross can be used to determine if a strain is *omega*$^+$ or *omega*$^-$ as follows.

(i) Cross the *rho*$^+$ strain in parallel with two tester strains (one *omega*$^+$ reference and one *omega*$^-$ reference) containing, at least, the two intron-linked mutations $chl^R{}_{321}$ and $ery^R{}_{221}$ (see *Table 6*).

(ii) For each cross analyse the ratio of the two reciprocal recombinants, that is $chl^R\ ery^S/chl^S ery^R$.

(iii) If this ratio is high (usually 30–50, at least) with the *omega*$^-$ tester and close to 1 with the *omega*$^+$ tester, then the strain is *omega*$^-$. If this ratio is close to 1 with the *omega*$^+$ tester and very low (usually 0.02–0.03 or lower) with the *omega*$^-$ tester, then the strain is *omega*$^+$. Rare instances may occur in which both ratios are close to 1 with the two testers. If this is the case the strain is a mutant of the *omega* system and it is not possible to deduce the presence/absence of that intron by this genetic test.

The presence of the intron can, obviously, also be determined by molecular analysis of mtDNA. In particular, one can use the minilysate procedure to screen large numbers of strains or clones. Detect the presence of the intron by hybridization with a specific intron probe (available upon request from B.Dujon).

8.3.5 *Isolation of mitochondrial recombinants*

Because genetic recombination between mtDNA molecules is an efficient process, the formation of recombinants, even between linked mutations, is easily obtained by crosses between haploid strains containing the mitochondrial mutations of interest. However, the screening and the isolation of such recombinants into strains able to mate (needed for subsequent crosses) may, depending upon their phenotype, become time consuming. The following cases should be distinguished.

(i) *Recombinants between anti^R mutations conferring different phenotype.* This is the simplest case since the recombinants are RC and phenotypically distinguishable. Use

Table 9. Sporulation of mitochondrial mutants of *Saccharomyces cerevisiae*.

Sporulation medium

0.25% (w/v) yeast extract (Difco)
0.1% (w/v) D-glucose
0.1 M potassium acetate
Prepare this medium and autoclave.

Sporulation and separation of ascospores

1. Inoculate desired homoplasmic diploid clone or mating mixture into 1–2 ml of sporulation medium placed in a 100 ml conical flask (sporulation is better if a very small volume is used). Incubate at 28°C for 2 or 3 days with shaking. Remember that only RC cells sporulate (see text).
2. Monitor tetrad formation under a microscope.
3. Take 0.1 ml of sporulated culture, add 0.1 ml (10 000 U) of β-glucuronidase from *Helix pomatia* (Industrie Biologique Française or Sigma) and incubate at 28°C for 60 min. Verify ascus disruption under microscope.
4. Either separate ascospores by micromanipulation on YPglu or dilute in 10 ml sterile water and vortex thoroughly.
5. In the second case, count the cell density with a haemocytometer and dilute to 10^4 and 10^3 cells/ml. Plate 0.1 ml aliquots of each dilution on YPglu. Incubate at 28°C for 2–3 days.
6. Replicate on W0 and either N3 or antibiotic-containing media (if homoplasmic RC diploids are sporulated) or YP10 and N3 (if heteroplasmic mit^+/mit^- zygotes are sporulated). Incubate at 28°C for 3 days.
7. Pick out either RC auxotrophic clones on N3 or antibiotic medium (for ant^R mutations) or RD auxotrophic clones on YP10 (for mit^- mutations). Dissociate in water and test auxotrophic requirements by dropping onto W0 plates supplemented with various combinations of nutrients (as required for the markers present in the diploid, see *Table 3*). Incubate for 3 days at 28°C.
8. Test the mating type of the clones with the desired nuclear genotype by crossing to **a** and **alpha** tester strains with complementary auxotrophy on W0 plates.
9. In the case of sporulation of heteroplasmic zygotes, check the mitochondrial genotype of RD haploid segregants by crossing to rho^- tester strains as in Section 8.3.1 (this step is necessary to eliminate undesirable spontaneous rho^- mutants that may appear during sporulation and subsequent steps and to distinguish between the required mit^- mutants from the mit^+ pet9 strains if pet9 is used).

the quantitative random cross method to obtain homoplasmic diploid clones, pick out the desired recombinant and sporulate as in *Table 9*. Since mitochondrial mutations do not sporulate during meiosis of a homoplasmic diploid, tetrad analysis is not necessary (except if you want to use it as a criterion of mitochondrial inheritance of your ant^R mutations). Haploid segregants can be used directly for subsequent experiments.

(ii) *Recombinants between ant^R mutations conferring the same phenotype*. This is a laborious procedure since the double mutant is indistinguishable from the parents. Use the quantitative random cross method. Pick out homoplasmic diploids of the resistant phenotype at random and mix them (pick out enough clones to ensure that at least one recombinant clone is present). Sporulate the mixture as in *Table 9*. Haploid segregants may be either single ant^R mutants of one type or the other or the desired double ant^R mutant. Check the mitochondrial genotypes by back-crosses of haploid segregants to each single mutant. Use the quantitative random cross method. Examine the appearance of sensitive recombinants in the progenies (the double mutant will not produce any sensitive recombinant in both crosses, score enough diploids for statistical significance, knowing the frequency of recombinants between the two ant^R/ant^S allelic pairs).

(iii) *Recombinants involving mit⁻ mutations.* The same difficulty exists in constructing double *mit*$^-$ recombinants or *mit*$^-$ *ant*^{R} recombinants, namely that the homoplasmic diploid is RD and does not sporulate. To circumvent this difficulty, zygotes are sporulated immediately after their formation before segregation of mtDNA molecules produces homoplasmic cells. Use the quantitative random cross Method *B* from step (i) to (vi). When zygotes are formed, inoculate into sporulation medium and proceed as in *Table 9*; it is essential to include step (ix). Verify the mitochondrial genotypes of segregants by the qualitative random cross method as follows. If a double *mit*$^-$ mutant is being constructed, cross to each single mutant and check for the absence of RC recombinants in both crosses by replica plating on N3 medium. If a *mit*$^-$ *ant*^{R} recombinant is being constructed, cross to a *mit*$^+$ *ant*^{S} strain and verify the formation of *mit*$^+$ *ant*^{R} recombinant by replica plating on the antibiotic-containing medium.

(iv) *Methods using the mating (MAT) system.* Mutations of the MAT system allow the formation of diploids able to mate, hence the possibility of constructing mitochondrial recombinants without the need for sporulation. To do this cross the mitochondrial mutant (an **a** strain wild-type for the MAT system) to a haploid strain with the following nuclear genotype: *hml a*$^-$ *mat a*$^-$ *HMRαmarl* (mate as an α strain). Use the quantitative random cross method and score phenotypes of diploid clones as normal. Diploids can now be used directly as an **a** parent in crosses. In particular, the qualitative replica cross method can be used to determine their mitochondrial genotype.

8.4 Complementation tests

Functional complementation between mitochondrial mutations can be examined during the transient heteroplasmic stage that follows zygote formation and precedes the appearance of recombinants. Complementation can be assayed either by measuring the oxygen consumption of cells arising from crosses between different *mit*$^-$ mutants or by examining the mitochondrial translation products of cells arising from crosses between a *rho*$^-$ mutant and a wild-type cell.

8.4.1 *Restoration of respiration in crosses between mit⁻ mutants*

(i) Grow the desired *mit*$^-$ mutants of opposite mating type and auxotrophic requirements in YP10 medium (note that mutants with mutations affecting the ATP synthetase genes cannot be used as they are themselves able to consume oxygen).
(ii) Centrifuge at 1000 *g* for 5 min to pellet the cells, wash and resuspend in YPgal at a cell density of 10^7 cells/ml.
(iii) Add 50 ml of each suspension to a 500 ml conical flask. Incubate at 28°C for 2 h with shaking.
(iv) Pellet the cells by centrifugation at 1000 *g* for 5 min. Let the tube stand for 15 min, then gently resuspend the pellet in the same medium.
(v) Incubate at 28°C for a further 1–3 h. Monitor zygote formation under the microscope.
(vi) Take 10 ml aliquots at regular time intervals (e.g. every 30 min) immediately after the first zygotes start to appear.
(vii) Immediately dilute 0.1 ml of each aliquot in 10 ml of sterile water, then dilute

again 1 ml in 10 ml. Plate 0.1 ml of each dilution on W0 and N3 media. Incubate the plates at 28°C for 5 days.

(viii) Immediately centrifuge the remaining 9.9 ml of culture, wash and resuspend the cells in 4 ml of 50 mM phthalate buffer, pH 4.0 containing 1% (v/v) ethanol. Incubate at 30°C for 5 min and vortex vigorously (for maximum oxygenation).

(ix) Place 1.8 ml of the aerated suspension into a Clark electrode. Add 5 μl of 5 mM dinitrophenol and record polarographic measurements of oxygen consumption at 30°C, using an automated oxygraph (e.g. Gilson).

(x) Repeat the same measurement on a second 1.8 ml sample after addition of dinitrophenol and of 40 μl of 0.4 M KCN. Subtract values from the previous measurement.

(xi) Count the number of colonies on W0 plates and calculate the number of prototrophic cells in the suspension placed in the Clark electrode. Plot as a function of time.

(xii) Count the number of colonies on N3 plates and calculate the number of cells able to give rise to RC colonies as before. This is a measure of the appearance of recombinants (a cell can form a colony on N3 if it contains recombined wild-type mtDNA molecules at the time of plating or if it is able to produce such molecules from its heteroplasmic pool during its survival time on the N3 medium). Plot the ratio of recombinants over the prototrophic cells as a function of time.

(xiii) Calculate respiration as nanomoles of KCN-sensitive oxygen consumption/h/10^6 prototrophic cells. Plot as a function of time. Complementation is positive if a significant rate of respiratory activity (approaching that of the wild-type) is observed during the time period before wild-type recombinants are found (~4–8 h after zygote formation in most cases). Partial complementation (with lower respiratory activity but significantly above background) is sometimes observed.

8.4.2 *Appearance of new translation products in rho$^-$ × rho$^+$ crosses*

(i) Grow the desired *rho*$^-$ mutant in YPglu and the *rho*$^+$ tester strain (with opposite mating type and auxotrophic requirements) in N3 medium. Note that the expected new translation product formed as a result of complementation should be distinguishable from the translation products of the *rho*$^+$ parent.

(ii)–(v) Same as in Section 8.4.1 but use a ratio of 3:1 of the *rho*$^-$ to *rho*$^+$ parental cells for mating.

(vi) Take 5 ml aliquots at regular time intervals and analyse the translation products as described in step (iii) of Section 10.

(vii) Dilute 0.1 ml of each aliquot in 10 ml of sterile water and dilute again 1 ml in 10 ml. Plate 0.1 ml aliquots of each dilution on W0 medium. Incubate at 28°C for 3 days. Score the plates and calculate the percentage of prototrophic cells relative to the *rho*$^+$ haploid parent.

(viii) Compare the frequency of appearance of new translation product with the previous ratio.

8.5 Quantitative and qualitative tests for suppressiveness

Suppressiveness was originally defined as the percentage of zygotic clones composed entirely of *rho*$^-$ cells arising from a cross between *rho*$^-$ mutant and a wild-type (53). This is a characteristic figure for a given *rho*$^-$ mutant and is quantitatively reproducible so long as the purity of culture is maintained.

8.5.1 *Quantitative determination of zygotic suppressiveness*

(i) Grow the desired *rho*$^-$ mutant as well as a *rho*^{0} (derived from the same strain) in YPglu or YP10 medium until stationary phase.
(ii) Grow the *rho*$^+$ tester strain in N3 medium until stationary phase.
(iii) Plate 0.1 ml aliquots of each culture on W0 medium to confirm the absence of prototrophic cells, incubate at 28°C for 3 days.
(iv) Dilute an aliquot of the *rho*$^-$ culture to 10^3 cells/ml and plate on YPglu to determine the purity of the culture as described in Section 8.2.2.
(v) Centrifuge the remainder of the culture at 1000 *g* for 5 min to pellet the cells, wash and resuspend in YP10 medium at a density of 10^7 cells/ml.
(vi) Prepare two sterile 100 ml conical flasks and place 5 ml of the *rho*$^+$ suspension in each. Add 5 ml of the *rho*$^-$ suspension in one flask and 5 ml of the *rho*^{0} suspension in the other. Incubate at 28°C for 90 min with shaking.
(vii) Pellet the cells by centrifugation at 1000 *g* for 5 min. Allow the tube to stand for 15 min.
(viii) Gently resuspend the pellet in the same medium taking care not to disrupt cell aggregates. Incubate at 28°C for a further 1–3 h. Monitor the appearance of zygotes under the microscope.
(ix) When the first zygotes appear, stop the incubation by placing the mating mixtures at 0°C. Never allow the first buds to separate from zygotes prior to plating.
(x) Dilute in sterile water to 10^3 zygotes/ml and plate 0.1 ml aliquots on W0 medium (if the percentage of visible zygotes is insufficient, prepare serial dilutions and plate all dilutions).
(xi) Incubate the plates at 28°C for 3–4 days.
(xii) Replicate on YPglu and N3 plates and incubate at 28°C for 2–3 days. [This step may be omitted since *rho*$^+$ *and rho*$^-$ clones can usually be distinguished on W0 medium by their size and colour (*rho*$^+$ clones are large, thick and white while *rho*$^-$ clones are small, flat and slightly translucent). The appearance of colonies on W0 is not, however, definite proof that the colony is *rho*$^-$.]
(xiii) Score the number of zygotic clones entirely composed of RD cells and compare with the total number of zygotic clones examined. Score at least 300 clones in total to obtain sufficient statistical significance.
(xiv) Score the purity of the culture. If purity is satisfactory proceed to step (xv), if not subclone and repeat the whole test as there are no means of predicting the effect of the secondary mutations on the suppressiveness.
(xv) The zygotic suppressiveness (S) of the *rho*$^-$ mutant is given by

$$S = \frac{S1 - S0}{100 - S0} \times 100$$

where S1 and S0 are the percentage of RD zygotic clones in crosses of the *rho*$^-$ mutant and of the *rho*^{0} mutant, respectively.

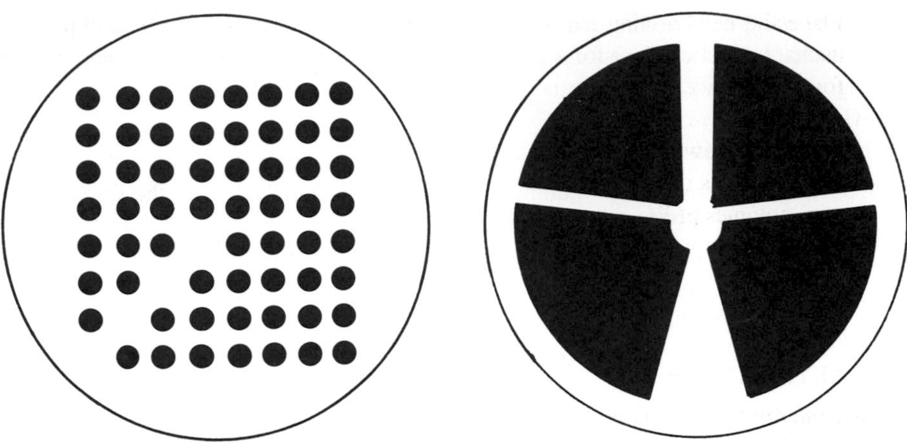

Figure 5. Popular dispositions of haploid clones for the qualitative replica cross. **Left:** 8 × 8 grid. **Right:** quadrant. Both dispositions are self-oriented.

8.5.2 *Qualitative determination of zygotic suppressiveness (54)*

This procedure allows one to distinguish rapidly between *rho*$^-$ mutants with extremely high zygotic suppressiveness (hypersuppressive mutants) and other *rho*$^-$ mutants. Hypersuppressive mutants ($S > 95\%$) form a distinct class of *rho*$^-$ mutants in which one origin of replication is retained and highly amplified.

(i) Place the *rho*$^-$ clones in grids (8 × 8 grids are convenient, see *Figure 5*) on YPglu medium and incubate at 28°C for 3 days.
(ii) Grow a *rho*$^+$ *pet9* strain of opposite mating type (and appropriate auxotrophic requirements) in YP10 medium.
(iii) Plate 0.1 ml of the YP10 culture on W0 medium (plating of the lawn should be very regular for good mating efficiency).
(iv) Replicate the YPglu grids on YPglu, W0 and N3 media first, then onto the lawn. Incubate all replica plates at 28°C for 3 days.
(v) Check that there is no growth of replicas on N3 and W0 plates. Check for the confluent growth of prototrophic diploids on each replica over the lawn and for the complete absence of growth on the N3 and W0 plates.
(vi) Replicate the diploids on N3 medium. Incubate at 28°C for *only* 2 days.
(vii) Score the growth of diploid colonies on N3 plates. Confluent growth indicates the presence of all types of *rho*$^-$ mutants except hypersuppressive. The appearance of discrete colonies of RC cells growing over the inhibited RD cells indicates a hypersuppressive *rho*$^-$ mutant.
(viii) Pick out the desired *rho*$^-$ clone on the YPglu replica plate.

8.5.3 *Semi-quantitative determination of zygotic suppressiveness (quadrant test)*

This test is based on the same principle as before except that the colonies of diploids are made larger making it easier to distinguish between the different classes of suppressiveness.

(i) Place the *rho*$^-$ clones on YPglu medium in quadrants (as shown in *Figure 5*) using sterile loop (ensure even inoculation over the quadrant). Incubate at 28°C for 3 days. Keep the original culture in the refrigerator at 5°C.
(ii)–(iv) Same as in Section 8.5.2.
(v) Score the growth of diploid colonies on N3 medium. Confluent growth indicates *rho*$^-$ mutants of low or moderate suppressiveness (lower than 50%); numerous mini-colonies of RC cells over the inhibited RD cells indicate high suppressiveness (~50–90%) while rare mini-colonies of RC cells indicate hypersuppressiveness (>95%).
(vi) Pick out the desired *rho*$^-$ clone from the original culture.

9. ANALYSIS OF YEAST MITOCHONDRIAL DNA

Because the mitochondrial genome of *S. cerevisiae* wild-type strains is relatively small (~70–80 kb, depending upon the strain) and almost entirely sequenced, direct examination of the mtDNA provides useful information on its genetic content. Furthermore, the minilysate and the colony hybridization methods are quick procedures permitting the simultaneous analysis of numerous clones.

9.1 **Purification of mitochondrial DNA (mtDNA)**

Several methods have been published for preparing mtDNA from yeast. The following method gives a good yield.

(i) Grow the cells of the desired strain until early stationary phase in 1–3 l of N3 medium or antibiotic-containing medium (if RC) or YP10 medium (if RD). Keep an aliquot to determine the purity of the culture by genetic tests.
(ii) Pellet the cells by centrifugation at 1000 *g* for 10 min, rinse the cells in water and centrifuge again. Weigh the pellet of cells. The following procedure is for 15 g of cells.
(iii) Resuspend the cells in 100 ml of 50 mM sodium potassium phosphate buffer, 25 mM EDTA, 1% (v/v) 2-mercaptoethanol, pH 7.5 and add 15 mg of Zymolyase 60 000 (Seikagaku Kogyo, Tokyo). Incubate at 37°C until cells lyse (medium becomes viscous, usually 5–15 min).
(iv) Immediately add 100 ml of 0.2 M Tris-HCl, 80 mM EDTA and 1% SDS, pH 9.5. Incubate at 65°C for 30 min.
(v) Cool in ice, then add 50 ml of 5 M potassium acetate. Incubate at 0°C for 45 min.
(vi) Centrifuge at 5000 *g* for 10 min at 5°C. Keep the supernatant and add 150 ml of isopropanol. Mix and freeze at –20°C for 60 min.
(vii) Centrifuge at 5000 *g* at 0°C for 15 min. Discard the supernatant. Add 20 ml of 1 mM EDTA, 10 mM Tris-HCl, pH 7.8 and dissolve the pellet (dissolution may be difficult, if so use a glass Teflon Potter-Elvehjem homogenizer if necessary).
(viii) Centrifuge at 15 000 *g* for 15 min at 5°C. Keep the supernatant, adjust the volume with buffer as necessary for ultracentrifugation tubes (usually 30 ml is convenient). For 1 ml of solution add 24 μl of a 10 mg/ml stock solution of bisbenzimide (Hoechst 33258) and 1.16 g of CsCl.

(ix) Centrifuge at 100 000 *g* at 15°C until equilibrium. Visualize the DNA bands under u.v. light (blue fluorescence). The upper band is the mtDNA, the lower intense band is nuclear DNA. Carefully pipette out the mtDNA. Place in a second (smaller) ultracentrifuge tube and centrifuge again at 100 000 *g* until equilibrium.

(x) Collect the pure mtDNA band (upper). Extract the bisbenzimide with three successive extractions with isopropanol (*previously equilibrated with CsCl solution*). Add 4 vols of 0.3 M sodium acetate. Mix and add 3 vols of ethanol. Freeze at −70°C for 5 min and centrifuge at 10 000 *g* for 10 min.

(xi) Discard the supernatant and dissolve the pellet in 300 μl of 0.3 M sodium acetate. Add 1 ml of ethanol, freeze and centrifuge.

(xii) Rinse the pellet with ethanol and dry under vacuum. Re-dissolve the pellet in 500 μl of 1 mM EDTA, 10 mM Tris-HCl, pH 7.8.

9.2 Quick, small-scale preparation of yeast mitochondrial DNA (minilysates)

Several methods have been published for the rapid preparation of total DNA out of which the mtDNA can be analysed directly by the Southern blotting technique using specific probes. The method described here is commonly used in the authors' laboratory and gives reliable results. It permits the analysis of numerous clones with limited effort (typically up to 100 clones can be analysed simultaneously), hence its usefulness in screening *rho*$^-$ mutants, polymorphic variants or recombinants.

(i) Grow the cells of the desired clone in a small volume (usually 3 ml are sufficient) of either YPglu medium (if RD) or N3 medium (if RC) until late log phase. It is not necessary to measure the optical density of the culture, an experienced eye is sufficient in this instance.

(ii) Take a 1.5 ml aliquot of the culture with a sterile pipette and transfer to an Eppendorf microcentrifuge tube. Store the remainder of the culture in the refrigerator until complete analysis of the mtDNA is finished.

(iii) Spin the tube at 12 000 *g* for about 10 sec (the pelleting of yeast cells in a microcentrifuge is almost immediate).

(iv) Discard the supernatant by inverting the tube and quickly drain on a clean tissue (do not leave tubes inverted for a long time as the pellet may start to drip down the side of the tube).

(v) Resuspend the pellet in 1 ml of water, vortex and centrifuge again as before.

(vi) Discard the water by inverting the tube and quickly drain on a clean tissue.

(vii) Resuspend the pellet in 0.2 ml of 25 mM EDTA, 1% (v/v) 2-mercaptoethanol, 50 mM sodium potassium phosphate buffer, pH 7.5. Vortex and add 0.1 ml of a 1 mg/ml Zymolyase 60 000 solution.

(viii) Incubate at 37°C until the cells lyse (usually a few minutes is sufficient). Lysis can be easily monitored as the medium becomes highly viscous and as a quick spin at 12 000 *g* no longer results in the formation of a pellet.

(ix) Immediately after the lysis, add 0.2 ml of 80 mM EDTA, 1% (w/v) SDS, 0.2 M Tris-HCl, pH 9.5. Mix and incubate at 65°C for 30 min.

(x) Add 0.1 ml of 5 M potassium acetate, mix and incubate at 0°C for 45 min.

(xi) Centrifuge at 10 000 *g* for 10 min.
(xii) Pipette out the supernatant and transfer to a fresh Eppendorf microcentrifuge tube.
(xiii) Add 0.25 ml of 7.5 M ammonium acetate and fill the tube with ethanol. Mix and freeze at −70°C for 5 min (do not use a longer precipitation step since impurities will also precipitate after longer periods).
(xiv) Centrifuge at 12 000 *g* for 5 min at 0°C. Discard the supernatant and re-dissolve the pellet in 0.2 ml of 0.3 M sodium acetate. If solubilization is difficult, use a vortex mixer or micropipette. Ensure that solubilization is complete (if not NH_4^+ ions may be trapped in the pellet and interfere with subsequent utilization of the DNA). After solubilization, fill the tube with ethanol and freeze at −70°C.
(xv) Centrifuge at 12 000 *g* for 5 min at 0°C, discard the supernatant and fill the tube with ethanol to rinse the pellet.
(xvi) Centrifuge, discard the ethanol supernatant and dry under vacuum.
(xvii) Re-dissolve the pellet in 100 μl of 1 mM EDTA, 10 mM Tris−HCl, pH 7.8. There should be enough DNA from such a preparation for 10 gel loadings.
(xviii) Restrict the DNA with appropriate restriction endonuclease(s) and analyse fragments by gel electrophoresis. If a minilysate is made from a *rho*$^+$ strain or from a *rho*$^-$ mutant containing a long fragment of the mitochondrial genome, blot onto nitrocellulose and hybridize using specific probes (if possible use recombinant plasmids with inserts of mtDNA as probes or use only highly purified mtDNA since any nuclear DNA contamination will result in a high background or even false hybridization). If a minilysate is made from a *rho*$^-$ mutant retaining only a short segment of the mitochondrial genome (e.g. <5000 bp), amplification is usually sufficient for the mtDNA bands to be clearly visible over the background of nuclear DNA after staining the gel with 0.1 μg/ml of ethidium bromide.

9.3 Colony hybridization

(i) Dilute appropriate culture to 10^3 cells/ml and plate 0.1 ml on YPglu medium (if RD) or N3 medium (if RC). Incubate at 28°C for 2 days (YPG) or 3 days (N3).
(ii) Place in grids on the same medium, picking colonies out one by one with sterile toothpicks. Prepare grids in duplicate in order to save viable colonies on one plate. Incubate at 28°C for 2 or 3 days.
(iii) Gently lay a Whatman 541 filter disk (9 cm diam.) on one plate of each duplicate. Store the other plate in the refrigerator.
(iv) Allow a few minutes contact with colonies, then slowly peel off the filter (only manipulate the filter with smooth forceps).
(v) Flip the filter disk onto Whatman 3MM sheets saturated with a solution of 25 mM EDTA, 1% (v/v) 2-mercaptoethanol, 50 mM sodium potassium phosphate, pH 7.5, containing 0.1 mg/ml of Zymolyase 60 000 (Seikagaku Kogyo, Tokyo). Incubate at 37°C for 1−2 h in a closed container to prevent desiccation.
(vi) Transfer the filter disk onto Whatman 3MM sheets saturated with 0.5 M NaOH and incubate for 10 min at room temperature.

(vii) Drain the filter disk on clean tissue and transfer onto Whatman 3MM sheets saturated with 1 M Tris-HCl, pH 7.5. Incubate for 10 min at room temperature. Repeat once.
(viii) Transfer the disk onto Whatman 3MM sheets saturated with 2.5 M NaCl, 0.5 mM Tris-HCl, pH 7.5. Incubate for 2 min.
(ix) Drain the filter on clean tissue and air dry.
(x) Bake at 80°C for 60 min.
(xi) Hybridize with a specific probe (See Section 2.3.1 of Chapter 8).

10. ANALYSIS OF YEAST MITOCHONDRIAL TRANSLATION PRODUCTS

This method is derived from the method of Douglas and Butow (55) which after modifications offers greater experimental convenience and allows one to analyse numerous clones at the same time. The method is based on the specific labelling of mitochondrial translation products when the cytoplasmic ribosomes are inhibited.

(i) Grow the desired strain in N3 medium (if RC) or YPglu medium (if RD). Remember that the strain used needs to be *rho*$^+$ since all *rho*$^-$ and *rho*^{0} mutants are completely deficient in mitochondrial protein synthesis.
(ii) Inoculate 0.2 ml of the culture in 5 ml of YPgal medium. Incubate at 28°C for 4 h.
(iii) Pellet the cells by centrifugation at 1000 *g* for 5 min, wash the cells in water and resuspend in 2 ml of Ggal medium without sulphate.
(iv) Incubate at 28°C for 30 min, then add 0.2 ml of a 6 mg/ml cycloheximide solution and further incubate at 28°C for 10 min.
(v) Add 400 μCi (14.8 MBq) of $H_2{}^{35}SO_4$ (at 20 Ci/mg, 740 GBq/mg) and incubate at 28°C for 30 min.
(vi) Add 0.2 ml of a solution containing 10% casamino acids (Difco) and 10 mg/ml of Na_2SO_4. Incubate at 28°C for 5 min.
(vii) Centrifuge, resuspend the cells in 1 ml of 1 mM EDTA, 1% (v/v) 2-mercaptoethanol, 50 mM Tris-HCl, pH 7.5. Transfer to an Eppendorf microcentrifuge tube.
(viii) Centrifuge for 1 min, resuspend in 1 ml of the same buffer and centrifuge again.
(ix) Resuspend in 0.2 ml of the same buffer, add 10 μl of a 10 mg/ml solution of Zymolyase 60 000 (Seikagaku Kogyo, Tokyo).
(x) Incubate at 37°C until the cells lyse (usually <15 min).
(xi) Add 0.2 ml of 65% (v/v) glycerol and 5 μl of 100 mM phenylmethylsulphonyl fluoride (dissolved in dimethyl sulphoxide). Mix and store at −20°C.
(xii) Proteins can be analysed by standard electrophoresis methods (see Section 3.2.4 of Chapter 9). For most mitochondrial translation products it is convenient to analyse samples on SDS−polyacrylamide gels according to Laemmli. In this case add an equal volume of the following loading buffer to the sample: 40% (v/v) glycerol, 2% (v/v) 2-mercaptoethanol, 2% (w/v) SDS, 125 mM Tris-HCl and 0.1% (w/v) bromophenol blue at pH 6.8). Heat denature the sample and load onto the gel.

Table 10. Strains of *Aspergillus nidulans*.

Strain	*Genotype*
R21	*pabaA1,yA2*
R153	*wA3;pyroA4*
O^R6	*pabaA1,yA2 (oli1)*
O^R31	*pabaA1,ya2;oliC31*
B3	*wA3;pyroA4 (oli1)*
B015	*wA3;pyroA4;oliC31*
O^R322	*pabaA1,ya2 (oli322)*
CS67-3	*pabaA1,yA2 (cs67)*
CS67-6	*wA3;pyroA4 (cs67)*
B6	*ya1;pabaA2 (cam112)*
B4	*wA3;pyroA4 (cam112)*
B24	*pabaA1,yA2 (cam51)*
B25	*wA3;pyroA4 (cam51)*
DJ.1-1	*blA4;yA2,pabaA1,hetA1,hetB1 (oli1)* (hybrid mitochondrial genome)
DJ.2-1	*pyroA5;hetA2;hetB2 (cam112)*

Extranuclear (mitochondrial genome) mutations are denoted by brackets: oli, oligomycin resistance; cs, cold sensitivity; cam, chloramphenicol resistance.
Nuclear markers: y, yellow conidia; w, white conidia; paba, *p*-aminobenzoic acid requirement; pyro, pyridoxine requirement; bl, blue ascospores; het, heterokaryon incompatibility.
Strains in this table and *Table 19* may be obtained from the Fungal Genetics Stock Center, Department of Microbiology, University of Kansas Medical School, Kansas City, MO 66103, USA.

11. MITOCHONDRIAL GENETICS OF ASPERGILLUS NIDULANS

11.1 **Introduction**

Since *Aspergillus nidulans* is an obligate aerobe, and mutations which block mitochondrial function would be lethal, the scope for mitochondrial genetics is much more limited than in *S. cerevisiae*. Nevertheless, a number of informative experiments can be easily carried out using the small number of mutants available. These mutants (*Table 10*) can be obtained from the Fungal Genetics Stock Centre (FGSC). Additional practical information about genetic experiments in *A. nidulans* can be found elsewhere (56,57).

Inheritance and recombination of the mtDNA can also be followed directly by isolation and restriction endonuclease digestion of the DNA, since certain closely related varieties of the *A. nidulans* species group have differences in mitochondrial genome structure, easily detectable as restriction fragment length polymorphisms. These subspecies can also be obtained from the FGSC.

Finally, almost the entire sequence of the mitochondrial genome is now known, and consequently the identity and location of most mitochondrial genes. Reviews of mitochondrial genetics and genome structure in *A. nidulans* should be consulted for further information (58–60).

11.2 **Isolation and identification of extranuclear mutants**

11.2.1 *Isolation of mutants*

One strategy for isolation of extranuclear mitochondrial mutants is to screen for resistance

Table 11. Growth media and supplements for *Aspergillus*.

A. Nutrient medium (NM)

Malt extract	20 g
Bactopeptone	1 g
Glucose	20 g
p-aminobenzoate	2 ml of 1 mg/ml autoclaved stock solution
Agar	20 g
Distilled water to 1 litre	

When replica plating of individual colonies is to be carried out, include sodium deoxycholate (0.08%) in the medium before autoclaving. This keeps the colonies compact.

B. Minimal medium (MM)

$NaNO_3$	6 g
KCl	0.52 g
KH_2PO_4	1.52 g
Trace elements solution	1 ml
Agar	15 g
Distilled water to 1 litre	
pH adjusted to pH 6.5 with KOH	

After autoclaving, add (per litre)	glucose	25 ml, 40% w/v sterile solution
	$MgSO_4$	2.5 ml, 20% w/v sterile solution

MM refers to the medium above, and supplements will be defined

and when required, add (per litre) to molten agar after cooling to ~50°C	*p*-aminobenzoic acid 2 ml, 1 mg/ml autoclaved stock solution
	pyridoxine HCl 1 ml, 1 mg/ml filter-sterilized stock solution (heat and light labile)
	oligomycin (Sigma), heat labile, 3 ml, 1 mg/ml stock solution in ethanol, stored in freezer
	chloramphenicol (Parke-Davies) 3 g added as non-sterile powder to hot agar, keep at 60°C until dissolved
	triethyltin (Merck, Darmstadt) 1 ml, 0.5 mg/ml stock aqueous solution added to the molten agar after cooling to ~50°C

Trace elements solution:

$ZnSO_4.7H_2O$	1.0 g
$FeSO_4.7H_2O$	8.8 g
$CuSO_4.5H_2O$	0.4 g
$MnSO_4.4H_2O$	0.15 g
$Na_2B_4O_7.10H_2O$	0.1 g
$(NH_4)6Mo_7O_24.4H_2O$	0.05 g
Distilled water to 1 litre	

Dissolve ingredients in 250 ml of distilled water, swirl to gather the crystals, then add approximately 0.2 ml of concentrated HCl directly onto the crystals to dissolve them. Make up to 1 litre.

to anti-mitochondrial antibiotics, then screen the mutants for extranuclear inheritance. Conditional lethal mutants, that is cold-sensitive or temperature-sensitive, can also be isolated in *A. nidulans*. Since these could affect many functions other than mitochondrial ones, screening for those which affect mitochondrial function can be laborious, but the heterokaryon test can be used to detect extranuclear inheritance. The mitochondrial mutation *cs67* was isolated in this way (61).

Table 12. Preparation of a conidial suspension of *Aspergillus*.

1. Inoculate a plate of NM with either a loopful of dry conidia from a colony or 0.1 ml of conidial suspension, and spread over the surface using a glass spreader. Incubate at 37°C for 2 days.
2. To harvest the conidia, add 10 ml of Tween 80/saline solution (0.25 ml Tween 80, Koch-Light Laboratories Ltd.; 8 g NaCl; distilled water to 1 litre; autoclave), and scrape the surface of the plate with a wire or glass spreader. Most of the non-wettable conidia will go into suspension and can be removed by pipette.
3. Stored in a suitable container at 4°C, the conidia remain viable for several months. Such conidial suspensions are the starting point for most experiments.
4. The suspension will contain about 10^8 conidia/ml and the numbers can be estimated using a haemocytometer or other counting chamber.

Table 13. Mutagenic treatment of *Aspergillus* conidia.

1. Standard mutagens such as u.v. light, ethane methane sulphonate or *N*-methyl-*N'*-nitro-*N*-nitrosoguanidine (MNNG) can be used. Use of u.v. is less effective but safer than MNNG. **MNNG must be handled with extreme care as it is a potential carcinogen.**
2. For u.v. mutagenesis, plates of conidiating mycelium (*Table 12*) can be exposed directly to a u.v. lamp for sufficient time to achieve 1 − 10% viability, and the conidia harvested as usual. The viability is checked by plating dilutions of conidia onto NM after different exposure times at a fixed distance from the lamp.
3. For MNNG mutagenesis:
   (a) Dissolve MNNG at 0.5 mg/ml in liquid MM (omit agar) lacking glucose, by shaking for 30 min at 37°C.
   (b) Add 10 ml of this solution to a pellet of conidia obtained by centrifuging (bench centrifuge) 10 ml of conidial suspension.
   (c) Transfer to a 250 ml conical flask, and incubate for 30 min at 37°C with shaking.
   (d) Re-pellet the conidia, and wash twice in liquid MM without glucose.
   (e) Resuspend in 1 ml of MM to obtain approximately 10^9 conidia/ml. For selection of drug-resistant mutants, spread 0.1 ml of the suspension onto NM containing an inhibitory concentration of the drug, and incubate at 37°C until resistant, conidiating colonies can be seen amidst a background of poorly growing, non-conidiating mycelium.

Mutant isolation is most easily done by selection for antibiotic resistance using either oligomycin, which inhibits the mitochondrial ATP synthetase, blocking oxidative phosphorylation and thus growth of an obligate aerobe, or chloramphenicol, an inhibitor of mitochondrial protein synthesis at high concentrations. While mutations occur spontaneously, the frequency can be increased by treatment of conidia with a mutagen (*Table 13*).

A yellow spored strain, such as R21, is preferable to white spored, since conidiating, resistant colonies can be clearly seen amongst a background of non-conidiating, white mycelium.

(i) Spread a dense suspension of conidia (10^8/plate in 0.1 − 0.2 ml) nutrient media (NM) agar containing either oligomycin or chloramphenicol (*Tables 11 − 13*).
(ii) Incubate at 37°C until resistant colonies are conidiating.
(iii) Using a bent, mounted, sterile wire as a needle, touch the conidia of putative mutants and stab or touch the surface of a fresh plate of NM plus drug, stabbing out wild-type conidia as a control (~25 strains/plate). Some of the putative mutants may just be areas of heavy inoculation which have grown and conidiated

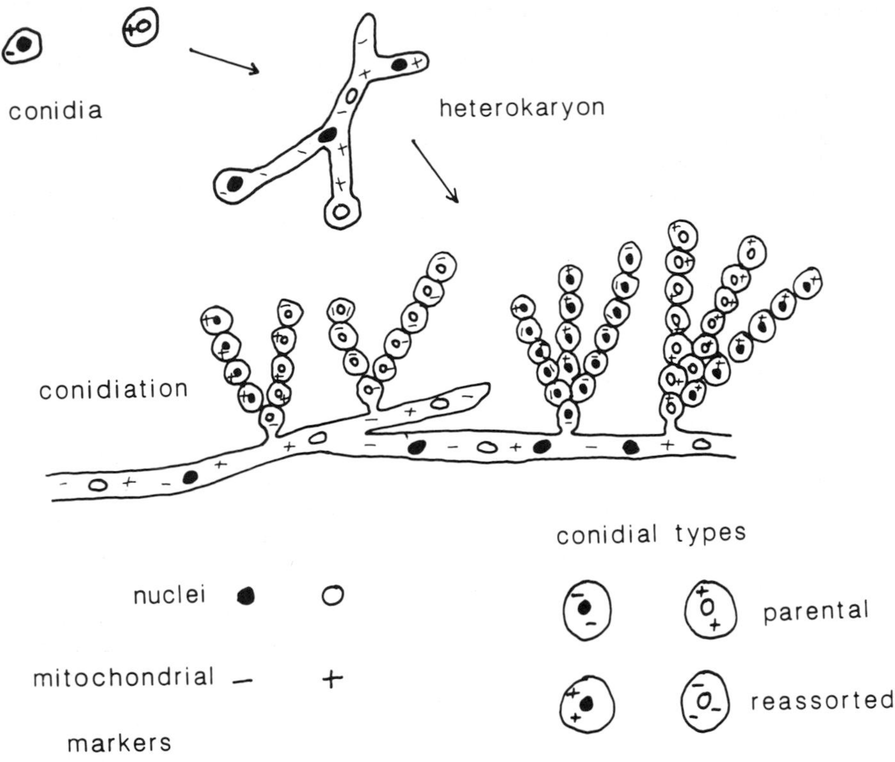

Figure 6. The heterokaryon test in *A. nidulans*. Conidia of mutant and wild-type strains are mixed and allowed to grow together to form a heterokaryon. Conidia taken from the heterokaryon are tested for 're-assortment', that is re-association of nuclear/extranuclear markers. No recombination is observed between nuclear markers.

despite the presence of the drug, and these will be eliminated by this re-testing procedure.

(iv) Streak the confirmed resistant mutants to single colonies on drug-free NM.

Stabbing conidia onto a fresh plate should be carried out with the plate inverted, since dry conidia tend to fall from the wire and can give badly contaminated plates. The static electricity of plastic Petri dishes sometimes causes cross-contamination problems when many mutants are stabbed out on a single plate. If need be, glass Petri dishes can be used to overcome this problem.

Another drug, mucidin, acts on cytochrome *b*, and is a good growth inhibitor of *A. nidulans*. Furthermore, we found that most of the mutants of *Aspergillus* resistant to mucidin were extranuclear. Unfortunately this useful drug is difficult to obtain, being sold only in Czechoslovakia as a topical anti-fungicide Mucidermin (Spofa, Prague).

11.2.2 *Testing for extranuclear inheritance*

Most of the mutants isolated will carry nuclear mutations. To distinguish between nuclear and extranuclear mutations, the heterokaryon test is used (59,60).

A heterokaryon (i.e. a cell containing mixed nuclei), also used for complementation

Table 14. Construction of an *Aspergillus* heterokaryon.

1. Pipette 0.1 ml of conidial suspension from each of the required strains onto the centre of a plate of NM and mix with a loop. Incubate upright overnight, by which time mixed mycelial growth will have occurred, with little apparent conidiation.
2. Using a sterile scalpel (dip in ethanol, flame and cool by stabbing in the agar) cut about five small blocks (2 mm^3) of agar containing mycelium from the growing edge and centre of the colony, and transfer each to the surface of a plate of MM. Since the starting strains used will have complementary nutritional requirements, no supplements are required in the MM other than glucose and magnesium.
3. Incubate for 2–3 days at 37°C, until conidia are seen on the surface of the heterokaryotic growth. It is likely that only some of the blocks will produced conidiating outgrowth.

Table 15. Testing for *Aspergillus* heterokaryons.

Set up a heterokaryon between O^R6 and R153 as in *Table 14*.

1. Rapid testing. With a wire loop carrying a drop of Tween 80/saline, scrape some conidia from the surface of a 2–3 day heterokaryon growing out from the agar block, and make a small streak directly onto a plate of MM supplemented with pyridoxine and oligomycin. Growth after 2 days at 37°C indicates a positive test result. As a control, try a heterokaryon between R153 and the nuclear oligomycin-resistant mutant O^R31.
2. Estimating the frequency of mitochondrial re-assortment:
   (a) Transfer a loopful of conidia from a heterokaryon to 1 ml of Tween 80/saline in a small bottle.
   (b) Vortex well to break up chains of conidia, dilute 1 in 10^4 in Tween 80/saline, and plate out 0.1 and 0.2 ml diluted suspension onto plates of NM containing deoxycholate. Aim for about 50 colonies/plate.
   (c) Incubate for 2–3 days (until conidiated), and replicate the colonies onto NM containing oligomycin. Aim for about 50 colonies/plate.

Replica plating

Use damp velveteen which has been washed in Tween 80/saline and squeezed well to remove excess liquid prior to autoclaving. This is essential to prevent scattering of the dry conidia

$$\text{\% Re-assorted conidia in heterokaryon} = \frac{\text{No. of white resistant + yellow sensitives}}{\text{Total}} \times 100$$

testing in *Aspergillus*, is formed by allowing two marked strains to grow together on the same plate. Anastomosis (joining of the hyphae) occurs and, in places, the hyphae carry mixed nuclei and mitochondria from both strains (*Figure 6*). Heterokaryons of *Aspergillus* are not very stable, and have to be 'forced' to establish and maintain them (*Table 14*). For this purpose, complementing auxotrophs are used, usually marked also with spore colour mutations for easy identification. Strains R153 and R21 and their derivatives (*Table 10*) are convenient pairs for this purpose. Neither strain can grow on unsupplemented minimal medium (MM) but, following mixing and growth on NM, a heterokaryon formed from them can grow indefinitely on MM because of complementation. The MM forces them to remain heterokaryons.

The heterokaryon test is best illustrated using O^R6, an extranuclear oligomycin-resistant mutant isolated from R21, and R153, a complementing strain with wild-type (oligomycin-sensitive) mitochondria (*Table 15*). As a negative control, carry out the test using the nuclear oligomycin-resistant mutant O^R31 with R153.

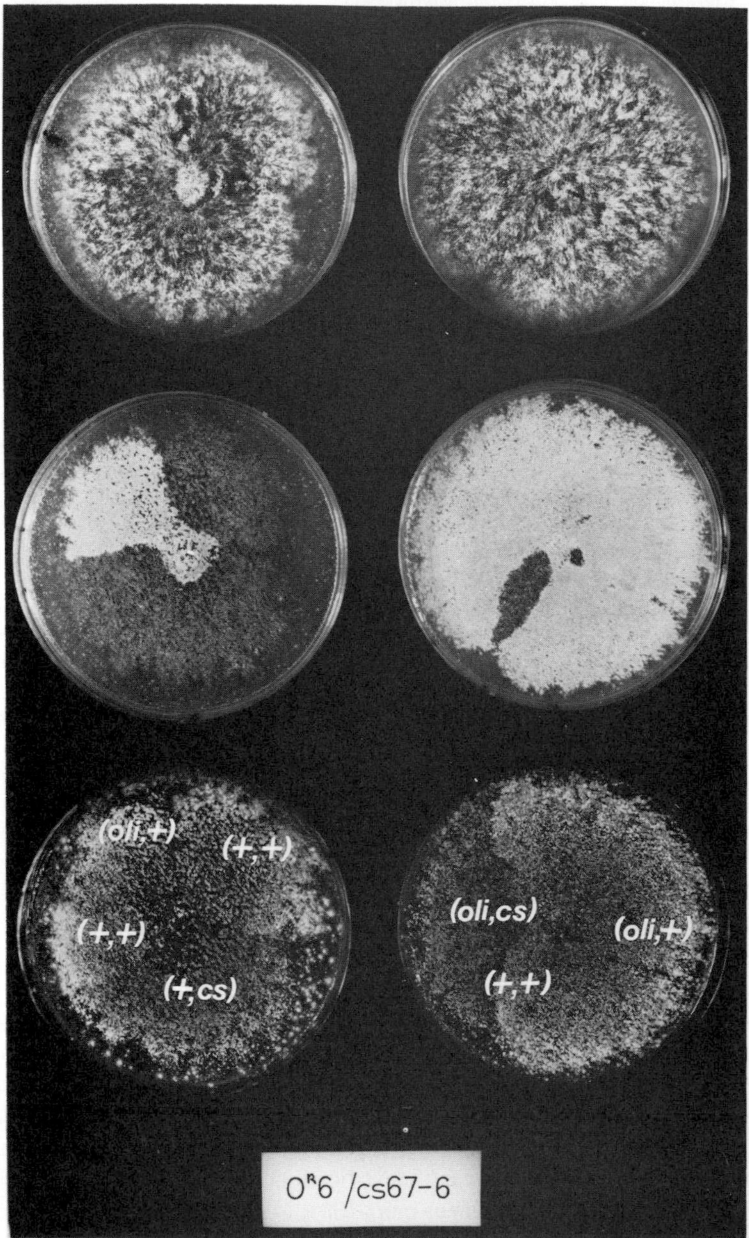

Figure 7. Sectoring-out of recombinants in heterokaryons. Each vertical set of plates consists of the original heterokaryon (top), its replica on oligomycin medium (middle) and its cold-grown replica (bottom). Both types of recombinant and both types of parental sectors are shown. The original genotypes were (*oliA1*, +)/ (+ ,*cs67*); the left side of the right-hand heterokaryon is oligomycin resistant (middle pair of plates) and is also cold-sensitive (bottom pair of plates), and thus is a recombinant sector with the genotype (*oliA1, cs67*). All the other sectors can be identified in the same way, but the existence of four degrees of growth at the cold temperature makes differentiation of the sectors more difficult than is the case with (*oli* A1, +). Reprinted from ref. 64 with permission.

Only at a very low frequency is there any genetic exchange between nuclei in a heterokaryon. The conidia which are formed by the heterokaryon carry the same combination of nuclear markers as did the original parent strains. Thus yellow conidia cannot grown on a medium lacking *p*-aminobenzoate, nor white conidia on a medium lacking a pyridoxine. If the spore suspension made from the heterokaryon is streaked on MM containing pyridoxine and oligomycin, no growth is possible for any parental conidia. This should be the case for O^R31/R153. However, if the oligomycin resistance is carried on the mitochondrial genome, then some of the white, pyridoxine-requiring spores will have acquired a mitochondrial genome(s) originating from O^R6, permitting them to grow on the selective medium.

If a large number of new drug-resistant mutants is made from R21, each one can be tested in this way. A similar approach is possible with cold/temperature-sensitive mutations.

The amount of mitochondrial exchange between the two parent strains can be easily assessed by plating out conidia on NM without drug, then replica plating onto selective media (*Table 15*). Each heterokaryon shows some difference in the amount of cytoplasmic mixing, and thus in the percentage of 're-assorted' conidia (conidia with a different combination of nuclear/mitochondrial markers from parents; *Table 16*).

11.3 Vegetative segregation of mitochondrial alleles

What happens to the two mitochondrial alleles, resistant (*r*) and sensitive (*s*), if the heterokaryon is allowed to grow for a longer period of time? Although the MM forces the two nucleotypes to remain evenly mixed, there is no such constraint on the mitochondriotypes. Continued growth results in a dramatic segregation of mitochondriotypes into 'homoplasmic' areas (*Figure 7*). This presumably results from the branching growth mode of the fungus, whereby a particular branch eventually receives only one mitochondrial genome, which multiplies during further growth to give a clonal sector.

11.4 Recombination of mitochondrial markers

Recombination of mitochondrial markers occurs readily when mitochondria are allowed to mix in the same cytoplasm, and this is achieved by forcing a heterokaryon between suitable strains, for example O^R6/CS67-6 or B3/B6. Analysis of the progeny can be carried out on a conidial suspension from a young heterokaryon (*Table 16*) or on an older, sectoring heterokaryon. *Figure 7* shows typical results obtained with the latter approach, where recombinant sectors can be detected.

Accurate and reproducible quantification of recombination frequency is complicated by the fact that each heterokaryotic colony may behave differently (*Table 16*). The extent of strain mixing in the heterokaryon may affect the amount of mitochondrial recombination. It is possible to attempt to quantify this by expressing mitochondrial recombinants as a percentage of re-assorted colonies, rather than as a percentage of total colonies (64,65). Even using this approach, the limited number of markers which have been used recombine with one another with similar frequencies in most cases.

An example of what appear to be tightly-linked mutations can be seen if recombination is examined between two *oli*^{r} alleles, (*oli1*) and (*oli322*). Set up a heterokaryon between B3 and O^R322. Make a suspension of conidia from a young heterokaryon,

Table 16. Recombination of mitochondrial markers in three *Aspergillus* heterokaryons.

	CS67-6/O^R6		
	HK1	*HK2*	*HK3*
Total genotypes scored	485	1466	1468
% Re-assorted types	7	37.5	30.5
RF as % total	2.5	10	6.5
RF as % re-assorted types	33	27	21
(+,+) as % re-assorted types	24	24	15
(oli,cs) as % re-assorted types	9	3	6
Heterokaryon balance (Y:W)	1:3.5	1:1	1:0.1

Re-assorted types: colonies with non-parental genotype.
RF = extranuclear recombination frequency.
Data from (64).

and search for a wild-type (oligomycin-sensitive) recombinant. Using replica plating, it will not be possible to find one. However, a finer screening can be carried out by plating conidia on NM containing 0.5 μg/ml triethyltin. While the wild-type strains are sensitive to this potent inhibitor of mitochondrial ATP synthetase, oligomycin-resistant mutants are hypersensitive. Amongst a background of very sparse growth, faster growing colonies can be detected which, if subcultured, are oligomycin sensitive. That these are not just revertants of either parent can be confirmed by plating out the pure parental conidia in appropriate numbers (66). Recombination between these alleles is about 50-fold lower than between (*oli1*) and either (*cam112*) or (*cs67*), and it is possible that they are both located in the same gene.

Similarly, (*cam51*), a mutation conferring chloramphenicol resistance, cold sensitivity and slow growth at all temperatures, is very tightly linked to (*cam112*) (66).

A second approach to the quantification of recombination frequency (67) is to count the frequency of recombinant sectors in a large number of heterokaryons.

11.5 Uniparental inheritance in the sexual cycle

Sexual crossing of a mitochondrial mutant with a wild-type strain demonstrates clearly the uniparental inheritance of the mitochondrial genome. Spore colour mutations provide easily scorable nuclear markers whose behaviour can be contrasted with that of the mitochondrial marker(s).

Set up a sexual cross between O^R6 and R153 (*Table 17*), and make ascospore suspensions from a few of the cleistothecia obtained. If a loopful of each suspension is streaked separately on a plate of NM and incubated for 3 days, hybrid cleistothecia can easily be detected by the mixture of coloured colonies obtained: yellow, white and green. Wild-type conidial colour is green, resulting from recombination of the unlinked *y* and *w* loci. Since *A. nidulans* is homothallic, identical nuclei can go through the sexual cycle, resulting in 'all-white' or 'all-yellow' cleistothecia (called 'selfed').

(i) Choose a hybrid suspension, dilute 1 in 10, 1 in 100 and plate out 0.1–0.2 ml on NM containing deoxycholate to obtain about 50 colonies per plate.
(ii) Replicate onto NM containing oligomycin, and incubate for 24 h.
(iii) Count the number of yellow, green and white colonies on the master plate.

Progeny genotype	*%*	*Spore colour*
$y^+;w^-$	25	White
$y^-;w^+$	25	Yellow
$y^+;w^+$	25	Green
$y^-;w^-$	25	White

Figure 8. Behaviour of the conidial colour markers in a sexual cross.
Cross: $y^+;w^-$ X $y^-;w^+$

Table 17. *Aspergillus* sexual cross.

1. Pipette 0.1 ml of conidial suspension from each of the required strains onto the centre of a plate of MM without supplements other than glucose and magnesium, and mix with a loop. Spread out the conidial suspension into a star or asterisk shape up to the edges of the plate.
2. Seal the plate with cellophane tape to prevent it drying out, and incubate upright for 10–14 days at 37°C, during which time the fruiting bodies or cleistothecia will develop. They can be observed under a low-power binocular microscope, using an overhead light, as black spheres (0.1 mm diameter) covered in grey mycelium and other vegetative cells. The larger cleistothecia are most likely to be hybrid, and very small ones are often sterile.
3. Use a mounted needle to scrape off mycelium containing cleistothecia, and transfer it to a plate of stiff agar (4% agar in water alone). Dissect out the cleistothecia using the needle, and roll a few of them across the plate until they are black and shiny, and free from contaminating mycelium and vegetative cells.
4. Transfer each cleistothecium to a small bottle containing 1 ml of Tween 80/saline, and crush it against the side using a needle. Vortex to resuspend the released ascospores. This suspension will remain viable at 4°C for some months.
5. To identify a hybrid cleistothecium, make a small streak of ascospore suspension from each bottle onto a plate of NM, and incubate at 37°C for 3 days (until the conidial colour can be determined).

Since *y* and *w* are unlinked, and should recombine freely, and *w* is epistatic to *y*, the colour ratio should be 1:1:2 (*Figure 8*). However, the mitochondrial marker will be that of one parent only, that is colonies are all sensitive or all resistant to oligomycin in this cross. For some unknown reason, oli^s tends to predominate. The experiment can be repeated with *B4* × *R153* strains and the cam^r gene (68).

Since sexual inheritance is uniparental, recombination of parental mitochondrial genomes cannot occur during the sexual cycle. However, since the strains of *Table 10* are all closely related and are 'heterokaryon compatible', a heterokaryon can form prior to development of the sexual stage, giving an opportunity for mitochondrial recombination. If you cross B4 × O^R6, you will find some cleistothecia which contain recombinant mitochondria. However, only one mitochondrial type, recombinant or parental, will be found in any single cleistothecium (68).

Although *A. nidulans* is homothallic, there is genetic evidence that a fertilization mechanism exists to bring together a pair of nuclei and initiate the sexual cycle. Sexual crossing is possible even between certain heterokaryon-incompatible strains of *A. nidulans*, emphasizing the fact that a heterokaryon is not a prerequisite for fertilization. This feature has made it possible to demonstrate that somehow in the process of fertilization, one of the pair of nuclei which are brought together acts as a 'maternal' nucleus, providing the genetic make-up of the cleistothecium and ascospore cell walls, and the cytoplasm associated with the maternal nucleus at the time of fertilization pro-

Table 18. Preparation of mitochondrial DNA from total *Aspergillus* DNA.

1. Dispense 800 ml of liquid MM + supplements into a 2 litre conical flask and inoculate with a conidial suspension to a final concentration of approximately 10^6 conidia/ml (conidia from 1 plate of NM).
2. Grow for 18–20 h at 37°C in an orbital incubator with vigorous shaking.
3. Harvest the mycelium by filtration through Miracloth (Calbiochem) or muslin (cheesecloth), wash well in cold distilled water and squeeze well to remove excess water.
4. Crumble into small pieces, and freeze-dry overnight. The dried mycelium can be stored in a deep freeze.
5. Grind to a fine powder with a pestle and mortar, weigh (yield 0.5–1 g), and transfer to a 250 ml conical flask.
6. Add DNA extraction buffer (0.15 M NaCl, 0.1 M EDTA, 2% SDS, 50 mM Tris-HCl, pH 8.0) at 25 ml/g mycelium, then 0.2 vol toluene.
7. Shake very slowly on a rotary shaker for 48–72 h.
8. Centrifuge at 1000 *g* for 15 min and carefully remove the supernatant with a wide bore pipette.
9. Shake the supernatant gently with an equal volume of chloroform–pentan-2-ol (24:1).
10. Centrifuge at 10 000 *g* for 30 min at 15°C, then remove and retain the aqueous layer, avoiding the protein at the interphase.
11. Add 10% (w/v) solid polyethylene glycol (PEG) to the aqueous layer and dissolve by gentle mixing. Leave overnight at 4°C.
12. Centrifuge at 2000 *g* for 5 min, and gently resuspend the PEG/DNA pellet in 5 ml of 0.5 mM EDTA, 10 mM Tris-HCl, pH 7.5.
13. Extract with chloroform–pentanol (step 9) and retain the aqueous layer after centrifugation (bench centrifuge, 5000 *g* for 5 min).
14. To each millilitre of aqueous phase, add 1.15 g of CsCl and 120 μg of bisbenzimide (Sigma), and centrifuge (10°C) at 100 000 *g* for 24 h.
15. Use a u.v. light source to visualize the DNA bands in the centrifuge tubes, and remove the upper, mitochondrial band (*Figure 13*) with a hypodermic syringe.
16. Pool the mitochondrial bands, and re-centrifuge as before to remove remaining nuclear DNA, and collect the pure mtDNA.
17. Remove the bisbenzimide by extracting four times with an equal volume of pentan-2-ol saturated with CsCl solution. After each extraction, allow the phases to separate at 4°C for 20 min (or centrifuge), and discard the upper layer.
18. Add three volumes of 0.5 mM EDTA, 10 mM Tris-HCl, pH 7.5, precipitate the DNA by addition of a tenth volume of 4 M sodium acetate, pH 6.0 and 0.6 vol propan-2-ol, and centrifuge at 10 000 *g* for 15 min.
19. Dry briefly, then dissolve the DNA in 0.5 mM EDTA, 10 mM Tris-HCl, pH 7.5.
20. Extract with an equal volume phenol–chloroform 1 to 1 (water-saturated phenol) (step 13) and re-precipitate the DNA (step 18, without buffer addition).
21. Dry the DNA, and re-dissolve it in 50–200 μl of 0.5 mM EDTA, 10 mM Tris-HCl, pH 7.5 (depending on the yield, which should be 50–200 μg). Store at −20°C.

vides the mitochondrial genome which ends up in all the asci. This can be illustrated using the strains described in Section 11.6.4.

11.6 Physical analysis of the mitochondrial genome

All of the experiments outlined above assume that 'extranuclear inheritance' reflects the behaviour of the mitochondrial genome, though they provide no direct proof of this. This section deals with isolation and analysis of mitochondrial DNA, and observation of its behaviour during the life cycle of the fungus.

Figure 9. Bisbenzimide/CsCl gradient of total *Aspergillus* DNA. DNA bands fluoresce under a u.v. light source.

11.6.1 *Isolation of mtDNA*

The mtDNA can be extracted from isolated mitochondria, or separated from total cell DNA, by density gradient centrifugation. Since the former method is more difficult, and prone to nuclease problems with some species of *Aspergillus*, the latter method is to be recommended (*Table 18*).

Mycelium is freeze-dried and total DNA extracted. As in the case of yeast (Section 9.1) this DNA is then purified on a CsCl gradient in the presence of the dye bisbenzimide, which binds preferentially to AT-rich DNA (such as fungal mtDNA), lowering its density relative to that of the chromosomal DNA. The DNA−dye complex is fluorescent, and the bands can be visualized under u.v. light (*Figure 9*).

11.6.2 *Restriction endonuclease digestion of the DNA*

A restriction map of the mtDNA of *A. nidulans* is presented in *Figure 10*. This also shows the location of genes obtained by sequencing. Note that the circular genome has been linearized for convenience.

Conventional techniques are used for digestion of the mtDNA and subsequent electrophoresis, and experimental details can be found elsewhere (Section 5 and ref. 11).

11.6.3 *Restriction fragment length polymorphism in the A. nidulans species group*

In the genetic experiments outlined above, it would be difficult to follow the fate of the DNA, since the strains used are virtually identical in DNA sequence and thus restriction pattern. However, we can take advantage of the existence of other members of the *A. nidulans* species group that are quite closely related to *A. nidulans* (Eidam) Winter, our original strain. These strains are listed in *Table 19*. DNA isolation is easier from *A. nidulans* var. *echinulatus* than from *A. quadrilineatus*. The latter species produces a large amount of viscous material in the growth medium which is difficult to remove completely from the mycelium.

Restriction endonuclease analysis of the mtDNA of these species reveals differences between them (*Figure 11*), which can be exploited to observe recombination of the mitochondrial genome at the level of the DNA. However, a problem remains, namely that these species are sufficiently different to prevent heterokaryons forming between

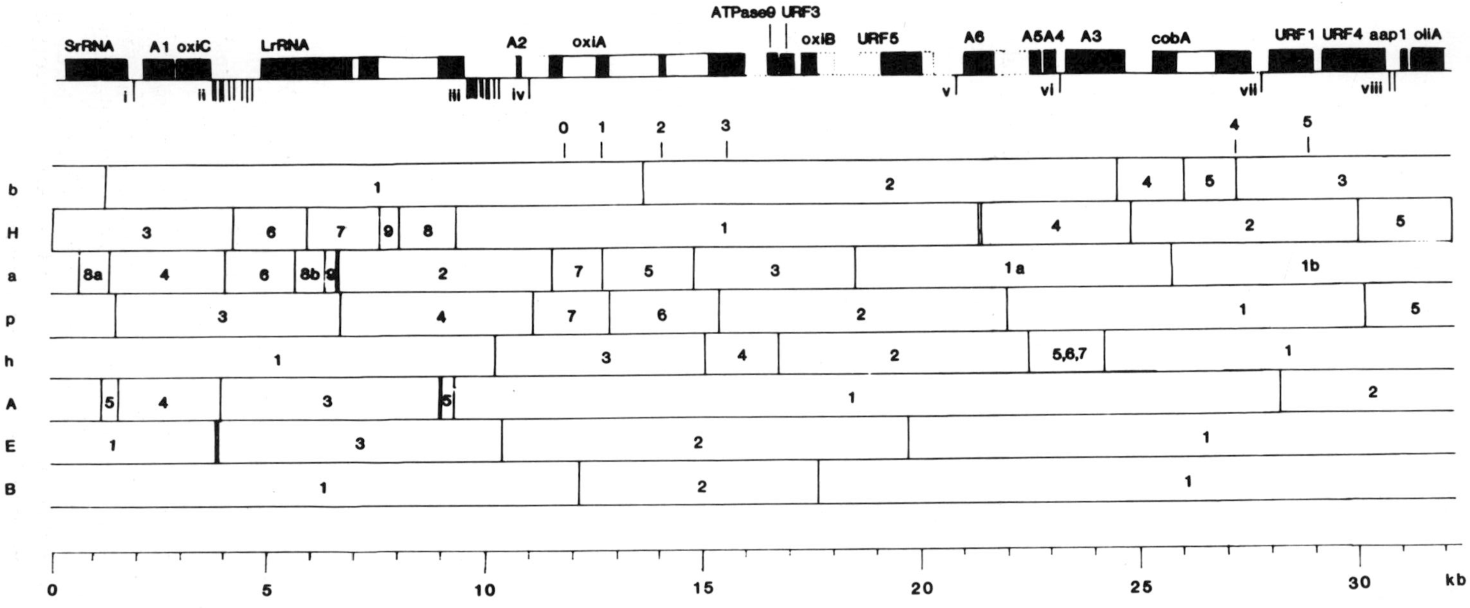

Figure 10. The mitochondrial genome of *A. nidulans*. The circular genome has been linearized for clarity. Black bars represent exons. Vertical lines below map indicate position of tRNA genes: (i) *tyr*; (ii) *lys*, *gly*1, *gly*2, *asp*, *ser*2, *trp*, *ile*, *ser*1, *pro*; (iii) *thr*, *glu*, *val*, *met*1, *met*3, *leu*1, *ala*, *phe*, *leu*2, *gln*, *met*2; (iv) *his*; (v) *asn*; (vi) *cys*; (vii) *arg*, *asn*; (viii) *cys*.
SrRNA, 16SRNA; LrRNA, 23SRNA.
*oxi*A, B, C, cytochrome oxidase subunits I, II, III.
*cob*A, apocytochrome b; *oli*A, ATPase subunit 6; *aap*1, ATPase-associated protein. URF, unidentified reading frames.
Vertical lines above restriction map indicate the position of additional sequences in *A. nidulans* var. *echinulatus:* 0 numerous small insertions (total 70 bp) within existing intron; 1–5 approx. 1 kb each, optional introns.
Enzyme recognition sites: **b** *Bgl*II, **H** *Hin*dII, **a** *Hae*III, **p** *Pvu*II, **h** *Hin*dII, **A** *Hha*I, **E** *Eco*RI, **B** *Bam*HI. Reprinted with permission from ref. 69.

Table 19. Strains of the *A. nidulans* species group.

Strains	*Genotype*
A. nidulans var. *echinulatus*	
25.2	*y25.2*
A. quadrilineatus	
12.1	*y12.1,pyro12.1*
12.22	*y12.1,pyro12.1 (oli12.7)*

The gene symbols are given in *Table 10*.

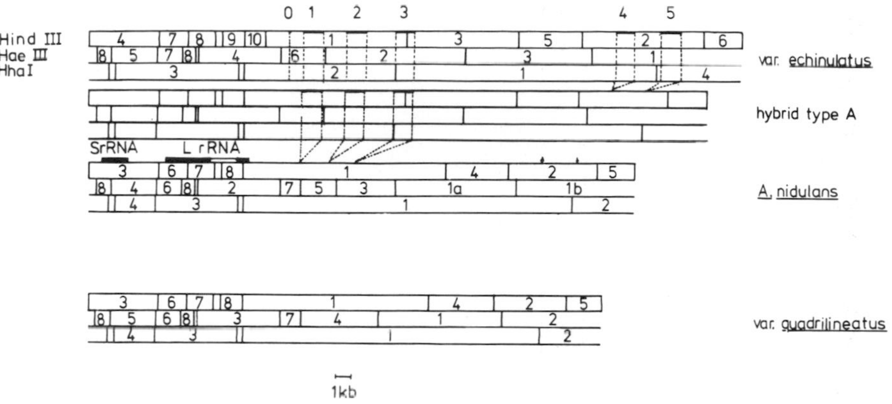

Figure 11. Restriction maps of *A. nidulans* species group mtDNA. A hybrid genome resulting from recombination between the mitochondrial genomes of *A. nidulans* and *A. nidulans* var. *echinulatus* is also shown and is present in strain DJ.1-1.

them, that is they are 'heterokaryon incompatible'. To overcome this problem it is necessary to remove the cell walls, producing protoplasts, and fuse protoplasts from different strains. Using this approach, it is possible to transfer the mitochondrial genome from one nuclear background to another. It is still necessary to have a selectable mitochondrial marker in at least one of the parents so that the desired nuclear/mitochondrial hybrid can be detected.

Protoplasts can be prepared from *Aspergillus* as follows.

(i) Inoculate 100 ml of YEG medium [2% yeast extract, 0.1% bactopeptone (Difco), 2% glucose, 1 mg/l *p*-aminobenzoic acid] in a 250 ml conical flask, with conidial suspension of the required strain to a final concentration of 10^6/ml, and incubate with shaking (orbital incubator) at 30°C for around 15–20 h. The aim is to obtain young mycelium which has barely aggregated into pellet growth.

(ii) Centrifuge the mycelium on a bench centrifuge to harvest (~5000 *g* for 5 min, but often does not pellet very well), and carefully remove most of the supernatant.

(iii) Wash the mycelium in sterile 0.6 M KCl, and centrifuge as before.

(iv) Combine the final pellets and suspend in 20 ml of 0.6 M KCl containing Novozym 234 (Novo Industri, Copenhagen) (5 mg/ml, filter sterilized).

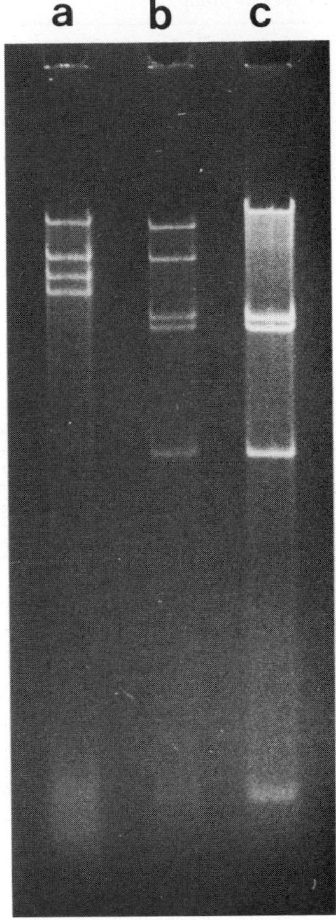

Figure 12. Recombination of mtDNA in *Aspergillus*. Restriction endonuclease digestion patterns of parental and hybrid mtDNA with *Hha*I. **a)** *A. nidulans* var. *echinulatus*, **b)** hybrid, **c)** *A. nidulans*.

(v) Shake slowly at 30°C for 90 min, gently pipetting the suspension a few times after 30 min to break up clumps of cells.

(vi) Check a sample under the microscope after 60 min to see whether the conversion to protoplasts is complete.

(vii) Filter through sterile Miracloth (Calbiochem), which has been pre-wetted with 0.6 M KCl, to remove larger pieces of remaining mycelium.

(viii) Harvest the protoplasts in a bench centrifuge (at 5000 *g* for ~2 min), then wash and re-centrifuge twice in 0.6 M KCl and resuspend the final pellet in 1 ml of 0.6 M KCl.

(ix) Take a small aliquot (50 μl) for viable counts, and use the remainder directly in fusion experiments.

(x) For viable counts, make parallel serial dilutions in water and 0.6 M KCl, and plate in a soft overlay of MM containing supplements in the medium with and without 0.6 M KCl, respectively. The protoplast viable count is the difference between the two types, that is the osmotically-sensitive count.

For plating protoplasts, dispense 5 ml aliquots of soft, molten MM (0.5% agar) with supplements and 0.6 M KCl in suitable bottles/tubes and keep in a water bath at 45°C until required. Remove from the bath, add protoplasts, mix and pour onto a plate of MM (1.5% agar) containing supplements and 0.6 M KCl. After the agar has set, incubate at 37°C.

Once protoplasts have been prepared protoplast fusion can be used to transfer the mitochondria. This can be carried out as follows.

(i) Prepare protoplasts from two strains with suitably marked mitochondria. As an example, a transfer between *A. nidulans* and *A. nidulans* var. *echinulatus* is described, using strains B3 and 25-2.

(ii) Mix approximately equal amounts of protoplasts from the two strains (numbers can be checked visually using a haemocytometer) for example 0.5 ml of each suspension, and centrifuge at 5000 *g* for 5 min.

(iii) Add 1 ml of 30% polyethylene glycol 6000 (PEG) in 0.1 M $CaCl_2$ and incubate for 10 min at room temperature.

(iv) Plate out aliquots (10 μl−0.5 ml) in MM containing 0.6 M KCl and 3 μg/ml oligomycin as described previously.

(v) Incubate at 37°C until prototrophic, oligomycin-resistant yellow colonies appear.

(vi) Pick off and streak to single colonies on suitable medium.

Isolation and analysis of DNA from a hybrid generated in this way often reveals a new restriction pattern which reflects the presence of DNA sequences from both parents (*Figures 11* and *12*). This kind of experiment demonstrates not only that mtDNA recombination occurs, but that certain introns predominate in the progeny (70).

11.6.4 *Uniparental inheritance of mtDNA*

Restriction fragment patterns can also be used to demonstrate uniparental inheritance of the mitochondrial genome (71), and the absence of recombination during the sexual cycle. To prevent the formation of a heterokaryon prior to crossing, heterokaryon incompatibility loci, *het*, have been introduced into the strains to be used for this purpose. While strain DJ.2-1 carries the normal *A. nidulans* mitochondrial genome, and (*cam112*) as a mitochondrial marker, DJ.1-1 has the *A. nidulans* nucleus, but carries a hybrid mitochondrial genome (*Figure 11*) (70) and (*oli1*). Thus the mtDNA is marked both genetically and physically. Crosses between these two strains produce hybrid and selfed cleistothecia with either mitochondrial type, but only one (parental) type is found in any single cleistothecium.

A further interesting point can be shown because of the presence of the *b1* mutation in DJ.1-1. This confers blue ascospore colour instead of the red wild-type. The *b1* mutation is a single nuclear mutation, but behaves rather strangely in the cleistothecium. If ascospore wall colour were determined by the genotype of each individual ascospore, then 50% of the ascospores in the cleistothecium would be blue, and 50% red. However, it turns out that the ascospores of a single hybrid cleistothecium are either all red or all blue, despite the fact that *b1* is a nuclear mutation. Although all the ascospores are one colour, if colonies from single ascospores are allowed to self, half of them give rise to blue cleistothecia, and half to red (72).

It appears that one of the pair of nuclei which take part in the unseen fertilization process acts as the 'maternal' nucleus, and determines the phenotype of both cleistothecial ascospore wall pigmentation. It is the cytoplasm around this nucleus which provides the mitochondriotype found in the cleistothecium (73).

11.7 A nuclear gene encoding a mitochondrial function

Strain O^R31 carries an oligomycin resistance mutation in the nucleus, on linkage group VII. The gene carrying this mutation has been isolated, and codes for subunit 9 of the mitochondrial ATP synthetase (74). By analogy with yeast, it is likely that (*oli1*) and (*oli322*) reside in the mitochondrially-encoded gene for subunit 6 of the same complex (*Figure 10*), but there is no direct proof of this. The *oliC31* (Oli^r) allele has been used as a selectable marker for the transformation of *A. nidulans* (74). If one makes oligomycin-resistant mutants, the heterokaryon test will show that most of them carry nuclear mutations. They will probably all be at the *oliC* locus, and it is possible to check this by sexual crosses between the nuclear mutants, isolated in R21, and B015. Absence or a very low frequency of wild-type (oligomycin-sensitive) recombinants amongst ascospores from a hybrid cleistothecium will indicate the occurrence of mutations at *oliC* (75).

12. ACKNOWLEDGEMENTS

Bernard Dujon wishes to thank his colleagues from the Centre de Génétique moléculaire du CNRS whose successful (or less successful) experiments resulted in this series of methods. He thanks M.Bolotin-Fukuhara, L.Colleaux, D.Hawthorne, A.Jacquier, F.Michel and P.Slonimski for their careful reading of the manuscript and fruitful comments and A.Jacquier for his photographic expertise.

13. REFERENCES

1. Brown,W.M. and Vinograd,J. (1974) *Proc. Natl. Acad. Sci. USA,* **79**, 4617.
2. Brown,W.M. (1981) *Ann. N.Y. Acad. Sci.*, **361**, 119.
3. Hauswirth,W.W. and Laipis,P.J. (1982) *Proc. Natl. Acad. Sci. USA,* **79**, 4686.
4. Denslow,N.D. and O'Brien,T.W. (1974) *Biochem. Biophys. Res. Commun.*, **57**, 9.
5. Piko,L. and Matsumoto,L. (1976) *Dev. Biol.*, **49**, 1.
6. Michaels,G.S., Hauswirth,W.W. and Laipis,P.J. (1982) *Dev. Biol.*, **99**, 246.
7. Clayton,D.A. (1982) *Cell,* **28**, 693.
8. Maxam,A.M. and Gilbert,W. (1981) In *Methods in Enzymology.* Grossman,L. and Moldave,K. (eds), Academic Press, New York, Vol. 65, p. 499.
9. Tu,C.D. and Cohen,S. (1980) *Gene,* **10**, 117.
10. McKay,S.D.L., Olivo,P.D., Laipis,P.J. and Hauswirth,W.W. (1986) *Mol. Cell Biol.*, **6**, 1261.
11. Maniatis,T., Fritsch,E.F. and Sambrook,J. (1982) *Molecular Cloning: A Laboratory Manual.* Cold Spring Harbor Laboratory Press, New York.
12. Tapper,D.P., Van Etten,R.A. and Clayton,D.A. (1983) In *Methods in Enzymology.* Fleischer,S. and Fleischer,B. (eds), Academic Press, New York, Vol. 97, p. 426.
13. Chang,D.D., Hauswirth,W.W. and Clayton,D.A. (1985) *EMBO J.*, **4**, 1559.
14. Leaver,C.J. and Gray,M. (1982) *Annu. Rev. Plant Physiol.*, **133**, 373.
15. Lonsdale,D.M., Hodge,T.P. and Fauron,C.M.-R. (1984) *Nucleic Acids Res.*, **12**, 9249.
16. Palmer,J.P. and Shields,C.R. (1984) *Nature,* **307**, 437.
17. Pring,D.R., Levings,C.S., Hu,W.W.L. and Timothy,D.H. (1977) *Proc. Natl. Acad. Sci. USA,* **74**, 2904.
18. Paillard,M., Sederoff,R.R. and Levings,C.S. (1985) *EMBO J.*, **4**, 1125.

19. Kemble,R.J., Gunn,R.E. and Flavell,R.B. (1980) *Genetics,* **95**, 451.
20. Sparks,R.B. and Dale,R.M.K. (1980) *Mol. Gen. Genet.*, **180**, 351.
21. Quetier,F. and Vedel,F. (1977) *Nature,* **268**, 365.
22. Levings,C.S. and Pring,D.R. (1977) *Science,* **193**, 158.
23. Masters,B.S., McCarty,D.M. and Hauswirth,W.W. (1983) *Plant Mol. Biol. Rep.*, **1**, 125.
24. Brown,W.M. (1980) *Proc. Natl. Acad. Sci. USA,* **77**, 3605.
25. Hauswirth,W.W. and Clayton,D.A. (1986) *Nucleic Acids Res.*, **13**, 8093.
26. Langer,P.R., Waldrop,A.A. and Ward,D.C. (1981) *Proc. Natl. Acad. Sci. USA,* **78**, 6633.
27. Leary,J.J., Brigati,D.J. and Ward,D.C. (1983) *Proc. Natl. Acad. Sci. USA,* **80**, 4045.
28. Singer,R.H. and Ward,D.C. (1982) *Proc. Natl. Acad. Sci. USA,* **79**, 7331.
29. Negro,F., Berninger,M., Chiaberge,E., Gugliotta,P., Bussolati,G., Actis,G.C., Rizzetto,M. and Bonino,F. (1985) *J. Med. Virol.*, **15**, 373.
30. Beckmann,A.M., Myerson,D., Daling,J.R., Kiviat,N.B., Fenoglio,C.M. and McDougall,J.J. (1985) *J. Med. Virol.*, **16**, 265.
31. Delius,H., van Heerikhuizen,H., Clarke,J. and Koller,B. (1985) *Nucleic Acids Res.*, **13**, 5457.
32. Southern,E.M. (1975) *J. Mol. Biol.*, **98**, 503.
33. Dujon,B. (1981) In *Molecular Biology of the Yeast Saccharomyces cerevisiae.* Strathern,J.N., Jones,E.W. and Broach,J.R. (eds), Cold Spring Harbor Laboratory Press, New York, Vol. 2, p. 505.
34. DeZamaroczy,M. and Bernardi,G. (1985) *Gene,* **37**, 1.
35. Stevens,B. (1981) In *Molecular Biology of the Yeast Saccharomyces cerevisiae.* Strathern,J.N., Jones, E.W. and Broach,J.R. (eds), Cold Spring Harbor Laboratory Press, New York, Vol. 2, p. 471.
36. Coen,D., Deutsch,J., Netter,P., Petrochilo,E. and Slonimski,P.P. (1970) In *Symposium of the Society for Experimental Biology.* Miller,P.L. (ed.), Cambridge University Press, Vol. 24, p. 449.
37. Netter,P., Petrochilo,E., Slonimski,P.P., Bolotin-Fukuhara,M., Coen,D., Deutsch,J. and Dujon,B. (1974) *Genetics,* **78**, 1063.
38. Wolf,K., Dujon,B. and Slonimski,P.P. (1973) *Mol. Gen. Genet.*, **125**, 53.
39. Avner,P.R., Coen,D., Dujon,B. and Slonimski,P.P. (1973) *Mol. Gen. Genet.*, **125**, 9.
40. Lancashire,W.E. and Griffiths,D.E. (1975) *Eur. J. Biochem.*, **51**, 403.
41. Lancashire,W.E. and Mattoon,J.R. (1979) *Mol. Gen. Genet.*, **176**, 255.
42. Pratje,E. and Michaelis,G. (1977) *Mol. Gen. Genet.*, **152**, 167.
43. Colson,A.M., Luu,T.V., Convent,B., Briquet,M. and Goffeau,A. (1977) *Eur. J. Biochem.*, **74**, 521.
44. Subik,J., Kovacova,V. and Takacsova,G. (1977) *Eur. J. Biochem.*, **73**, 275.
45. Thierbach,G. and Michaelis,G. (1982) *Mol. Gen. Genet.*, **186**, 501.
46. Galzy,P. and Slonimski,P.P. (1957) *C.R. Acad. Sci. Paris,* **245**, 2423.
47. Deutsch,J., Dujon,B., Netter,P., Petrochilo,E., Slonimski,P.P., Bolotin-Fukuhara,M. and Coen,D. (1974) *Genetics,* **76**, 195.
48. Putrament,A., Baranowska,H. and Prazmo,W. (1973) *Mol. Gen. Genet.*, **126**, 357.
49. Kocavova,V., Irmlerova,J. and Kovac,L. (1968) *Biochim. Biophys. Acta,* **162**, 157.
50. Kotylak,Z. and Slonimski,P.P. (1977) In *Mitochondria 1977.* Bandlow,W., Schweyen,R.J., Wolf,K. and Kaudewitz,F. (eds), DeGruyter, Berlin, p. 83.
51. Michaelis,G., Douglass,S., Tsai,M.J. and Criddle,R. (1971) *Biochem. Genet.*, **5**, 487.
52. Slonimski,P.P. and Tzagoloff,A. (1976) *Eur. J. Biochem.*, **61**, 27.
53. Ephrussi,B. and Grandchamp,S. (1965) *Heredity,* **20**, 1.
54. Fangman,W.L. and Dujon,B. (1984) *Proc. Natl. Acad. Sci. USA,* **81**, 7156.
55. Douglas,M.G. and Butow,R.A. (1976) *Proc. Natl. Acad. Sci. USA,* **73**, 1083.
56. Clowes,R.C. and Hayes,W., eds. (1968) *Experiments in Microbial Genetics.* Blackwell Scientific Publications, Oxford and Edinburgh.
57. Clutterbuck,A.J. (1974) In *Handbook of Genetics.* King,R.C. (ed.), Plenum Press, New York and London, Vol. 1, p. 447.
58. Turner,G. and Rowlands,R.T. (1977) In *Genetics and Physiology of Aspergillus.* Smith,J.E. and Pateman, J.A. (eds), Academic Press, London, New York and San Francisco. British Mycological Society Symposium Series No. 1, p. 319.
59. Scazzocchio,C., Brown,T.A., Waring,R.B., Ray,J.A. and Davies,R.W. (1983) In *Mitochondria 1983.* Schweyen,K. and Kaudewitz,F. (eds), Walter de Gruyter & Co., Berlin and New York, p. 303.
60. Brown,T.A., Waring,R.B., Scazzochio,C. and Davies,R.W. (1985) *Curr. Genet.*, **9**, 113.
61. Waldron,C. and Roberts,C.F. (1973) *J. Gen. Microbiol.*, **78**, 379.
62. Jinks,J.L. (1963) In *Methodology in Basic Genetics.* Burdette,W.J. (ed.), Holden-Day Inc., San Francisco, p. 325.
63. Rowlands,R.T. and Turner,G. (1973) *Mol. Gen. Genet.*, **126**, 201.
64. Rowlands,R.T. and Turner,G. (1974) *Mol. Gen. Genet.*, **133**, 151.
65. Rowlands,R.T. and Turner,G. (1975) *Mol. Gen. Genet.*, **141**, 69.

66. Lazarus,C.M. and Turner,G. (1977) *Mol. Gen. Genet.*, **156**, 303.
67. Waring,R.B. and Scazzocchio,C. (1983) *Genetics*, **103**, 409.
68. Mason,J.R. and Turner,G. (1975) *Mol. Gen. Genet.*, **143**, 93.
69. Spooner,R.A. and Turner,G. (1984) In *Genetic Maps 1984.* O'Brien,S. (ed.), Cold Spring Harbor Laboratory Press, New York, Vol. 3, p. 262.
70. Earl,A.J., Turner,G., Croft,J.H., Dales,R.B.J., Lazarus,C.M., Lünsdorf,H. and Kuntzel,H. (1981) *Curr. Genet.*, **3**, 221.
71. Jemayel,D. (1986) Ph.D. Thesis, University of Birmingham.
72. Apirion,D. (1963) *Genet. Res., Camb.*, **4**, 276.
73. Rowlands,R.T. and Turner,G. (1976) *Genet. Res., Camb.*, **28**, 281.
74. Ward,M., Wilkinson,B. and Turner,G. (1986) *Mol. Gen. Genet.*, **202**, 265.
75. Rowlands,R.T. and Turner,G. (1977) *Mol. Gen. Genet.*, **154**, 311.
76. Blanc,H. and Dujon,B. (1986) *Proc. Natl. Acad. Sci. USA*, **77**, 3942.
77. Tzagoloff,A., Akein,A. and Foury,F. (1976) *FEBS Lett.*, **65**, 391.
78. Kruszewska,A. and Szczesniak,B. (1980) *Genet. Res., Camb.*, **35**, 225.
79. Bolotin-Fukuhara,M. and Fukuhara,H. (1976) *Proc. Natl. Acad. Sci. USA*, **73**, 4608.
80. Murphy,M., Choo,K.B., McReadie,I., Marzuki,S., Lukins,H.B., Nagley,P. and Linnane,A.W. (1980) *Arch. Biochem. Biophys.*, **203**, 260.
81. Alexander,N.J., Vincent,R.D., Perlman,P.S., Miller,D.H., Hanson,D.K. and Mahler,H.R. (1979) *J. Biol. Chem.*, **254**, 2471.
82. Dujon,B., Bolotin-Fukuhara,M., Coen,D., Deutsch,J., Netter,P., Slonimski,P.P. and Weill,L. (1976) *Mol. Gen. Genet.*, **143**, 131.
83. Conde,J. and Fink,G.R. (1976) *Proc. Natl. Acad. Sci. USA*, **73**, 3651.

CHAPTER 8

DNA replication and transcription

M.BARAT-GUERIDE, R.DOCHERTY and D.RICKWOOD

1. INTRODUCTION

As can be gauged from the previous chapter on the genetics of mitochondria, the biogenesis of mitochondria involves complex interactions between the nuclear and mitochondrial genomes. The mitochondrial DNA exhibits both regulated transcription and a pattern of replication which is quite clearly distinct from that occurring in the nucleus. It is both beyond the scope of this book and inappropriate to review the complexities of the transcription and replication of the mitochondrial DNA (mtDNA), however this topic has been reviewed elsewhere (1,2). In this chapter an attempt will be made to give the reader an insight into the methods involved in studying these two processes in mitochondria. Although any such attempt is invariably limited by our knowledge of the precise enzymes and molecular mechanisms involved, since it is not at all clear at the present time as to the exact degree of complexities involved in the processes of transcription and replication.

2. REPLICATION OF MITOCHONDRIAL DNA

2.1 **Introduction**

The mtDNA of a cell usually represents only a few percent of the total DNA but, in some cases, the mtDNA is much more abundant as, for example, in *Drosophila* or mouse eggs, or even preponderant as in the case of *Xenopus laevis* oocytes. While the synthesis of nuclear DNA is restricted to a particular period, the S phase, of the cell cycle, the replication of mtDNA appears to be continuous throughout the cell cycle in most types of cells. However the relative rates of synthesis of nuclear and mtDNA in tissue culture cells can be affected by culture conditions in that it has been claimed that in cells synchronized by selective detachment the mtDNA only replicates in the S and G2 phases of the cell cycle, while in thymidine-block synchronized cells replication of mtDNA proceeds at a constant rate throughout the cell cycle. The discrepancy between these results could arise from a number of sources, either from variations in experimental conditions or from the use of different types of cells.

Essentially two types of mitochondrial genome can be defined, namely circular and linear molecules. Linear molecules are found in some lower eukaryotes, notably protozoa; while it has clearly been indicated that most other genomes are circular. Another notable source of diversity of the mitochondrial genome is its size, it can be from as small as 16.5 kb (human) to as large as 570 kb in plants; as a general rule vertebrates have a small tightly compacted genome while plants have large genomes that contain repeats, introns and even chloroplast DNA sequences. Most of the studies of mtDNA replication have been done on a fairly limited range of cell types such as fungi (e.g. *Saccharomyces cerevisiae*), ciliated protozoa (e.g. *Tetrahymena* and *Paramecium*), *Drosophila* and a number of vertebrate cells (usually in tissue culture). There is no universal pattern of replication of mtDNA. However, in all vertebrate cells so far studied the same distinctive replication process has been found, it is characterized by the asymmetric replication of each DNA strand starting from origins at different sites. In the circular mtDNA from vertebrates one also finds that the origin of replication of the heavy strand (H-strand) is associated with a displacement loop (D-loop) structure which consists of a short length of H-strand paired with the parental L-strand. The frequency of these D-loop structures is very variable depending on the organism, cell type and replicating activity, from less than 1% to 75%. The structure and organization of mitochondrial genomes has been reviewed elsewhere (3,4), this part of the chapter will concentrate on the methods used to study replication in mitochondria.

2.2 **Systems for studying DNA replication**

2.2.1 *In vivo studies of mtDNA replication*

A great deal of work has been done on the mechanism of replication of DNA in mitochondria by studying the patterns of incorporation of [^{3}H]thymidine *in vivo*. Such an approach has been widely used to study replication but the use of *in vivo* labelling methods for kinetic studies or even for comparative studies with nuclear DNA replication are hampered by the different pools of deoxynucleoside triphosphates for the nucleus and mitochondria; an added complication is that these pools can vary independently of each other.

2.2.2 *Systems for studying mtDNA replication based on purified mitochondria*

DNA replication can be studied in isolated mitochondria purified as described in Section 2 of Chapter 1. However most of the work with isolated mitochondria has been carried out some time ago and predominantly with the aim of proving the replication of the DNA inside the mitochondrion (e.g. see refs. 5,6); although isolated mitochondria have been used to study the process of replication in the absence of cytoplasmic pools of deoxynucleoside triphosphates (7). A typical incubation medium for the *in vitro* synthesis of DNA in isolated mitochondria is given in *Table 1*. Following incubation it is necessary to treat the assay mixture with 1% SDS and 1 mg/ml nuclease-free protease to avoid artefactual binding of the labelled deoxynucleoside triphosphates (dNTPs) by the mitochondrial proteins (7). The incorporation of radioactivity into DNA can then be measured by acid precipitation with trichloroacetic acid. An alternative approach is to study DNA replication in mitochondrial lysates prepared by treating mitochondria with 0.5% Triton X-100; the incubation conditions used in this case (8) are notable in that there is 0.1 M KCl and manganese rather than magnesium in the assay mixture (*Table 2*). One important aspect of this type of assay is to ensure that the mitochondria are not contaminated with nuclear chromatin and that neither the mitochondrial preparation nor any of the solutions are contaminated by bacteria since in both cases artefactual incorporation of radioactive dNTPs could occur. Density-labelling studies

Table 1. Incubation conditions for DNA synthesis in isolated liver mitochondria (7).

Sucrose	80 mM
Tris-HCl, pH 7.4	25 mM
Sodium phosphate, pH 7.4	10 mM
Sodium succinate	10 mM
Nicotinamide	2 mM
Malate	1 mM
KCl	2 mM
$MgCl_2$	10 mM
ATP	2 mM
dATP, dCTP, dGTP	15 μM of each
[^{3}H]dTTP (19 Ci/mmol, 0.7 TBq/mmol)	4 μM
Mitochondrial protein	0.2–0.4 mg/ml

Table 2. Incubation conditions for DNA synthesis in lysed mitochondria (8,9).

Tris-HCl, pH 8.3	20 mM
KCl	100 mM
$MnCl_2$	2 mM
Dithioerythritol	1 mM
Bovine serum albumin	1 mg/ml
ATP	5 mM
dATP, dCTP, dGTP	50 μM
[^{3}H]-dTTP (25 Ci/mmol,0.9 TBq/mmol)	10 μM

To 100 μl of this incubation medium add 20 μl of a 0.5% Triton X-100 mitochondrial lysate containing approximately 100 μg of protein and incubate at 37°C for 20 min.

of the DNA synthesized *in vitro* using the techniques described in Section 2.3.3 reveal that the newly synthesized DNA is from true replication rather than repair synthesis.

2.2.3 *Use of cloning techniques*

With the advent of cloning techniques there has been increasing interest in cloning the origins of replication of the mtDNA. Here as an example, the procedure for cloning a 1903-bp fragment containing the origin of replication of *X. laevis* into pUC18 is described.

Cloning of mtDNA into plasmids. The plasmid pUC18 contains genes for ampicillin resistance and *Escherichia coli* β-galactosidase, the latter containing a polylinker sequence into which restriction fragments can be inserted.

(i) Prepare purified *X. laevis* mtDNA using CsCl–ethidium bromide gradients as described in *Table 3*; oocytes are a good source of mtDNA because it is preponderant in this type of cell.
(ii) Digest approximately 2 μg of mtDNA with *Hpa*I and *Eco*RI and mix with approximately 0.6 μg of plasmid pUC18 digested with *Sma*I and *Eco*RI. Both digestions are carried out using the optimum conditions as given by the suppliers.
(iii) Heat the DNA mixture for 5 min at 65°C to inactivate the enzymes and precipitate the DNA by the addition of a tenth volume of 3 M sodium acetate and 2 vols of ethanol, store at −20°C for 30–60 min. Pellet the DNA by centrifugation

Table 3. Preparation of mtDNA on CsCl–ethidium bromide gradients.

1.	Resuspend the pellet of purified mitochondria in 4 ml of 20 mM Tris-HCl, pH 8.0, 10 mM EDTA, 100 mM NaCl.
2.	Add SDS to 1% (0.45 ml of a 10% stock solution).
3.	Add 7.44 g of solid CsCl and dissolve.
4.	Centrifuge at 10 000 *g* and discard the 'protein–SDS cake' on top.
5.	Add 0.5 ml of a 10 mg/ml ethidium bromide solution.
6.	Adjust the volume to 10 ml and, if necessary, adjust the density to 1.55 g/cm^3.
7.	Centrifuge in a fixed-angle Beckman R50Ti or equivalent rotor at 100 000 *g* for 36–48 h at 20°C.

Table 4. Preparation of competent bacteria.

1.	Inoculate 250 ml of nutrient broth medium with 2.5 ml of overnight culture of *E. coli* strain JM109[a] and incubate for ~2 h to give a final optical density of 0.4 at 500 nm.
2.	Pellet the bacteria by centrifugation at 6000 *g* for 10 min and wash them in 100 ml of cold 0.1 M $MgCl_2$ and pellet the bacteria as before.
3.	Resuspend the bacteria in 100 ml of cold 0.1 M $CaCl_2$ and incubate in ice for 20 min, then pellet the bacteria by centrifugation at 6000 *g* for 10 min.
4.	Resuspend the bacteria in 10 ml of a mixture containing 14% glycerol and 86% of 0.1 M $CaCl_2$; freeze 1 ml aliquots in liquid nitrogen and store at −70°C. Before use thaw the sample slowly in ice.

[a]This strain is derived from strain K12 and has the following genotype: *recA*1, *endA*1, *gyrA*96, *thi, hsdR*17, *supE*44, *relA*1, λ^-, Δ(*lac-proAB*), [F′, *traD*36, *proAB, lac*I^{q}Z ΔM15].

at 12 000 *g* for 10 min at 4°C and dry under vacuum.

(iv) Dissolve the DNA in 9 μl of ligation buffer containing 10 mM $MgCl_2$, 10 mM dithiothreitol (DTT) and 20 mM Tris-HCl, pH 7.6 then add a tenth volume of 10 mM ATP and 1 unit of bacteriophage T4 ligase. Incubate the mixture at 4°C for 16 h.

(v) Add 5 μl of the ligated DNA to 100 μl of competent *E. coli* strain JM109 (*Table 4*). Incubate the mixture for 30 min at 4°C and then 5 min at 37°C to complete the transformation. Add 2 ml of nutrient broth and continue the incubation at 37°C for 60–90 min with gentle shaking. Any unused transformation mixture can be kept at 5°C until you are sure that you have enough transformed clones; if required the mixture can be centrifuged, the bacteria resuspended in a smaller volume of nutrient broth and spread on a single plate.

(vi) Spread 0.1 ml of the bacterial suspension on nutrient agar plates containing 25 μg of ampicillin and previously spread with 30 μl of 10 mg/ml of Xgal (5-bromo-4-chloro-3-indolyl β-D-galactopyranoside dissolved in *N,N′*-dimethylformamide). Spread between two and five plates together with a control plate of non-transformed bacteria.

(vii) Incubate the plates overnight at 37°C. Ampicillin-resistant colonies which are white amongst the blue ones as a result of the insertion of a DNA fragment into the β-galactosidase gene are picked out and streaked out on the same medium to ensure that they are resistant to ampicillin and do not contain galactosidase activity.

(viii) Isolate plasmid DNA by the rapid small-scale alkali lysate method (*Table 5*) and carry out restriction digests on the plasmids of the various clones. The insert

Table 5. Mini-preparation of plasmid DNA.

1. Centrifuge 2 ml of an overnight culture grown in the presence of the appropriate antibiotics in a microcentrifuge to pellet the bacteria.
2. Remove the supernatant using a Pasteur pipette, the end of which has been drawn out into a capillary, and resuspend the bacteria in 100 μl of solution containing 50 mM glucose, 10 mM EDTA, 10 mM Tris-HCl, pH 8.0, then add freshly prepared lysozyme solution to a final concentration of 2 mg/ml, mix and incubate in ice for 30 min.
3. Add 150 μl of freshly prepared 1% SDS in 0.2 M NaOH to the bacterial suspension, mix gently and leave to stand for 5 min in ice to allow the lysis to go to completion.
4. Add 150 μl of 3 M sodium acetate, pH 4.8 to the lysate, mix gently and leave to stand in ice for 30 min prior to centrifugation at 12 000 *g* for 5 min.
5. Remove 400 μl of the supernatant and add it to 1 ml of absolute ethanol at −20°C; allow the nucleic acids to precipitate at −20°C for 20 min then pellet the precipitated material by centrifugation at 12 000 *g* for 5 min.
6. Remove the supernatant using a capillary Pasteur pipette, add 100 μl of 0.3 M sodium acetate, pH 7 and mix by vortexing prior to adding 300 μl of absolute ethanol at −20°C; leave at −20°C for 20 min and then centrifuge as before.
7. Remove the supernatant using a Pasteur pipette, add 100 μl of 0.3 M sodium acetate, pH 7 and mix by vortexing prior to adding 300 μl of absolute ethanol at −20°C; leave at −20°C for 15 min and centrifuge as before.
8. Carefully remove all of the supernatant, dry under vacuum and dissolve the pellet in 40 μl of sterile distilled water. Dissolve the plasmid preparation by vortexing, heat to 65°C for 10 min to inactivate any nucleases present and then cool.
9. Use 5−10 μl of the plasmid preparation for restriction enzyme analysis which should be carried out after the addition of 1 μl of 10 mg/ml of pancreatic RNase boiled before use to inactivate any contaminating DNases.

can be excised from the plasmid by digestion with *Pst*I and *Eco*RI. Alternatively, the plasmids can be digested with *Pst*I and *Eco*RI together with either *Hinc*II or *Hind*III to ensure the absence of *Hinc*II and *Hind*III sites in the inserted DNA restriction fragments.

Once prepared the plasmids can be transfected into cells to monitor whether or not the inserted fragments display autonomous replication sequence (ARS) activity. High frequency yeast transformation can be used as a positive selection method for mtDNA fragments exhibiting possible functions as origins of replication. As an example of this technique, a *ura*3 auxotrophic mutant of yeast can be transformed to a prototrophic cell by transfection with a recombinant plasmid vector such as YIP5 that contains the *ura*3 gene, and if a sequence of mtDNA with ARS activity is placed in the plasmid then it will be maintained as an extra-chromosomal element. The method used for transforming yeast cells is given below.

Transformation of yeast cells.

(i) Grow up 250 ml of yeast in glucose medium containing 2% glucose, 1% yeast extract, 1% bactopeptone and 5 mM sodium phosphate, pH 6.5.

(ii) Harvest the cells at mid-log phase ($\sim 2 \times 10^7$ cells/ml) by centrifugation at 1000 *g* for 5 min and wash the pelleted cells in distilled water.

(iii) Resuspend the cells in 25 ml of 1.0 M sorbitol, 10 mM $CaCl_2$, 10 mM Tris-HCl, pH 7.5 and add 500 μg of zymolyase. Incubate at 37°C for 30−45 min without shaking to produce protoplasts. Note that the length of incubation required to convert the cells to protoplasts will vary from strain to strain but to ensure good regeneration of the protoplasts the incubation time should not exceed 45 min; if this is insufficient try adding more zymolyase.

(iv) When osmotic lysis of the cells gives a reduction in the optical density of 50%, stop the incubation by cooling the cell suspension in ice, the protoplasts are very delicate and must be handled with care.

(v) Wash the protoplasts twice in the protoplasting buffer without zymolyase by centrifugation at 1000 *g* for 5 min at 5°C and gently resuspend the washed protoplasts in 0.5 ml of the same buffer.

(vi) Mix 150 μl of the protoplast suspension with 1 μg of plasmid DNA in 5 μl of 1 mM EDTA, 10 mM Tris-HCl, pH 8.0 and incubate at room temperature for 15 min.

(vii) Add 1.0 ml of 20% polyethylene glycol 4000 (PEG, mol. wt 4000, Koch-Light) in 10 mM $CaCl_2$, 10 mM Tris-HCl, pH 7.5 and incubate at room temperature for a further 20 min. Note that it is important to run a parallel control incubation without added plasmid DNA at the same time.

(viii) Sediment the protoplasts by centrifugation at 1000 *g* for 5 min and resuspend them in 100 μl of 20% PEG 4000, 10 mM $CaCl_2$, 10 mM Tris-HCl, pH 7.5 and gently mix the suspension with 5 ml of regeneration agar containing 3% bacto agar, 0.6% yeast nitrogen base, 2% glucose and 1.0 M sorbitol kept at 45°C.

(ix) Immediately pour the protoplasts suspended in the regeneration agar onto a minimal agar plate (3% bacto agar, 2% glucose, 0.6% yeast nitrogen base without amino acids, 1.0 M sorbitol together with any required amino acids) pre-warmed to 36°C.

(x) Incubate the plates for 5–7 days at 28°C and pick out the transformed colonies.

Using this procedure segments of mtDNA from a wide range of organisms containing a known or putative origin of replication have been able to transform yeast at a high frequency (*Table 6*). Further discussion of the use of cloning techniques to study the origin of replication is given in Section 2.4.3.

At the present time gene banks containing the whole of the mitochondrial genome of essentially all types of cell under active investigation have been prepared by various groups around the world. These clones should be freely available from one source or another. In addition, the significant level of conservation of the coding regions of the mtDNA often allows one to use a clone of a region of mtDNA from one organism

Table 6. Cloning of ARS sequences from different organisms.

Organism	*Vector*	*Plasmid*	*Ref.*
Xenopus laevis	YIp5	pXEY26	10
Paramecium aurelia	YIp5	YPaM	11
Saccharomyces cerevisiae	pBR322	pSCM237	
		pGT20	12
Saccharomyces cerevisiae	Y8p5	YRMpl etc.	13
Candida utilis	Cosmid pHC79		14

Table 7. Proteins associated with mtDNA.

Enzyme activity	*Organisms*	*Sedimentation coefficient*	*Mol. wt (kd)*	*Template specificity*	*Divalent ion requirement*	*Inhibitors*	*Ref.*
mt DNA polymerase[a]	rat liver	9.2 S[b]	150[c]				16
	rat liver	4 S, 11.3 S[b]		poly(A).$(dT)_{12-18}$ activated DNA	Mn^{2+} Mg^{2+}	NEM, EB	17
	rat liver	9.2 S		idem	idem	NEM, EB	18
	rat liver	9.2–9.4 S		idem	idem	NEM, ddTTP	19
	chick embryo	7–7.5 S		idem	idem	NEM	20
	chick embryo	7.5 S	180[c], 47[d]	poly(A).$(dT)_{12-18}$	Mn^{2+}		21
	HeLa cells (human)	8.1 S		poly(A).$(dT)_{12-18}$ activated DNA	Mn^{2+}	NEM, EB	17
	Xenopus laevis oocytes		70[d]	poly(A).$(dT)_{12-18}$ activated DNA	Mn^{2+} Mg^{2+}	NEM, EB, ddTTP	9
	Drosophila melanogaster	7.6 S	160[c] 125 and 35[d]	poly(A).$(dT)_{10}$ activated DNA	Mg^{2+}	NEM, ddTTP	22
	Saccharomyces cerevisiae		60[d]	activated DNA poly(dA).$(dT)_{10}$	Mg^{2+}	Mn^{2+} (5 mM) PCMB	23
	wheat embryo			activated DNA poly(dA).$(dT)_{12}$		ddTTP	24
primase	KB cells (human)	30 S (+RNA)				NaCl (250 mM)	25
Single-strand DNA Binding protein	*Xenopus laevis* oocytes	4 S	15.5 [d]	SS DNA			26
	rat liver		15.2[d]	SS DNA			27
Topoisomerase I	*Xenopus laevis* oocytes	4.4 S	65–70[d]	positive and negative supercoiled DNA		EB, Berenil	28
	HALL cells (human)	7.1 S	132[c], 60[d]			ATP (90 mM)	29, 30
	rat liver		44[c]	positive and negative supercoiled DNA		EB, NEM Berenil	31

[a]Purified from mitochondria except in ref. 19 where it was purified from mtDNA–membrane complexes. [b]Determined in low salt. [c]Determined by gel permeation chromatography. [d]Determined by SDS–polyacrylamide gel electrophoresis. ddTTP = 2′3′ dideoxythymidine triphosphate; EB = ethidium bromide; NEM = N-ethylmaleimide; PCMB = p-chloromercuribenzoate.

as a probe for the same gene in another organism. In exceptional cases where appropriate clones are not available then it will be necessary to clone the required pieces of DNA following the procedures described in one of the specialized manuals on cloning (15).

2.2.4 *Purification of proteins associated with DNA replication*

Work on a wide range of organisms has shown that a variety of enzyme activities that may be associated with the replication of mtDNA are present in the mitochondrion, these include primase, DNA polymerase and topoisomerase activities as well as single-strand DNA-binding proteins. *Table 7* summarizes the characteristics of a number of these proteins. However, of these various proteins only the mitochondrial DNA polymerase (γ-DNA polymerase) has been fully characterized as it is relatively easy to prepare pure enzyme from a variety of tissues using standard protein purification procedures. DNA polymerase can be purified from cells or mitochondria, but in the authors' experience it is most convenient to isolate the enzyme from purified mitochondria or mitoplasts, particularly since a very similar activity of unknown function appears to be present in the nucleus of some cells (17,19,20). The following method has been used for the purification of γ-DNA polymerase from chick embryo mitochondria as described elsewhere (20).

Purification of γ-DNA polymerase from chick embryos.

(i) Prepare purified mitochondria by differential pelleting and isopycnic sucrose gradient centrifugation (see Section 2 of Chapter 1).

(ii) Suspend the pellet of purified mitochondria (1.5–2 g) in 4 vols of 0.5 M KCl, 2 mM DTT, 0.5% Triton X-100, 20 mM Tris-HCl, pH 7.6 by Dounce homogenization.

(iii) Sonicate the mitochondrial suspension in ice 2–4 times each for 20 sec with cooling periods of 30 sec between each sonication (take care to avoid foaming of the solution at this stage).

(iv) Centrifuge the lysed mitochondria at 180 000 *g* for 60 min at 4°C and dialyse the supernatant against 2 mM 2-mercaptoethanol, 20 mM potassium phosphate, pH 7.5, overnight.

(v) If necessary clear the dialysates by centrifugation at 180 000 *g* for 60 min prior to loading the mitochondrial extract onto a DEAE–cellulose column (28 cm × 2.5 cm) equilibrated with 20 mM potassium phosphate, pH 7.5.

(vi) Wash the column with two column volumes of the equilibration buffer and elute the bound proteins using a 600-ml gradient of 20–500 mM potassium phosphate, pH 7.5, the γ-DNA polymerase usually elutes as a single peak at 150 mM potassium phosphate. The exact elution position should be determined by DNA polymerase assays (*Table 8*).

(vii) Pool the fractions containing the DNA polymerase activity and dialyse them against 20% glycerol, 2 mM 2-mercaptoethanol, 20 mM potassium phosphate, pH 7.5 prior to loading them onto a phosphocellulose column (15 cm × 1.8 cm) equilibrated with the same buffer.

(viii) Wash the column with two column volumes of equilibration buffer and elute the bound proteins using a 120-ml gradient of 20–600 mM potassium phosphate,

Table 8. Assay for mitochondrial DNA polymerase γ.

	Activated DNA 180 $\mu g/ml$	*Poly(A). (dT)$_{12-18}$* 50 $\mu g/ml$
Tris-HCl, pH 8.3	25 mM	25 mM
Potassium phosphate, pH 8.4	–	50 mM[b]
KCl	150 mM[a]	100 mM
$MnCl_2$	–	0.5 mM
$MgCl_2$	10 mM	–
Dithiothreitol	2 mM	2 mM
BSA	200 μg/ml	200 μg/ml
dATP, dCTP, dGTP	0.05 mM	
[Me-^{3}H]dTTP (1 Ci/mmol 37 GBq/mmol)	0.05 mM	0.05 mM

Total assay volume is 50 μl.

1. Incubate the assay mixture at 30°C for 30 min.
2. To measure the incorporation of [^{3}H]dTTP spot 45 μl of the assay mixture on a Whatman GF/C filter.
3. Immerse the filters in cold 5% trichloroacetic acid, 1% sodium pyrophosphate (10 ml/filter) with occasional stirring for 10 min.
4. Discard the trichloroacetic acid solution and add fresh 5% trichloroacetic acid solution, wash for 10 min. Repeat this step twice.
5. Rinse the filter twice in cold ethanol.
6. Dry the filter in an oven at 45°C or under an infra-red lamp and measure the radioactivity of each in a liquid scintillation counter.

[a]Can be replaced by 200 mM NaCl

[b]The presence of potassium phosphate makes this assay specific for DNA polymerase γ activity because poly(A).(dT)$_{12-18}$ does not act as a template for DNA polymerase β in the presence of this concentration of potassium phosphate.

pH 7.5, the DNA polymerase activity usually elutes at 450 mM potassium phosphate. Perform DNA polymerase assays to check the exact elution position of the enzyme.

(ix) Pool the fractions containing the DNA polymerase activity and dialyse them against 20% glycerol, 2 mM 2-mercaptoethanol, 20 mM potassium phosphate, pH 7.5 and then load onto a column of hydroxyapatite (16 cm × 1.5 cm) equilibrated with the same buffer.

(x) Wash the column with a column volume of equilibration buffer and elute the bound proteins using a 200-ml linear gradient of 20–600 mM potassium phosphate, pH 7.5. The DNA polymerase activity usually elutes at about 400 mM potassium phosphate.

This type of procedure usually gives a purification of several hundred fold with a yield in the region of 30%. The enzyme can be further purified if required by passing the purified extract in 20% glycerol, 2 mM 2-mercaptoethanol, 1 mM EDTA, 20 mM potassium phosphate, pH 7.5 over a double-stranded DNA–cellulose column prepared as described in *Table 9*. The DNA polymerase activity is eluted using a 20–600 mM potassium phosphate gradient at about 450 mM potassium phosphate.

This type of protocol, with minor variations, can be used for purifying γ-DNA

Table 9. Preparation of native DNA–cellulose.

1. Wash 50 g of cellulose (Whatman CF11) with gentle stirring successively with 500 ml of 0.1 M NaOH, 1000 ml of 1 mM EDTA, pH 7, 600 ml of 10 mM HCl and then double-distilled water until neutral pH.
2. Allow the cellulose to settle, discard the supernatant and add approximately 2 vols of absolute ethanol. Mix gently.
3. Discard the ethanol and allow the cellulose to dry for approximately 24 h at 36°C or for 3 days at room temperature.
4. Dissolve 300 mg of commercial calf thymus DNA (for example Sigma type I) in 100 ml of 1 mM EDTA, 100 mM Tris-HCl, pH 7.5. Then add 0.6 g of NaCl (100 mM). Shear the DNA by passing the solution 4–5 times through a 23-gauge needle.
5. Gradually add the cellulose powder to the DNA solution until the mixture is a thick paste (use ~1 g of powder for each 3 ml of DNA solution). Cover the beaker with gauze and allow it to dry at room temperature (cold air can be blown over the mixture to hasten the drying process).
6. Reduce the dry mixture to powder in a mortar and resuspend it in approximately 5 vols of absolute ethanol. The depth of the suspension should be approximately 1 cm.
7. Expose to 100 000 ergs/mm^2 of u.v. light (254 nm) and keep the suspension gently stirred with a magnetic stirrer.
8. Remove the ethanol by washing three times with 1 mM EDTA, 100 mM Tris-HCl, pH 7.5. Then add NaCl to 100 mM and pour the suspension into the column at 4°C.

The DNA cellulose suspension can be kept as a slurry frozen at −20°C.

polymerase from a wide range of animal cells although there may be some differences in terms of the elution profiles of the polymerase activities from the various columns used to purify the enzyme. However, the properties of mitochondrial DNA polymerases from plants appear to be significantly different from the γ-DNA polymerase of animal cells (24) and hence it may be necessary to make significant changes to the method used for animals cells.

The role(s) of the other activities associated with DNA replication remain unclear. It is known that *in vitro* inhibitors of topoisomerase I activity such as berenil also inhibit DNA replication *in vivo* but inhibitor studies are limited by the specificity of the inhibitor, for example, inhibitors of gyrase activity can also affect respiration. Thus the characterization of these activities must await the development of efficient protocols and assay procedures.

2.3 Methods for studying DNA replication

2.3.1 *Determination of the amount of mtDNA in cells*

The amount of mtDNA in cells can in some instances vary depending on the environment of the cell as well as other physiological factors. Thus it can be very useful to measure the mtDNA content of a cell relative to that of the nuclear DNA in order to gauge the level of replicative activity in cells. The following dot–blot method is very simple to do and has been used extensively in the authors' laboratory for cells containing only a few percent of mtDNA relative to the nuclear DNA.

(i) Homogenize the cells and prepare total cell DNA using the standard phenol–chloroform extraction procedure (see Section 2 of Chapter 7). Dissolve the DNA in 1 mM EDTA, 10 mM Tris-HCl, pH 7.5 and dilute the DNA to a final concentration of 5–10 μg/ml with the same buffer.

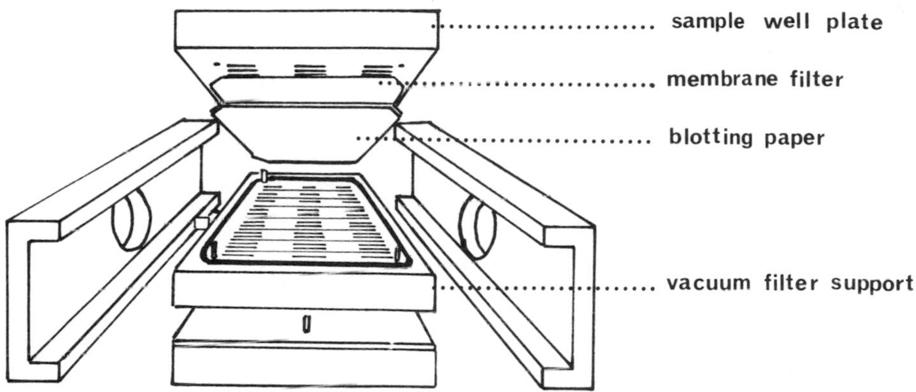

Figure 1. Minifold II slot blotter from Schleicher and Schuell.

(ii) Denature the DNA by adding a tenth volume of 3 M NaOH and incubate at 65°C for 45 min before neutralizing the solution by the addition of an equal volume of 2 M ammonium acetate.

(iii) Dot triplicate samples of the DNA solution containing about 500 ng of DNA onto a nitrocellulose membrane using the Minifold apparatus (*Figure 1*) from Schleicher and Schuell (Dassel, FRG).

(iv) At the same time dot in triplicate standard mixtures of CsCl-purified mtDNA (1−25 ng) and nuclear DNA (~500 ng).

(v) Bake the filters in a vacuum oven at 80°C for 2 h.

(vi) Float the baked filters on 6 × standard saline citrate (SSC) until they are thoroughly wetted, then submerge the filter in this solution for 5 min.

(vii) Pre-hybridize the filter in a plastic bag using 0.5 ml/cm^2 of filter in pre-hybridization medium containing 50% de-ionized formamide, 0.1% SDS, 5 × Denhardt's solution, 5 × SSPE and 100 μg/ml of heat-denatured sonicated calf thymus DNA. Incubate the filter with gentle agitation at 42°C for 2 h. Denhardt's solution is prepared as a 50-fold concentrated solution containing 1% Ficoll, 1% polyvinylpyrrolidone (PVP) and 1% bovine serum albumin (BSA) while SSPE is prepared as a 20-fold concentrated solution containing 3 M NaCl, 20 mM EDTA and 0.2 M sodium phosphate buffer, pH 7.4; both of these stock solutions are made in advance and can be stored at −20°C. Formamide is de-ionized by stirring it with Biorad AG501x8(D) mixed-bed resin (10 g/100 ml) for 4 h at room temperature and then it is also stored at −20°C.

(viii) Add the denatured mtDNA probe, labelled with [^{32}P]phosphate by nick-translation either by using one of the commercially available kits (e.g. from Amersham International) or the procedure outlined in *Table 10*. The specific activity of the probe should be in the range of $5 \times 10^7 - 5 \times 10^8$ c.p.m./μg and the final concentration in the hybridization solution should be approximately 5 ng/ml.

(ix) Re-seal the bag and mix the probe into the solution by gentle squeezing and inversion of the bag and then incubate the filter for 18 h at 42°C.

(x) After hybridization, carefully remove the filter from the plastic bag and wash in approximately 200 ml of SSC containing 0.1% SDS at room temperature for

Table 10. Nick-translation of mtDNA.

1.	Dissolve 200−250 ng of DNA in 10 μl of 1 mM EDTA, 10 mM Tris-HCl, pH 8.0.
2.	Add 4 μl of nucleotides/buffer solution containing 3 dNTPs 100 μM of each in 50 mM Tris-HCl, pH 7.8, 10 mM $MgSO_4$, 0.1 mM DTT.
3.	Add 2−5 μl of the other [α-^{32}P]dNTP (10 mCi/ml, 370 MBq/ml).
4.	Add 2 μl of enzyme solution (1 unit of DNA polymerase I, 20 pg of DNase I in a buffer containing 20 mM Tris-HCl, pH 7.5, 10 mM $MgSO_4$, 10% glycerol and 50 μg/ml BSA).
5.	Make up to 20 μl with distilled water, mix gently and incubate at 15°C for 75 min.
6.	Stop the reaction by adding 2 μl of 0.5 M EDTA, 0.25% bromophenol blue (the dye makes it possible to follow the purification of the labelled DNA by gel filtration).
7.	Load the reaction mixture onto a Sephadex G-75 column (prepared in a siliconized Pasteur pipette) equilibrated with 1 mM EDTA, 10 mM Tris-HCl, pH 8.0, elute with the same buffer, collect four-drop fractions and measure the radioactivity of each fraction by Cerenkov counting. DNA is eluted first then the peak of deoxynucleotides and the dye.

15 min. Then wash the filter three or four times in the same solution at 65°C each for 20−30 min.

(xi) Dry the filter on a piece of Whatman 3MM filter paper, cover the filter with Saran wrap and press the covered, dried filter onto X-ray film (Hyperfilm MP Amersham International or equivalent).

(xii) After development scan the spots corresponding to the standards and the unknown samples using a spectrophotometer and deduce the relative concentrations in the unknown samples from the relative areas under the peaks.

In the case of yeast cells, an isotope dilution method has been used to estimate the mtDNA content (32). In this technique the DNA of the reference cell type is labelled by growth in [^{3}H]adenine and then some of the reference cells are mixed with yeast cells to be examined. Following extraction and purification of the nuclear and mtDNA from the reference cells and the mixture of cells calculation of the ratios of the specific activity of the DNA reveals whether the two types of cells have the same ratio of nuclear to mtDNA.

In special cases it is possible to quantitate the amount of mtDNA in cells by Feulgen staining. This method has been used for measuring the amount of mtDNA in very young oocytes of *X. laevis* where mitochondria accumulate in one region of the oocyte (33).

2.3.2 *Autoradiographic analysis of mtDNA replication*

To the authors' knowledge very little work has been done on using the autoradiography of mitochondria for studying the replication of mtDNA and none has been done using autoradiography for the analysis of the mechanism of replication of mtDNA molecules. Autoradiography has been used to study mitochondrial DNA replication in *Tetrahymena* and *Xenopus* oocytes (34,35) where it is easy to distinguish between the nuclear and mtDNA. As an example the labelling and analysis of *Xenopus* oocyte mitochondria by autoradiography as has been carried out in the authors' laboratory is given here.

(i) Dissect out the ovaries of very young females and incubate the ovarian lobes in Barth's medium (36) containing 100 μCi/ml (3.7 MBq/ml) of [Me-^{3}H]thymidine (35 Ci/mmol, 1.3 TBq/mmol) for 48 h at 20°C.

(ii) Wash the lobes free of unincorporated radioisotope with Barth's medium and cut into small pieces.
(iii) Fix the specimens in 2.5% glutaraldehyde, 0.15 M sodium phosphate buffer, pH 7.3 and then post-fix in 1% osmium tetroxide prior to dehydration in ethanol (see Section 2 of Chapter 2).
(iv) For light microscopy cut sections (1 μm thick) systematically from different areas of the lobes of the ovaries and mount them on glass slides.
(v) Cover the slides with Ilford L4 emulsion diluted with an equal volume of distilled water, dry and store at 4°C in a sealed slide box for 2 weeks.
(vi) Develop these preparations in Kodak Microdol X for 8 min at 18°C, rinse in distilled water and fix prior to a final wash in running water for 20 min. At a later stage these slides can be stained with toluidine blue.
(vii) For electron microscopy obtain ultra-thin sections of the fixed ovaries using a diamond knife on an ultra-microtome. Collect the sections onto collodion-coated grids and then stain and treat with emulsion as described elsewhere (37).
(viii) After 6 weeks develop the grids in Phenidon (Kodak), rinse in distilled water and fix in 30% sodium thiosulphate.

As a control, treat some of the oocytes with 50 μg/ml of DNase I in sodium phosphate buffer, pH 6.8 for 4 h at 37°C after fixation with 10% formaldehyde. Post-fix the samples with osmium tetroxide and embed in Epon 812 (see Chapter 2). As a guide you will need to examine at least 20 electron autoradiographs in order to obtain a reliable quantitative analysis of the distribution of silver grains.

2.3.3 *Centrifugal analysis*

Isopycnic centrifugation. The base composition of the mtDNA of some organisms, particularly lower eukaryotes (e.g. yeast), is sometimes sufficiently different from that of the nuclear DNA to allow the two types of DNA to be separated in an isopycnic CsCl gradient. Moreover, in some cases it is possible to enhance the differences in density by adding compounds such as bis-benzimide or 4,6-diamidine-2-phenylindole hydrochloride (DAPI) which bind selectively to AT-rich DNA. However, the fact that in most cells the mtDNA is the only supercoiled circular DNA present enables one to separate the mtDNA on the basis of its density in CsCl−ethidium bromide gradients (see *Table 3*) even if it has the same base composition because the supercoiled DNA binds less ethidium bromide and so bands at a higher density than the linear nuclear DNA. Using this method it is possible to examine the incorporation of isotopically-labelled precursors into the mtDNA. However, if the mtDNA has been cut by nucleases or physical shearing, as often happens in the case of the larger molecules (e.g. plant mtDNA), then this method is of much less use since the linearized DNA will band with the nuclear DNA. Similarly this method will not work with organisms that have a linear mtDNA (e.g. ciliates). The conformation of the mtDNA does vary during replication and so in the case of vertebrate cells it is possible to isolate replicative intermediates on the basis of their buoyant density in gradients of CsCl containing ethidium bromide or propidium iodide. A typical example of the type of separation that can be obtained is shown in *Figure 2*.

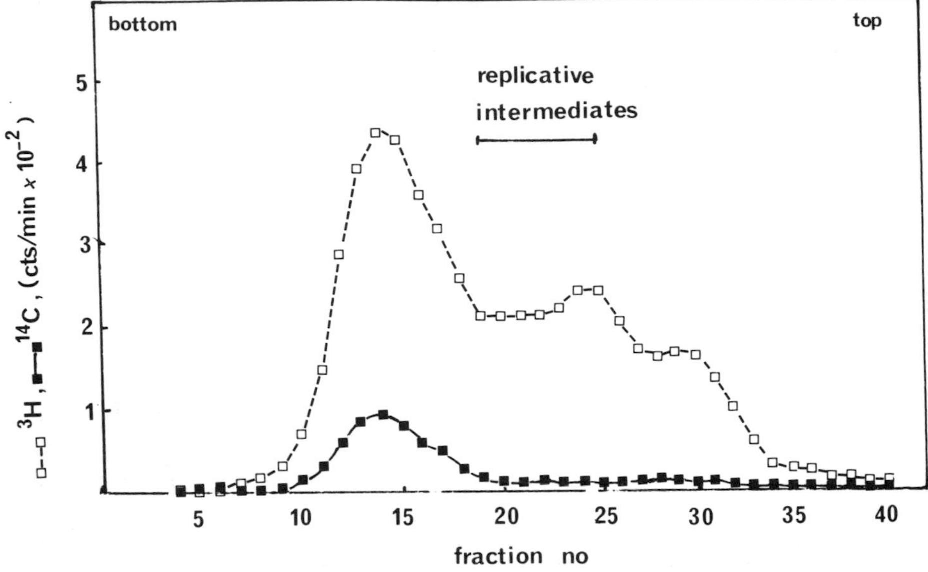

Figure 2. Separation of mtDNA and replicative intermediates on CsCl–ethidium bromide gradients. Mouse L cells grown in [^{14}C]thymidine to label the DNA uniformly were pulse-labelled with [^{3}H]thymidine for 10 min and the mtDNA was fractionated on CsCl–ethidium bromide gradients essentially as described in *Table 3*. Data derived from ref. 38.

Rate-zonal centrifugation. Rate-zonal gradients, usually of sucrose, can be used to separate DNA on the basis of its size and conformation; molecules which have a higher molecular weight or a more compact conformation sediment faster in a centrifugal field. An indication as to the usefulness of the method is given in *Figure 3* which shows the release of single-stranded DNA from the D-loop after denaturation of the mtDNA. This method is useful in that it is possible to use the appropriate size of gradient depending on the amount of material available and it is very easy to recover the DNA by precipitation after fractionation. However the degree of resolution obtainable with these gradients is rather limited and if a high degree of resolution is required it is better to use gel electrophoresis.

2.3.4 *Electrophoretic analysis*

Gel electrophoresis separates DNA molecules on the basis of their molecular weight and conformation; molecules with larger molecular weights or extended conformations move more slowly. This method is ideal for separating whole mtDNA molecules, restriction fragments of different sizes as well as fragments of the same length with or without D-loop structures. This last application is of particular interest in that the separated fragments of DNA containing the D-loop can be used for determining the origin of replication (see Section 2.4). Electrophoresis is carried out in 1% agarose gel in buffer containing 0.16 M Tris base, 0.08 M sodium acetate, 0.08 M NaCl, 8 mM EDTA, pH 8.0.

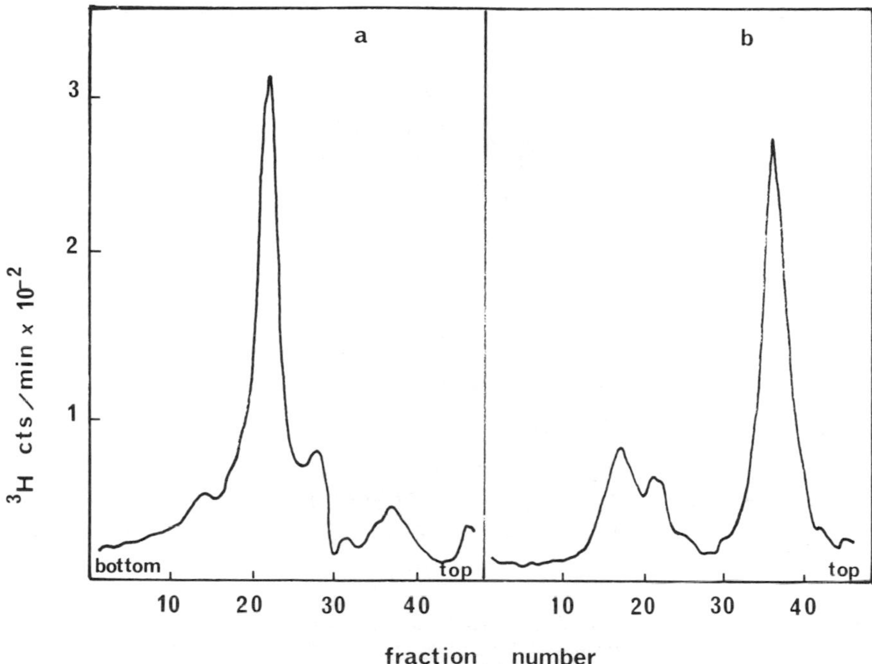

Figure 3. Separation of mtDNA on rate-zonal sucrose gradients. *X. laevis* oocytes treated with human chorionic gonadotrophin were labelled with [^{3}H]thymidine for 24 h. The mtDNA purified on a CsCl–ethidium bromide gradient was dialysed into 10 mM NaCl, 1 mM EDTA and 10 mM Tris-HCl, pH 8.0 and loaded either directly (**a**) or after heat denaturation at 95°C for 2 min (**b**) onto 5–20% (w/w) sucrose gradients. The gradients were centrifuged at 120 000 *g* for 2 h at 20°C. After heat denaturation a peak of radioactivity corresponding to the single strand H-fragment of the D loop appears at the top of the gradient. Data from ref. 39 with permission.

(i) Digest the purified mtDNA with the appropriate restriction nuclease using the recommended incubation conditions and then add a fifth volume of 1% SDS, 50 mM EDTA, 0.025% xylene cyanol, 0.025% bromophenol blue, 50% glycerol to stop the reaction.
(ii) Load about 1 μg of the digested DNA onto the gel using about 1.5 V/cm and run overnight (~18 h) at room temperature using a submarine gel apparatus.
(iii) After electrophoresis stain the gel by immersion in a solution of 1 μg/ml of ethidium bromide, rinse the gel with distilled water and visualize the bands using a long wave u.v. transilluminator.
(iv) If the fragments are small and you do not obtain a good separation using these conditions then try increasing the concentration of agarose.

A typical separation of *X. laevis* mtDNA restriction fragments is shown in *Figure 4*.

2.3.5 *Microscopic analysis of mtDNA replication*

Light microscopy. The limits of resolution of light microscopy preclude any detailed analysis of DNA replication in mitochondria. However, as mentioned in Section 2.3.1,

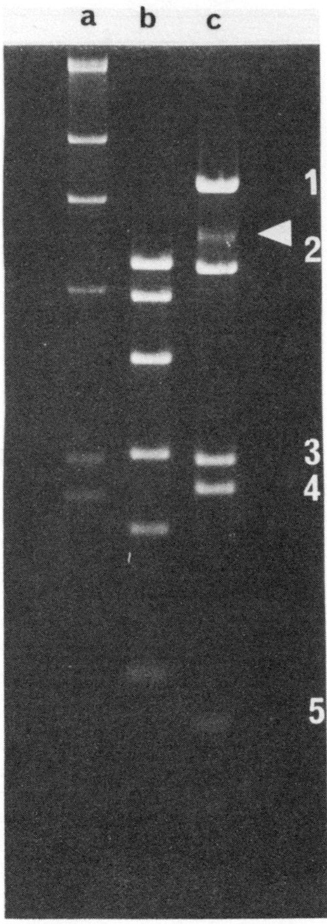

Figure 4. Analysis of mtDNA by gel electrophoresis. Purified mtDNA of *X. laevis* oocytes containing a high percentage of D-loop structures was digested with *Bcl*I (4 units/μg of DNA) for 60 min at 37°C. The digested DNA fragments were separated on a 1% agarose gel, stained by soaking the gel in 1 μg/ml ethidium bromide and photographed. The arrow indicates the fragment containing the D-loop. **Lanes a** and **b**: markers.

in situations where the mitochondria are concentrated in one part of the cell, for example during oogenesis in *X. laevis*, it is possible to use cytophotometric analysis to examine the accumulation of mtDNA after Feulgen staining (33). The other approach has been to use microscopic fluorimetric methods to study the division of mitochondria.

Electron microscopy. For the novice, expectations of the great resolving power of the electron microscope should be tempered by the knowledge that in order to obtain good results a great deal of patience and practice in this art is required. As described in Chapter 2, electron microscopy has played an important role in elucidating the structure and function of many features of mitochondria. Similarly, electron microscopy has proved to be a very powerful tool for studying DNA replication. The restriction of space only allows the authors to describe the most frequently used methods, another volume in the Practical Approach series (40) is devoted to the electron microscopy of nucleic acids and nucleoproteins and should be consulted if you wish to familiarize yourself with other methods.

Essentially all of the electron microscopic studies of mtDNA have used the method of spreading the DNA in the presence of formamide and a basic protein, cytochrome *c*, devised by Inman (41) as a modification of the original method of Kleinschmidt *et al.* (42). In this method the DNA sample is spread over the outside of a drop of either water or a formamide–water mixture and then the DNA is adsorbed from the surface onto an electron microscopy grid. To do this use either a Teflon block with small indentations (~20 mm diameter and 1 mm deep) or a convenient alternative is to use a piece of clean Parafilm.

(i) Prepare the DNA solution (~8 μg/ml) in 8 mM EDTA, 80 mM Tris-HCl, pH 7.2, 40% formamide, incubate in this solution for 10 min in ice and then

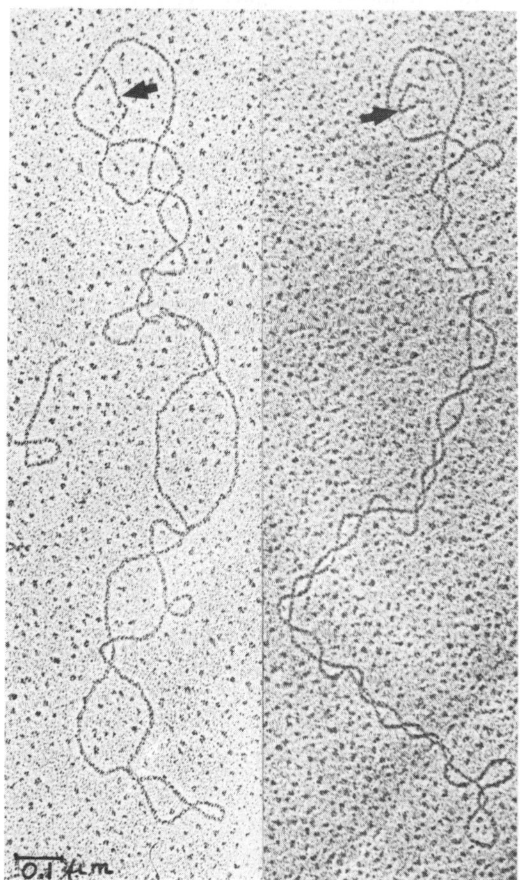

Figure 5. Electron micrograph of mtDNA. Electron micrograph of *X. laevis* mtDNA spread using the formamide technique (41) and shadowed as described in the text. The arrow indicates the position of the D-loop. (Kindly supplied by J-C.Callen.)

add cytochrome *c* to a final concentration of 0.01%.

(ii) Using a fine glass capillary apply about 5 μl of the DNA solution to the outside of a 1.2 ml drop of double-distilled water over a period of 5−10 sec, to do this it is convenient to insert a clean fine glass rod into the water droplet and to allow the solution to flow from the capillary down the rod and onto the surface of the droplet.

(iii) Compress the film on the outside by removing 0.1 ml of the water from the inside of the drop and pick up the sample by touching a carbon-coated mica disk on the surface of the drop.

(iv) Immerse the disk together with any adhering droplets of water in ethanol and then dry it in a vacuum desiccator.

(v) Rotary shadow the sample with platinum−palladium prior to floating off the samples and picking them up on electron microscopy grids. An example of this type of preparation is shown in *Figure 5*.

An alternative is to pick the samples off from the drop of water using a parlodion-coated 300-mesh copper electron microscopy grid, in this case the sample can be dehydrated and shadowed directly on the grid.

2.4 **Methods for identifying the origin of replication**

2.4.1 *Introduction*

One of the keys to our understanding of the replication of mtDNA has been the study of the origins of replication in terms of their location in the genome, their function during replication and the nucleotide sequence around the origin of replication.

2.4.2 *Electron microscopy of origins of replication*

Electron microscopy has been used to analyse replicative structures either after partial denaturation or after restriction nuclease digestion. In studying the replication structures it is usually necessary to treat the DNA with formaldehyde, glyoxal or glutaraldehyde prior to enzyme digestion to prevent branch migration. As an example, glutaraldehyde can be added to samples of DNA in 50% formamide, 0.1 M potassium phosphate, pH 7.3 to a final concentration of 10 mM and the solution is incubated for 15 min at 23°C. After fixation the DNA sample is dialysed into the appropriate solution for restriction nuclease digestion. The partially denatured fragments are then analysed as described in Section 2.3.5.

2.4.3 *Cloning of origins of replication*

As has already been described in Section 2.2.3, the advent of gene cloning techniques has been extremely important in allowing one to prepare large amounts of DNA containing the sequence around the putative origin of replication and, in addition, it is possible to assay the ARS activity of such sequences in yeast. However, although a wide range of putative origins of replication have been cloned, in some cases difficulties do arise, for example the D-loop region of human mtDNA is very refractory to cloning when cloned in *E. coli* and some parts of the mouse mtDNA genome have been shown to decrease the plating efficiency of *E. coli* under selective conditions. Hence it is possible that in cloning putative origins of replication some problems may arise.

2.4.4 *Sequence analysis of origins of replication*

A number of approaches have been made to study the nucleotide sequence(s) of origins of replication depending on the type of cell. In the particular case of the yeast *S. cerevisiae* it has been possible to study their activity in spontaneous petite mutants that arise as a result of deletion of large parts of the DNA (see Section 8 of Chapter 7). In this work the 5′ ends of DNA and RNA molecules from a super-suppressive petite strain were mapped to the repeat *ori* sequences on the mtDNA (43).

In vertebrate cells the origin of replication for the H-strand is associated with the D-loop structure and the short length of DNA in the displacement loop has the same sequence as the new daughter H-strand. Thus the new H-segment can be labelled at the 5′ end by polynucleotide kinase and its sequence compared with that of the fragment of DNA known to contain the origin of replication for the H-strand. To minimize the risk of introducing artefacts during its isolation and purification the 5′ end of the D-loop H-strand fragment is dephosphorylated while it is still hybridized to the L-strand. The protocol is as follows.

(i) Incubate 2.5 μg of purified mtDNA containing a high proportion of D-loops in 20 mM Tris-HCl, pH 8.0 with 25 units of bacterial phosphatase at 60°C for 30 min.
(ii) Terminate the incubation by heating to 90°C and cool rapidly in ice.
(iii) Deproteinize the DNA by extracting it with an equal volume of re-distilled phenol–chloroform saturated with 20 mM Tris-HCl, pH 8.0 at 5°C, add a tenth volume of 3 M sodium acetate and precipitate the DNA by the addition of 2 vols of absolute ethanol at −20°C for a least 2 h. Pellet the DNA (10 000 *g* for 10 min) and dissolve in distilled water.
(iv) Mix the dephosphorylated DNA with at least 50 pmol (150 μCi, 5.6 MBq) of [γ-^{32}P]ATP (3000 Ci/mmol, 111 TBq/μmol) and 10–20 units of T4 polynucleotide kinase in 50 μl of solution to give a final concentration of 10 mM $MgCl_2$, 5 mM DTT, 0.1 mM EDTA, 0.1 mM spermidine and 50 mM Tris-HCl, pH 7.6.
(v) Incubate the mixture at 37°C for 30 min.
(vi) Terminate the incubation by the addition of 2 μl of 0.5 M EDTA and then extract with an equal volume of phenol–chloroform.
(vii) After centrifugation precipitate the DNA from the aqueous phase by the addition of 2 vols of ethanol and after 2 h at −20°C pellet the DNA by centrifugation and re-dissolve the DNA in 50 μl of 1 mM EDTA, 10 mM Tris-HCl, pH 7.9.
(viii) Remove unincorporated [^{32}P]ATP from the labelled DNA by precipitation or, if necessary, by centrifugation through a Sephadex G-50 column (see *Table 12*).

It should be noted that the D-loop H-strand usually consists of a family of different species of different lengths. Depending on the type of organism the heterogeneity can be at the 5′ or 3′ ends and so it is necessary to analyse the labelled D-loop H-strand fragments by polyacrylamide gel electrophoresis to determine the number of species and the location of the length heterogeneity. To do this cut the 5′ end-labelled single-stranded fragments isolated by gel electrophoresis under denaturing conditions with various appropriate restriction nucleases (e.g. *Hae*III, *Msp*I, *Hin*fI and *Dde*I) and run

Table 11. Spin column chromatography with Sephadex G-50.

1.	Swell Sephadex G-50 (F) in 10 mM Tris-HCl, pH 7.4, 1 mM EDTA, 0.1% SDS and autoclave at 110°C for 20 min.
2.	Take a 1 ml sterile syringe and remove the plunger. Place a small quantity of baked glass wool in the bottom of the syringe.
3.	Pipette a quantity of the swollen Sephadex G-50 (F) into the syringe and place it inside a 15 ml Corex glass tube. Centrifuge at 1000 *g* for 5 min at 4°C in a swing-out rotor.
4.	Sephadex G-50 (F) will pack down in the syringe. Add more Sephadex G-50 (F) and repeat the centrifugation in step 3 until the column reaches the 1 ml calibration mark.
5.	Wash the column twice with 100 μl of 10 mM Tris-HCl, pH 7.4, 1 mM EDTA, 0.1% SDS.
6.	After the final wash remove the syringe and fix a sterile MCC tube to the end of the syringe to collect the nucleic acids.
7.	Replace the syringe and add 100 μl[a] of the nucleic acids sample in 10 mM Tris-HCl, pH 7.4, 1 mM EDTA, 0.1% SDS to the column.
8.	Repeat the centrifugation as in step 3. Either precipitate the nucleic acids with 3 vols of absolute ethanol and store at −20°C or cap the microcentrifuge tube and store at −70°C.

[a]Sample volume should not exceed 0.1 vol of the Sephadex G-50 (F) bed volume, i.e. 100 μl of sample per 1 ml column.

Table 12. Denaturing 6% polyacrylamide gels.

1.	Mix together 10 ml of 30% acrylamide stock solution (29% acrylamide, 1% *N,N′*-methylenebis-acrylamide), 5 ml of 10 × TBE buffer (25 mM EDTA, 0.9 M Tris-borate buffer, pH 8.3) and 17 ml of distilled water.
2.	Dissolve 21 g of urea in the solution with stirring and adjust the volume to 50 ml.
3.	Filter and de-gas the solution.
4.	Add 0.25 ml of freshly made 10% ammonium persulphate solution and 50 μl of *N,N,N′,N′*-tetra-methylethylenediamine (TEMED), mix and pour the gel immediately.
5.	Overlayer the gel with water and allow to set.
6.	Electrophorese the gel using TBE as buffer at approximately 5 V/cm for 15 h at room temperature.

the fragments on a 6% polyacrylamide gel containing 7 M urea (*Table 12*). In the case of *X. laevis*, digestion with *Hae*III produces two 5′ end-labelled fragments of 80 and 115 nucleotides. These bands can be cut out of the gel and sequenced as described in another volume of the Practical Approach series (45).

In vertebrates the origin of replication for the L-strand is quite separate from that of the H-strand and in order to localize this origin it is necessary to prepare labelled replicating mtDNA. To do this pulse-label the mtDNA *in vivo* and centrifuge the mtDNA on an ethidium bromide−CsCl gradient (see *Table 3*). Most of the replicating DNA intermediates band in the gradient between the upper band of linear DNA and the band of supercoiled DNA. Recover the DNA from between these two bands, extract the ethidium bromide with CsCl-saturated isopropanol, dialyse to remove the CsCl and digest first with appropriate restriction nucleases and then S1 nuclease to remove single-stranded ends. Digest uniformly labelled, essentially non-replicating mtDNA in the same way and compare the electrophoretic patterns of both digested samples. The appearance of a new band in the replicating DNA sample allows you to map the location of the L-strand origin of replication between restriction enzyme sites. As in the case of the H-strand fragment, the L-strand fragment can be end labelled using polynucleotide kinase and sequenced.

3. TRANSCRIPTION IN MITOCHONDRIA

3.1 Introduction

There are a number of approaches that can be taken in the study of transcription in mitochondria, both in terms of the systems used and the methods used to analyse the transcriptional products. In the following sections of this chapter detailed descriptions of the different systems and the methods of analysis will be given based not only on the authors' experience with yeast and animal mitochondria but also, where relevant, commenting on the results found in other species.

3.2 Systems used to study transcription

A number of systems can be used to study transcription in mitochondria. A significant amount of work has been carried out using isolated mitochondria purified free of nuclear and other cellular contaminants by differential centrifugation and, usually, isopycnic gradient centrifugation (see Chapter 1, Section 2). Isolated, intact mitochondria incubated under optimal conditions (see Section 3.3) will incorporate radioactive nucleoside triphosphates (NTPs) into RNA. The extent of transcription can be measured either in terms of the rate of synthesis or, after isolation of the RNA, by assessing the nature of the transcripts.

The advantage of *in organello* studies is that, as there is little disruption of the interior organization of the mitochondrion, including the nucleoid structure of the mtDNA, one can expect that the pattern of transcription will accurately reflect that occurring *in vivo*. The difficulty with using this system is that the mitochondrial membranes may be selectively permeable to components needed for transcription. Similarly, the presence of intact mitochondrial membranes may lead to the formation of endogenous pools of ions and other compounds within the mitochondrion which introduce difficulties, not only in defining exactly what are the precise requirements of the mitochondrial RNA polymerase, but also in interpreting other experimental results.

One possible solution to this problem is to 'permeabilize' the mitochondrial membranes; however the method used must be chosen with care. Triton X-100 and other non-ionic detergents can be used to lyse mitochondria at a concentration of 1–2% but, although this method works well in terms of solubilizing membranes, detergent lysis often inhibits transcription in mitochondria; a similar inhibition has been found if one uses macrolide antibiotics to permeabilize yeast mitochondrial membranes. Ultrasonication does disrupt mitochondria but overall it is not advisable to use this approach in that it may disrupt the integrity of mtDNA which is loosely packaged inside the mitochondrion. The method used by the authors and which could have a wide range of applications is to permeabilize the mitochondria by osmotic shock. This method involves gently resuspending the mitochondria in a very hypotonic buffer (e.g. 1 mM EDTA, 20 mM Tris-HCl, pH 7.9) and the subsequent swelling of the mitochondria permeabilizes the membranes.

In an attempt to define the transcriptional mechanisms and its controls better, other systems have been devised. The obvious system, that of analysing the transcriptional pattern of the nucleoid, has been relatively neglected because of the difficulties in obtaining purified nucleoids and the likelihood of losing transcriptional factors during

Table 13. Variations in conditions used for transcription *in organello* or *in vitro* in a variety of species.

Species	*Assay tube*	*NTPs (mM)*	*Divalent cation*	*Ionic strength*	*Buffer*	*pH*	*Tempera-ture*	*Ref.*
S. cerevisiae	*in organello*	0.30	10 mM $MgCl_2$	150 mM KCl	20 mM Tris	6.7	30°C	50
S. cerevisiae	*in organello*	0.20	10 mM $MgCl_2$	50 mM NaCl	20 mM Tris	7.9	28°C	51
S. cerevisiae	*in organello*	0.05	10 mM $MgCl_2$	50 mM KCl	50 mM Bicine	7.4	28°C	52
S. cerevisiae	*in vitro*	0.15	10 mM $MgCl_2$	–	10 mM Tris	7.9	37°C	53
S. cerevisiae	*in vitro*	0.15	10 mM $MgCl_2$	50 mM KCl	10 mM Tris	7.9	20°C	54
S. cerevisiae	*in vitro*	0.04	10 mM $MgCl_2$	–	50 mM Tris	7.9	35°C	55
S. cerevisiae			1 mM $MnCl_2$	–				
N. crassa	*in vitro*	1.00	3 mM $MgCl_2$ 1.6 mM $MnCl_2$	150 mM KCl	30 mM Tris	7.9	37°C	56
X. laevis	*in organello* *in vitro*	0.50	10 mM $MgCl_2$	–	50 mM Tris	8.0	30°C	57
Rat	*in organello*	0.50	1 mM $MgCl_2$ 3 mM $MnCl_2$	60 mM KCl	50 mM Hepes	7.0	30°C	58
Rat	*in vitro*	0.16	4 mM $MgCl_2$ 1 mM $MnCl_2$	–	100 mM Tris	7.9	37°C	59
Mouse	*in organello*	2mM ATP, 1 mM GTP[a]	6 mM $MgCl_2$	60 mM KCl	5 mM Hepes	7.4	35°C	60
Plant (*Vignia sinensis*)	*in organello*	5 mM ATP[b]	8 mM $MgCl_2$	–	50 mM Tris	7.4	37°C	61
Human	*in vitro*		10 mM $MgCl_2$	–	10 mM Tris	8.0	28°C	62
Human	*in organello*	2 mM ATP[b]	5 mM $MgCl_2$	20 mM NaCl	35 mM Tris	7.8	37°C	63
Human	*in vitro*	0.5	7.5 mM $MgCl_2$	–	10 mM Tris	8.1	30°C	64

[a]No added CTP or UTP.
[b]No added CTP, UTP or GTP.

the isolation procedure. Instead studies have tended to concentrate on the isolation of mitochondrial protein fractions that exhibit RNA polymerase activity using either artificial templates or cloned fragments of mtDNA containing known transcription promoters (46–49). The cloned DNA can be transcribed *in situ* in the plasmid or after excision using restriction nucleases. However, the major problem in this case is in knowing the exact protein subunit composition of the purified mitochondrial polymerases of the different species since these enzymes and their likely transcriptional factors remain almost completely uncharacterized at the present time.

3.3 **Characteristics of mitochondrial RNA polymerases**

The properties of many different types of RNA polymerases have been studied using a number of different systems (*Table 13*). The following sections provide only a general outline of the characteristics of these enzymes in a number of systems. They do, however, provide a guide to the reader in devising transcriptional assays for their own particular requirements.

3.3.1 *Nucleoside triphosphates*

There is general recognition that transcription requires the presence of all four nucleoside triphosphates (NTPs). Intact mitochondria seem to be permeable to NTPs although osmotic shocking does appear to increase the permeability of the membranes. The actual concentration of NTPs used has varied between 0.04 and 1 mM. Usually higher concentrations of NTPs (0.5–1 mM) have been used for transcriptional assays for mitochondria of higher eukaryotes. The authors have found that, in the case of osmotically-shocked yeast mitochondria, 0.2 mM of each NTP gives maximal transcription and, at this concentration of ATP, there is no need to include an ATP-generating system in the assay mixture. However, care is required since it has been reported that the concentration of ATP can affect the pattern of transcription of mtDNA (65).

3.3.2 *Effects of divalent cations and ionic strength*

All mitochondrial RNA polymerases need divalent ions for activity and the activities of most are also affected by the presence of potassium or sodium ions. However, because of the diversity of systems and species, there is considerable variation in the ionic requirements for mitochondrial RNA polymerases. Animal mitochondrial RNA polymerases also show variations in their ionic requirements. In *X. laevis* and HeLa cells Mg^{2+}-dependent RNA polymerase activities have been identified with optima of 5–10 mM Mg^{2+} but, while the *X. laevis* enzyme is inhibited by all concentrations of monovalent ions (57), the HeLa cell enzyme is stimulated by NaCl or KCl concentrations up to a maximum of 20 mM (63), higher concentrations inhibit. Rat mitochondria appear to be unusual in that the mitochondrial RNA polymerase appears to require both Mg^{2+} and Mn^{2+} at unusually low concentrations of 1 mM and 3 mM, respectively (58). Some plant mitochondrial RNA polymerases studied show some similarities with the HeLa cell enzyme in that they are stimulated by low concentrations of KCl (66). It should be emphasized that, while this information provides the reader with a guide as to the optimum divalent ion concentrations for RNA polymerase assays, there is no guarantee that this reflects the actual environment of these enzymes *in vivo*.

3.3.3 *Effects of pH and buffers*

The mitochondrial RNA polymerases of most organisms are maximally active between pH 7.5 and pH 8 although, exceptionally, in the case of yeast, studies of transcription *in vitro* have been carried out at pH 6.7. The buffer of choice is usually Tris-HCl. In the case of yeast mitochondria the authors have found that the presence of Bicine as a buffer is inhibitory as compared with Tris-HCl. Low concentrations of Tris-HCl (~20 mM) should be used in order to avoid possible inhibitory effects that have been noted at higher concentrations (67).

3.3.4 *Effect of incubation temperature on transcription in vitro*

The usual guideline is that the incubation temperature should reflect the 'normal' temperature of the organism, for example, 37°C for mammalian mitochondria with lower temperatures being used for mitochondria from fungi and cold-blooded vertebrates. However, on occasions it may be appropriate to use a lower temperature, in that it has been found in some systems that lowering the incubation temperature preferentially inactivates the endogenous nuclease activity as compared with the RNA polymerase activity (68).

3.4 **Isolation of RNA from mitochondria**

For a number of techniques it is necessary to isolate the RNA prior to analysis. During the isolation procedure it is of paramount importance to ensure that the RNA is not degraded. Contamination with exogenous nucleases can be minimized by baking all glassware at 200°C for 2 h and by autoclaving all plasticware and solutions at 120°C for 20 min. Solutions that do not contain compounds with free amino groups can also be treated with diethylpyrocarbonate (DEPC, 20 μl/100 ml) prior to autoclaving; this inactivates proteins and when heated it decomposes into CO_2 and ethanol. Often endogenous nucleases are also present in the transcription mixture, especially if one is using a system derived from whole mitochondria, and so it is very important to inactivate

Table 14. Isolation of mitochondrial RNA.

1. Add 0.1 vol. of 20% (w/v) SDS to the mitochondrial suspension. Note that potassium ions precipitate SDS and should be omitted if possible, otherwise use 2% laurylsarcosine instead of SDS.
2. Add an equal volume of redistilled phenol–chloroform–isoamyl alcohol (50:50:2 by vol.) equilibrated with 10 mM Tris-HCl, pH 7.4, 1 mM EDTA. Extract nucleic acids for 10 min with mixing of the phenol and aqueous phases every minute for 10 sec.
3. Separate the phenol–chloroform and aqueous phases by centrifugation at 10 000 *g* for 5 min at 8°C in a swing-out rotor.
4. Immediately remove the aqueous phase (using a Pasteur pipette taking care not to disturb the interface) and mix it with a half volume of phenol–chloroform–isoamyl alcohol. Repeat the extraction as in step 2.
5. Repeat the separation in step 3 and remove the aqueous phase. Add 0.1 vol of 25% (w/v) sodium acetate, pH 5.0.
6. Add 2.5 vols of absolute ethanol chilled to −20°C. Leave the nucleic acids to precipitate overnight at −20°C.
7. Pellet the precipitated RNA by centrifugation at 20 000 *g* for 10 min at 0°C. For very small amounts of RNA it will be necessary to centrifuge at 60 000 *g* for 30 min to ensure a good recovery of RNA.

Table 15. RNA synthesis in isolated yeast mitochondria.

1. In a sterile conical glass test tube prepare the following assay solution (total volume 200 μl).

   0.6 M sorbitol
   20 mM Tris-HCl, pH 7.9
   10 mM $MgCl_2$
   50 mM NaCl
   0.5 mM ATP
   0.2 mM GTP and CTP

   To each assay solution add 25 μl of mitochondrial suspension (~2 mg/ml).
2. Pre-incubate the assay at 28°C for 2 min to pre-equilibrate the mixtures.
3. Initiate transcription by adding 25 μl of either 2 mM [α-^{32}P]UTP or [^{3}H]UTP (50 Ci/mmol, 1.85 TBq/mmol) and continue the incubation for another 10 min[a].
4. Add 1 ml of 0.8 M perchloric acid, 100 mM sodium pyrophosphate and leave on ice for a minimum of 10 min.
5. Pipette the solution onto a GF/C filter (25 mm diameter) on a vacuum filtration apparatus.
6. Wash filters twice with 5 ml of 10% trichloroacetic acid containing 10 mM sodium pyrophosphate and once with 5 ml of 5% trichloroacetic acid, 10 mM sodium pyrophosphate. Finally wash each filter with 2 ml of an ethanol–ether mixture (1:1, v/v).
7. Place each filter in a scintillation vial and dry for at least 2 h at 45°C. Cool the vials and add 4.5 ml of toluene-based scintillator.

[a]Background radioactivity is estimated by preparing identical non-incubated assays kept in ice. The isotope is added followed directly by 0.8 M perchloric acid, 0.1 M sodium pyrophosphate. Acid-precipitable material is collected as in steps 5 and 6.

all of the proteins and to separate them from the RNA after *in vitro* incubation. The main types of deproteinization procedure involve extraction using phenol–chloroform, or phenol alone (e.g. see Section 3.1 of Chapter 7), or by centrifugation in either CsCl (e.g. see Section 3.2 of Chapter 7) or guanidine thiocyanate solutions (69). In the authors' experience the most consistent results are obtained by extracting the RNA with phenol–chloroform as described in *Table 14*. Variations of this procedure, aimed at improving the yield of RNA, have included carrying out the phenol–chloroform extraction at 65°C and digestion with 100 μg/ml of protease K for 30 min at 37°C prior to phenol extraction.

The RNA obtained by this procedure is contaminated by DNA which may interfere with subsequent analytical procedures. The DNA can be removed by ultracentrifugation at 100 000 *g* for 24 h in 50% (w/w) CsCl gradients with an initial density of 1.55 g/ml, this allows the RNA to pellet while the DNA bands in the solution; this method is only recommended for high molecular weight RNA. An alternative approach is to digest the total nucleic acids with 200 μg/ml of RNase-free DNase I at 37°C for 5 min in 10 mM $MgCl_2$, 1 mM EDTA, 10 mM vanadyl ribonucleoside, 10 mM Tris-HCl, pH 7.5, followed by re-extraction with phenol–chloroform (*Table 14*); RNase-free DNase I is commercially available. Another problem is that often the RNA obtained is contaminated with large amounts of unincorporated isotope. This acid-soluble material is not readily removed by simple washing of the precipitated RNA with 70% ethanol, especially if the RNA is contaminated with [^{32}P]phosphate. The most effective way to remove this material is to pass the RNA dissolved in 0.1% SDS, 10 mM NaCl, 10 mM Tris-HCl, pH 7.4, over a column of Sephadex G-50(F) as described in *Table 11*. The concentration of the final RNA solution can be determined spec-

Table 16. Denaturing formaldehyde–agarose gels.

1.	Denature the sample by mixing 4.5 μl of the sample solution containing up to 20 μg of DNA or RNA with 3.5 μl of formaldehyde, 10 μl of formamide and 2 μl of 5 × gel buffer containing 50 mM sodium acetate, 5 mM EDTA and 0.2 M Mops–NaOH, pH 7.0. Heat the sample at 65°C for 15 min.
2.	Prepare an agarose gel, usually the concentration should be between 0.8% and 1.5% agarose depending on the nature of the sample, in gel buffer (10 mM sodium acetate, 1 mM EDTA and 40 mM Mops–NaOH, pH 7.0) containing 6.6% formaldehyde.
3.	Mix the sample with a tenth its volume of 50% glycerol, 1 mM EDTA, 0.5% bromophenol blue and 0.5% xylene cyanol.
4.	Load the sample onto the gel and run overnight at 2 V/cm.
5.	Rinse the gel in several changes of distilled water for 2 h.
6.	If required stain the gel by soaking it in 5 μg/ml of ethidium bromide for 30 min, then wash the gel in distilled water for 60 min and view under u.v. light.
7.	Radioactive samples can be visualized by autoradiography or fluorography.

Both formamide and formaldehyde are oxidized in air, the former should be de-ionized by mixing it with a mixed-bed resin until it is neutral while the formaldehyde should be neutralized before use.

trophotometrically at 260 nm; a 1 mg/ml solution of RNA has an optical density of 25 at 260 nm.

3.5 Methods to study the rates of synthesis and initiation in mitochondria

3.5.1 *Measurement of the rate of transcription*

The 'rate' of transcription measures the rate of initiation and elongation of transcripts as well as, in some cases, nucleolytic degradation of the RNA. Measurements of the rate of RNA synthesis are readily made simply by measuring the amount of radioactively labelled NTP incorporated into acid-precipitable material. For this type of assay mitochondria are incubated in the required incubation medium (see Section 3.3); *Table 15* gives the conditions used by the authors for yeast mitochondria. After an appropriate time interval the incubation is terminated by the addition of 0.8 M perchloric acid containing 100 mM sodium pyrophosphate (*Table 15*); this precipitates all polynucleotides of five nucleotides or longer.

3.5.2 *Measurement of initiation by capping of primary transcripts using* [α-^{32}*P*]*GTP and guanyltransferase*

RNA initiation and processing produce two different types of 5′ termini. Initiation sites are unique in that they retain their 5′ triphosphate terminus. Those 5′ termini that are the product of processing lack this distinctive feature. Mitochondrial RNA is not capped *in vivo* and so capping is particularly suitable for measuring the level of initiation using this method first developed by Levens *et al.* (70). Guanyltransferase from *Vaccinia virions* can be used to catalyse the capping of 5′ triphosphate termini with [α-^{32}P]GTP thereby identifying unique sites of RNA initiation. The guanyltransferase required is now commercially available; the procedure originally described is a slight modification of the method developed by Monroy *et al.* (71).

The protocol for capping the mitochondrial RNA is as follows.

(i) Dissolve 3–7.5 μg of mitochondrial RNA in 2.5 μl of double-distilled water.

Add methyl mercuric hydroxide to a final concentration of 5 mM and incubate at room temperature for 10 min to denature the RNA.

(ii) Adjust to a final volume of 20–40 μl with 50 mM Tris-HCl, pH 7.5, 1 mM $MgCl_2$, 20–40 μM [α-^{32}P]GTP (~2500 Ci/mmol, 100 TBq/mmol), 1 mM dithiothreitol (DTT).

(iii) Add 10–12 units of guanyltransferase and incubate for 15 min at 37°C.

(iv) Terminate the incubation by addition of an equal volume of 2% SDS, 20 mM EDTA, 20 mM Tris-HCl, pH 7.5, containing 40–100 μg of *E. coli* tRNA.

(v) Add an equal volume of re-distilled phenol equilibrated with 1% SDS, 10 mM EDTA, 10 mM Tris-HCl, pH 7.5.

(vi) Separate the phenol and aqueous phases by centrifugation at 13 000 *g* in a microcentrifuge for 10 min at 4°C.

(vii) Remove the aqueous phase and apply the capped RNA to a 5 ml (bed volume) Sephadex G-50 column equilibrated with 0.2% SDS, 2 mM EDTA, 10 mM Tris-HCl, pH 7.5.

(viii) Wash the column with equilibration buffer and locate the eluted RNA either with a Geiger counter or by acid precipitating aliquots followed by filtration and liquid scintillation counting.

(ix) Pool the peak fractions and adjust to 1 M ammonium acetate. Precipitate the RNA by the addition of 2.5 vols of absolute ethanol chilled to −20°C and leave overnight at −20°C.

(x) Pellet the precipitated RNA by centrifugation in a microcentrifuge for 10 min at 13 000 *g* at 4°C.

(xi) Re dissolve the RNA in a small volume of sterile double-distilled water and store at −80°C until required.

One possible problem in the use of this technique is that the triphosphate terminus may be degraded by phosphatase activity and hence these transcripts will not be detected. It is also necessary to denature the sample RNA completely before carrying out the capping reaction because RNA secondary structure has to be kept to a minimum to ensure that the 5′ termini are freely available to the guanyltransferase.

This technique does have considerable advantages if transcripts of mitochondrial origin are to be identified because the cytoplasmic mRNA is already capped and the cytoplasmic rRNA does not have a triphosphate terminus, thus neither of these likely contaminants interferes with this method. The capped mitochondrial RNAs can be analysed by a number of methods. In the absence of pre-existing transcripts determination of the incorporation of [α-^{32}P]GTP, that is the number of transcripts capped, gives a direct measure of the rate of initiation; whereas capping of RNA in *in organello* assays measures the steady-state levels of transcripts. However, more sophisticated analyses can also be carried out. The isolated RNA can be simply electrophoresed on a denaturing formaldehyde–agarose gel (*Table 16*) and the capped species visualized by autoradiography. A useful method for the identification of the number of initiation sites in the case of yeast is to digest the capped RNA with the RNase T1 which is specific for GMP residues. *S. cerevisiae* mtDNA has an extremely low G+C content (18% G+C) and so the digestion of capped RNA with RNase T1 generates a number of short

Table 17. Northern blotting of RNA.

1. Dissolve 10 μg of mitochondrial RNA in 8 μl of 1 M de-ionized glyoxal[a], 50% (v/v) dimethyl sulphoxide, 10 mM sodium phosphate, pH 7.0. Incubate for 60 min at 50°C.
2. Cool the mixture in ice and add 2 μl of 50% (v/v) glycerol, 10 mM sodium phosphate buffer, pH 7.0, 0.01% bromophenol blue.
3. Separate the RNA on a 1.5% agarose gel in 10 mM sodium phosphate buffer, pH 7.0 at 90 V for 6 h. Circulate the buffer between electrodes to maintain the pH at neutrality.
4. Place gel on two sheets of Whatman 3MM paper saturated with 20 × SSC.[b]
5. Soak nitrocellulose first in distilled water then in 20 × SSC. Place the nitrocellulose over the gel followed by two sheets of Whatman 3MM paper. Place several paper towels and a weight on top of the filter paper. Transfer is complete after 12–15 h.
6. Air dry the RNA blots and bake in a vacuum oven at 80°C for 2 h.
7. Soak the filter in 20 mM Tris-HCl, pH 8.0 at 100°C and allow it to cool to room temperature to remove any remaining glyoxal.
8. Pre-hybridize RNA blots for 8–20 h at 42°C in 50% (v/v) formamide, 5 × SSC, 50 mM sodium phosphate, pH 6.5, 250 μg/ml sonicated salmon sperm DNA, 10 × Denhardt's solution[c]. Incubate with gentle agitation at 42°C for 8–20 h.
9. Replace the pre-hybridization buffer with hybridization buffer containing four parts pre-hybridization buffer with one part 50% (w/v) dextran sulphate.
10. Heat the labelled probe DNA ($>10^8$ c.p.m./μg) to 100°C for 5 min, then cool in ice before adding it to the hybridization buffer in the bag; re-seal the bag. Incubate the filters for 20 h at 42°C. The optimum hybridization temperature will vary depending on the base composition of the probe.
11. Wash the filters with four changes of 2 × SSC containing 0.1% SDS for 5 min each at room temperature.
12. Wash the filters with two changes of 0.1 × SSC containing 0.1% SDS for 15 min each at 50°C.
13. Expose blots to pre-flashed Kodak X-Omat R X-ray film or equivalent at −70°C.
14. Hybridized probe can be removed by washing the filters in 0.1–0.05 times wash buffer (1 times wash buffer is 50 mM Tris-HCl, pH 8.0, 2 mM EDTA, 0.5% sodium pyrophosphate, 10 × Denhardt's solution for 1–2 h at 65°C.

[a]De-ionize glyoxal by mixing 20 ml of 40% glyoxal with 20 g of AG501-X8 ion-exchange resin. Stir for 30 min and then decant the supernatant and mix with fresh ion-exchange resin for 30 min, continue until the final pH is between pH 5.5 and pH 6.0. Store at −20°C.
[b]SSC is 0.15 M NaCl, 15 mM sodium citrate, pH 7.0.
[c]100 × Denhardt's solution contains 2% bovine serum albumin, 2% Ficoll, 2% polyvinylpyrrolidone and can be stored at −20°C.

RNA species which can be readily sequenced. By comparing the RNA sequence with known DNA sequences it is then possible to identify those sequences at the site of the initiation of transcription.

It is very important to appreciate that the guanyltransferase will cap the 5′ termini of pre-existing as well as newly initiated RNA as long as the terminal triphosphate has been retained. Thus this method cannot be used for measuring *in vitro* initiation in whole mitochondria and other methods must be used.

3.5.3 *5′ End-labelling transcripts with [^{32}P]phosphate or [^{35}S]sulphate nucleoside triphosphates*

The retention of the triphosphate group at the 5′ terminus of RNA molecules enables one to end-label the 5′ terminus of transcripts initiated *in vitro* in the presence of β- or γ-phosphate-labelled ATP or GTP; unlike guanyltransferase capping, this assay for the initiation of RNA is not affected by the presence of pre-existing transcripts. However,

β-labelled nucleoside triphosphates are not generally available and so γ-position labelled nucleoside triphosphates have been used; but there are a number of problems associated with their use. The γ-phosphate is incorporated from ATP into protein and lipid molecules by kinases. Measurement of the incorporation of this label into RNA therefore necessitates the deproteinization and purification of RNA. The use of [γ-^{32}P]ATP or GTP may give erroneous results because the 5′ triphosphate termini can be degraded by phosphatase activity. This problem can be reduced by using either [γ-^{35}S]-adenosine-5′-*O*-(3-thiotriphosphate) (ATP-γ-S) or guanosine-5′-*O*-(3-thiotriphosphate) (GTP-γ-S). These nucleotides are more resistant to phosphatase cleavage and should be used in preference to the ^{32}P-labelled ATP and GTP. Work in the authors' laboratory has successfully used [γ-^{35}S]ATP-γ-S to study the rates of initiation *in vitro*. In this case the *in vitro* assay is terminated by addition of SDS to a final concentration of 2% and the RNA is extracted as described in *Table 14*. The incorporation of radioactivity into RNA is quantified by acid-precipitating an aliquot of the RNA solution (steps 4–7 of *Table 15*) followed by liquid scintillation counting. Alternatively, the extracted RNA can be analysed on a denaturing gel (*Table 16*). The ^{35}S-labelled transcripts are visualized by fluorography. To do this soak the gel in 10 vols of 1.0 M sodium salicylate solution, pH 7 for 30 min, transfer the gel onto wetted Whatman 3 MM paper and dry down under vacuum. Place the dried gel in contact with Kodak X-omat R film or equivalent at −70°C. The incorporation of radioactivity into RNA is an indication that the RNA polymerase is initiating transcription *in vitro* and that the incorporation of labelled NTPs is not just the result of 'run-off' synthesis. A possible disadvantage of the use of γ-labelled NTPs to assess rates of initiation is the fact that the label is only incorporated at one position in the RNA molecule, thus limiting the specific activity of the RNA.

3.5.4 *Isolation of newly-initiated transcripts*

All correctly initiated transcripts have either ATP or GTP at the 5′ terminus. Hence, if ATP or GTP is completely substituted in the *in vitro* reaction by their γ-thiol-containing analogues all newly-initiated RNA molecules will have a thiol at their 5′ terminus. The RNA can then be extracted, purified and passed through a mercury–agarose or mercury–cellulose column. The thiol-initiated transcripts bind to the mercury column and those transcripts that are the result of run-off synthesis are removed in the flow-through wash. The bound RNA is then eluted from the column using either 10 mM DTT or 2-mercaptoethanol. In the case of yeast, results in our laboratory found this method to yield somewhat variable results; the reasons for this remain unclear. A typical protocol for the isolation of thiol-containing RNA is as follows.

Isolation of RNA.

(i) Synthesize RNA under the appropriate conditions substituting ATP-γ-S and GTP-γ-S completely for ATP and GTP, respectively.

(ii) Terminate transcription by the addition of 0.2 vol. of 5 × TNES buffer (TNES is 10 mM Tris-HCl, pH 7.9, 50 mM NaCl, 50 mM EDTA, 0.1% SDS).

(iii) Add an equal volume of phenol–chloroform (1:1, v/v) equilibrated with TNES buffer and vortex for 2–3 min.

(iv) Separate the phases by centrifugation at 10 000 *g* for 10 min at 8°C.
(v) Remove the aqueous phase and add an equal volume of phenol–chloroform (1:1, v/v).
(vi) Briefly mix the phases and repeat the separation in step (iv).
(vii) Remove the aqueous phase and add an equal volume of chloroform–isoamyl alcohol (48:2, v/v). Mix the phases and centrifuge as in step (iv) but for only 5 min.
(viii) Precipitate nucleic acids from the aqueous phase with 3 vols of absolute ethanol chilled to −20°C. Leave the nucleic acids to precipitate overnight.
(ix) Pellet the precipitated nucleic acids by centrifugation at 10 000 *g* for 10 min at 0°C.
(x) Dry the nucleic acid pellet and dissolve in 100 μl of TNES buffer. Store at −70°C or apply directly to the mercury–agarose column.

Affinity chromatography.

(i) Pipette 1–2 ml of mercury–agarose (Sigma Chemical Co.) into a sterile Pasteur pipette that has been plugged with a small quantity of sterile glass wool.
(ii) Wash the column with five column volumes of double-distilled water that has been treated with DEPC (Section 3.4) and then with 5 vols of TNES buffer.
(iii) Apply the RNA sample in 100 μl of TNES buffer to the column and leave for 10 min.
(iv) Wash the column with 10 vols of TNES buffer to remove the unbound RNA. Collect 1 ml aliquots.
(v) Add 5 vols of TNES buffer containing 10 mM DTT or mercaptoethanol to elute the bound RNA. Collect 1 ml aliquots.
(vi) Wash the column with 10 vols of TNES buffer and collect 1 ml aliquots.
(vii) Remove 100 μl aliquots from each fraction and add 20 μg of carrier tRNA and 4 vols of ice-cold 10% (w/v) trichloroacetic acid. Leave samples on ice for 15 min.
(viii) Collect the acid-precipitable material by vacuum filtration as in *Table 15* steps 4–7.
(ix) Count samples in a scintillation counter using a toluene-based scintillator to identify the fractions containing the RNA that bound to the column.
(x) After use, store the columns at 4°C in 10 mM Tris-HCl, pH 7.4, 1 mM EDTA, 0.02% sodium azide.

Column regeneration.

(i) Wash the column with 10 vols of 50 mM sodium acetate, pH 5.0.
(ii) Wash the column with 5 vols of 50 mM sodium acetate, pH 5.0, 4 mM mercuric chloride or mercuric acetate.
(iii) Wash the column with 10 vols of 50 mM sodium acetate, pH 5.0 to remove all excess mercuric ions.
(iv) Either wash column with 5 vols of TNES buffer and use directly or store at 4°C in 10 mM Tris-HCl, pH 7.4, 1 mM EDTA, 0.02% sodium azide.

An alternative, but much less frequently used technique, is to density label the newly-initiated RNA by including density-labelled NTP analogues; the most frequently used analogues are labelled with mercury (72).

3.6 **Determination of the accuracy of initiation of transcripts**

As described in Section 3.7.3, hybridization analysis is a very powerful tool for studying the accuracy of initiation of transcripts using S1 nuclease mapping and primer extension methods. However, other relatively simple methods to show that the *in vitro* pattern of transcription reflects the *in vivo* pattern also exist. One simple approach is to isolate the *in vivo* labelled RNA and compare the relative sizes of the transcripts synthesized with those products of the *in vitro* assay. Molecular weight markers are run on a denaturing formaldehyde−agarose gel (*Table 16*) for accurate size determination (e.g. cytoplasmic rRNA, *E. coli* 23S and 16S rRNA or a restriction digest of lambda bacteriophage DNA if the gel is completely denaturing). This will only reveal those RNAs synthesized at a relatively high rate, usually the rRNAs and tRNAs, but it is still a good indicator of the accuracy of the initiation of transcription *in vitro* for all transcripts. However, often transcripts are subjected to extensive post-transcriptional processing and in this case the *in vitro* synthesized products may well be much larger as a result of the lack of accurate processing of the RNA *in vitro*.

The 'run-off assay' is, in principle also a very simple method for the determination of specific initiation in an *in vitro* assay. A cloned restriction fragment of mtDNA is chosen that is known to contain the site of initiation and to terminate within the gene. This restriction fragment is then used as a template to assay the specificity of protein extracts that exhibit mitochondrial RNA polymerase activity. *In vitro* transcription should generate RNA molecules of known length that can be accurately sized by electrophoresis on a denaturing gel by comparison with RNA molecules of known size and composition. Alternatively, the RNA molecules synthesized *in vitro* can be sequenced and compared with the DNA sequence of the restriction fragment. A method that has been used to generate RNA molecules of known size is to omit a specific NTP from the *in vitro* incubation mixture. In the absence of the particular NTP the mitochondrial RNA polymerase terminates transcription prematurely thus synthesizing a short RNA molecule of defined size. The size of the RNA can then be determined by gel electrophoresis.

3.7 **Hybridization analysis of transcription**

3.7.1 *Introduction*

The technique of DNA−RNA hybridization has been used extensively to analyse the patterns of transcription of mtDNA. However, the type of analysis depends on the nature of the transcription assay. In the case of *in organello* assays most of the RNA represents pre-existing transcripts which can interfere with some types of hybridization analyses. In some instances it is possible to isolate *in vitro* synthesized transcripts by labelling with sulphydryl analogues (Section 3.5.4) or by density labelling (72). In contrast, *in vitro* transcription assays of cloned mtDNA avoid this problem.

There are two strategies that can be used for hybridization analysis of *in vitro* transcribed RNA. The classical approach is solution hybridization in which the amount of hybrid formed is measured by acid precipitation or isolation of the hybrid followed by liquid scintillation counting of the amount of radioactive hybrid formed. However, the development of blotting techniques has led to the development of a range of semi-quantitative methods based on autoradiographic techniques.

For blot hybridization either the DNA or RNA can be immobilized by adsorption onto nitrocellulose or covalently attached to diazobenzyloxymethyl paper (DBM-paper) and then incubated with radioactively labelled DNA or RNA. The separation of DNA restriction fragments on agarose gels and the transfer of fragments to nitrocellulose or DBM-paper is termed Southern blotting. However, the converse approach is usually more appropriate for transcription studies and it involves separating the RNA by gel electrophoresis and then transferring the RNA to either DBM-paper or nitrocellulose. This is known as the Northern blot technique (*Table 17*). A variation of these techniques is to immobilize the RNA or DNA directly onto nitrocellulose in a series of dots. After hybridization the amount of radioactive hybrid associated with the dots can be determined by autoradiography or the dots cut out and the amount of radioactivity of each quantitated by liquid scintillation counting. Autoradiographic analysis of the Southern or Northern blots after hybridization can be used to analyse RNA transcribed *in vitro* both semi-quantitatively and qualitatively. By scanning the developed films with a densitometer it is possible to obtain a good estimate as to the relative rates of transcription of specific genes of the mtDNA (73).

Within the very limited space available in this chapter it is impossible to describe in detail the whole range of hybridization techniques used to analyse RNA transcripts. For such an overview the reader is strongly recommended to consult the much more detailed descriptions of hybridization protocols given in *Nucleic Acid Hybridisation*, another volume in the Practical Approach series (74) which covers this topic in depth. This section of the chapter will confine itself instead to the techniques most frequently used for the characterization of mitochondrial RNA.

3.7.2 *Choice of mtDNA probe for hybridization analysis of* in vitro *transcribed mitochondrial RNA*

The products of *in vitro* transcription can be probed with a restriction digest of mtDNA immobilized on nitrocellulose or DBM-paper. There are many published physical maps of the mtDNA from a variety of species throughout the scientific literature. An alternative, and usually more convenient, approach is to clone defined fragments of the mtDNA into double-stranded plasmid DNA or the single-stranded DNA of M13 bacteriophage. Gene banks derived from the mtDNA of a wide range of organisms have been established by a number of groups around the world and clones of a wide range of genes can usually be obtained from one source or another. In the absence of the availability of such clones the required mtDNA restriction fragments can be cloned using a protocol very similar to that given in Section 2.2.3.

Restriction digests of the cloned DNA can then be used to probe the RNA synthesized *in vitro* in the same way as the restriction digest of the mtDNA but with greater specificity. In the particular case of the yeast *S. cerevisiae,* a whole range of petite deletion mutants is available (see Section 8 of Chapter 7); petites covering most of the genome have been mapped and can be used as an alternative source of specific DNA probes.

A final aspect of the use of mtDNA sequences as probes is that one can take advan-

Table 18. Mapping with nuclease S1.

1. Hybridize 10–70 ng of S1 probes to 50 ng of mtRNA in 10 μl of 80% (v/v) formamide (de-ionized), 500 mM NaCl, 60 mM Hepes–NaOH, pH 7.2, 2.5 mM EDTA. Heat at 100°C for 90 sec, bring to 46°C and incubate for 2–10 h.
2. Dilute the mixture with 100 μl of 250 mM NaCl, 30 mM sodium acetate, pH 4.8, 1 mM $ZnCl_2$, 20 μg/ml denatured salmon sperm DNA and 50–100 Vogt units of nuclease S1.
3. Incubate the mixture at 30°C for 40 min.
4. To stop the reaction add a tenth volume of 3 M sodium acetate, pH 5, 3 vols of ethanol and 10 μg of carrier tRNA.
5. Pellet nuclease S1-resistant hybrids and resuspend in de-ionized formamide with 0.05% bromophenol blue and xylene cyanol as marker dyes. Heat at 100°C for 90 sec.
6. Separate the protected fragments by electrophoresis through 4% (w/v) polyacrylamide, 7 M urea (see *Table 12*).

tage of the remarkable sequence conservation of some coding sequences of the mtDNA. There is enough homology in the coding sequences of mtDNA genes of lower and higher eukaryotes to allow mtDNA genes to be used in cross-hybridization experiments between totally unrelated species (e.g. maize and yeast mitochondrial genes).

3.7.3 *Use of hybridization analysis to study the initiation of transcription in vitro*

(i) *S1 nuclease mapping*. The single strand-specific nuclease S1 has been used extensively by workers studying transcription. The principle of this technique involves solution hybridization of *in vitro* transcribed RNA to a DNA probe known to contain the transcriptional unit including the site of RNA initiation (75). After hybridization, the DNA–RNA hybrid is digested with nuclease S1 to remove non-specific DNA and regions of the specific RNA not hybridizing with the DNA probe. The size of the hybrid is determined on a non-denaturing gel and compared with the protected fragment generated between the probe DNA and the *in vivo* transcribed RNA. If they have identical sizes one can conclude that the mitochondrial RNA polymerase is initiating at the correct site *in vitro*.This method can also be used to map the 3′ termini of RNA synthesized *in vitro* to establish the accuracy of termination. A protocol for nuclease S1 mapping is given in *Table 18*, however because of differences in the base composition of the DNA the hybridization conditions will vary slightly according to the species of mtDNA.

(ii) *Primer extension analysis*. This method involves the hybridization of the RNA transcript to a DNA probe that is known to be entirely contained within the gene of interest but *not* including the site of initiation. The DNA probe is end-labelled (see Section 2.4.4) and hybridized in solution to the complementary RNA synthesized *in vitro* (76). The molar ratio of DNA probe to RNA should be in the region of 5- to 10-fold, for example 0.04 μg of a 50-nucleotide DNA primer hybridized to 0.2 μg of RNA 1 kb long; large excesses of primer should be avoided otherwise non-specific priming may occur. The DNA primer is then extended to the 5′ (i.e. initiating) nucleotide of the RNA using reverse transcriptase and unlabelled deoxyribonucleotide triphosphates. The products are then run on a denaturing gel which, after autoradiography, will show the primer extension product and the DNA probe. The product is accurately sized by

Table 19. Primer extension analysis of transcripts.

1. Hybridize the end-labelled DNA primer (see Section 2.4.4) to 1 μg of mitochondrial RNA in 120 mM KCl, 10 mM Tris-HCl, pH 8.5 in a final volume of 10 μl. Incubate the mixture at 65°C for 20 min then slowly cool to 42°C over 60–90 min[a].
2. Add 10 μl of 120 mM KCl, 100 mM Tris-HCl, pH 8.5, 20 mM $MgCl_2$, 10 mM dithiothreitol, 1.6 mM of each dNTP and 5–20 units of avian myeloblastosis virus reverse transcriptase.
3. Incubate the extension mixture for 2 h at 42°C.
4. At the end of the incubation add 3 vols of absolute ethanol chilled to −20°C. Leave the nucleic acids to precipitate for 2 h at −20°C.
5. Collect nucleic acids by centrifugation in a microcentrifuge at 13 000 *g* for 10 min.
6. Briefly dry the pellet under vacuum and resuspend it in 10–20 μl of 80% formamide, 20 mM EDTA, 0.04% bromophenol blue, 0.04% xylene cyanol.
7. Heat the sample at 100°C for 2 min followed by cooling on ice.
8. Separate the samples by electrophoresis in 6% polyacrylamide, 7 M urea, 90 mM Tris-borate, pH 8.3 (*Table 12*). Include suitable DNA size markers for accurate size determination of the primer extension product.
9. Visualize the primer extension product by autoradiogaphy.

[a]This suggested re-annealing temperature is for yeast RNA which has a very low content of G+C. For mitochondrial DNA with G+C content of 40–60% use a higher temperature, typically 55–60°C.

Table 20. Solution hybridization analysis.

1. End-label the chosen DNA probe using polynucleotide kinase (Section 2.4.4) the probe should be small, typically 100–200 bp.
2. Mix together the DNA probe with the purified mtDNA both in distilled water and lyophilize.
3. Re-dissolve the sample in 20 μl of 0.3 M NaCl, 0.03 M sodium citrate, 8 mM Pipes–NaOH, pH 6.4 containing, if desired, 1% SDS.
4. Incubate the solution at 65°C for 10 min in a sealed capillary.
5. Cool the solution to the hybridization temperature, 45°C for yeast, and incubate for 30 min.
6. *Either*
   Mix the hybridization solution with 4 μl of 35% Ficoll, 0.5% SDS, 0.1% bromophenol blue and 0.1% xylene cyanol and separate the mixture on a non-denaturing agarose or polyacrylamide gel.
   or
   Digest the non-hybridized RNA by digestion with S1 nuclease or RNase A, acid precipitate the hybrid with 10% trichloroacetic acid, collect on filters and measure the amount of radioactivity using a scintillation counter.

comparison with molecular weight markers to identify the exact position of the 5′ terminus of the transcript. Alternatively, the end-labelled primer extension product can be sequenced and compared with the DNA sequence in the coding region of the gene. A method for primer extension is given in *Table 19*. Methods for sequencing DNA are described in detail in another volume of the Practical Approach series (45).

3.7.4 *Determination of the rate of transcription of specific genes*

(i) *Solution hybridization*. A very useful and rapid method for the hybridization of DNA and RNA in solution is described in *Table 20*. While typically, the hybridization of filter-bound DNA to labelled RNA can take from 8 to 48 h, the hybridization of a short DNA probe to RNA in solution is complete in 15–30 min (77). The hybrid molecule can then be digested with RNase A or nuclease S1 to remove non-hybridized, single-stranded RNA. The hybrid can then be separated on a non-denaturing gel followed by autoradiography or fluorography to identify the hybrid. Alternatively, the hybrid can be precipitated with cold trichloroacetic acid and the amount of RNA hybridizing to the probe DNA quantitated by liquid scintillation counting of the acid-precipitable material.

(ii) *Dot–blot methods*. These methods allow the measurement of the level of specific transcripts depending on the exact nature of the DNA probe used. Labelled RNA can be hybridized to DNA on filters or, alternatively, the RNA can be dotted onto the filters and incubated with an appropriately labelled DNA probe. In the former case the DNA is blotted onto the filter as described in Section 2.3.1 and then, after pre-hybridization, the filter is incubated with either labelled mRNA or, more usually, its cDNA copy: note that not all transcripts have a poly(A) tail making the preparation of cDNA from these transcripts significantly more difficult. Thus it is more convenient to bind RNA to the filter and to hybridize it to a radioactive DNA probe. Prior to dotting the RNA onto the filter the RNA (~5 mg/ml) is denatured by adding an equal volume of 34% de-ionized glyoxal (see *Table 17*), 20 mM sodium phosphate, pH 6.5 and incubating the mixture at 50°C for 60 min in a small sealed tube. Prepare a range of dilutions in 0.1% SDS and apply 4 μl aliquots to a dried nitrocellulose filter previously soaked in 20 × SSC. Bake the filter at 80°C for 2 h and then wash it in 20 mM Tris-HCl, pH 8.0 to 100°C to remove any residual glyoxal. Transfer the filter to a plastic bag for pre-hybridization prior to hybridizing it to the labelled DNA probe (final concentration ~5 μg/ml, 10^8 c.p.m./μg) using essentially the same procedure as described in *Table 17*.

3.8 **Inhibitors of transcription in mitochondria**

There are a number of drugs and antibiotics that can be used to inhibit mitochondrial transcription. These inhibitors can be broadly divided into two categories based on whether they bind to the DNA or RNA polymerase. Template-binding drugs are numerous and this section deals only with those drugs that have been shown to inhibit RNA synthesis by binding preferentially to the mtDNA. Drugs that bind specifically to the mitochondrial RNA polymerase are relatively few but they do exist and can be used for some applications.

3.8.1 *Inhibitors that bind to DNA*

Ethidium bromide and acriflavine have long been known to bind preferentially to mtDNA. The exact features that make the mtDNA particularly sensitive to these drugs are not clear. The authors have found that, in the case of yeast, acriflavine is a more potent inhibitor of mitochondrial transcription than ethidium bromide. Both drugs, however, result in over 90% inhibition of transcription at relatively low concentrations (5–10 μg/ml). Rat mtDNA transcription is also sensitive to these drugs but the inhibition is not as complete as is found with yeast mitochondrial transcription. Acriflavine at 10 μg/ml inhibits transcription by only 26%. Daunomycin, an anthrocycline antibiotic, has been shown to inhibit yeast mitochondrial RNA synthesis preferentially *in vivo*. Work in the authors' laboratory has shown this antibiotic to be a very effective inhibitor of RNA synthesis in an *in vitro* assay using osmotically-shocked yeast mitochondria. Rat-liver mitochondrial transcription is also sensitive to this drug. Actinomycin D is a frequently used inhibitor of RNA synthesis. It acts by binding specifically to the GpC sequences of the DNA. Yeast mitochondrial transcription is very sensitive to the addition of this drug in an *in vitro* assay.

3.8.2 *Inhibitors binding to RNA polymerase*

The resistance of mitochondrial RNA polymerases from a variety of species of α-amanitin is well established; concentrations as high as 1 mg/ml lead to no significant inhibition of mitochondrial RNA polymerase activity. However, this insensitivity has allowed this antibiotic to be used to assess the degree of contamination of isolated mitochondria or purified fractions of mitochondrial RNA polymerase with nuclear RNA polymerases. Rifampicin and its derivatives have been used extensively to characterize the RNA polymerase activity in mitochondria. Rifampicin is an extremely effective inhibitor of prokaryote RNA polymerases. However, it has been found that some species appear to contain rifampicin-sensitive mitochondrial RNA polymerase activity while others do not. Rat-liver mitochondria and the purified rat mitochondrial RNA polymerase have been shown to be sensitive to this antibiotic by some workers (78). A rifampicin-insensitive activity has been isolated from rabbit-liver mitochondria (55). The *X. laevis* mitochondrial RNA polymerase is sensitive to only some of the rifampicin derivatives.

Fungal mitochondrial RNA polymerases have been shown to be both sensitive (79,80) and insensitive (81) to rifampicin. Finally, plant mitochondrial RNA polymerases are the most sensitive to the addition of rifampicin. There is a 50% decrease in plant mitochondrial RNA polymerase activity at a concentration of 5 μg/ml (61). In conclusion, there is a general response to rifampicin. Sensitivity appears to be species specific and largely dependent on the procedure used to isolate either the mitochondria or the mitochondrial RNA polymerase.

Heparin is a sulphated polysaccharide that has been used to study the kinetics of *E. coli* RNA polymerase initiation and elongation in prokaryote *in vitro* assays. Transcription in osmotically-shocked yeast mitochondria is sensitive to heparin in the presence of $MgCl_2$ and NaCl. Results from this laboratory suggest that heparin does have a complex inhibitory effect on the yeast mitochondrial RNA polymerase in that the presence of heparin does not inhibit initiation of new transcripts as it does in prokaryotic systems and the inhibition by heparin is not complete.

4. REFERENCES

1. Clayton,D.A. (1982) *Cell,* **28**, 693.
2. Attardi,G. (1985) *Int. Rev. Cytol.*, **93**, 93.
3. Tzagoloff,A. and Meyers,A.M. (1986) *Annu. Rev. Biochem.*, **55**, 249.
4. Quetier,F., Lejeune,B., Delorme,S. and Falconet,D. (1986) *Encyclop. Plant Physiol.*, **13**, 25.
5. Ter Schegett,J. and Borst,P. (1971) *Biochim. Biophys. Acta,* **254**, 239.
6. Parsons,R. and Simpson,M.V. (1973) *J. Biol. Chem.*, **248**, 1912.
7. D'Agostino,M.A. and Nass,M.M.K. (1976) *J. Cell Biol.*, **71**, 781.
8. Dunon-Bluteau,D.C., Cordonnier,A. and Brun,G.M. (1987) *J. Mol. Biol.*, in press.
9. Cordonnier,A. (1982) These de Doctorat 3eme cycle Universite de Paris *VI.*
10. Zakian,V.A. (1981) *Proc. Natl. Acad. Sci. USA,* **78**, 3128.
11. Lazdins,I. and Cummins,D. (1984) *Curr. Genet.*, **8**, 483.

12. Blanc,H. (1984) *Gene,* **30**, 47.
13. Hyman,B.C., Cramer,J.H. and Rownd,R.H. (1982) *Proc. Natl. Acad. Sci. USA,* **79**, 1578.
14. Tikomirova,L.P., Kryukov,V.M., Strizhov,N.I. and Bayev,A.A. (1983) *Mol. Gen. Genet.*, **189**, 479.
15. Glover,D.M. (ed.) (1985) *DNA Cloning: A Practical Approach*. IRL Press, Oxford.
16. Probst,G.S. and Meyer,R.R. (1973) *Biochem. Biophys. Res. Commun.*, **50**, 111.
17. Bolden,A., Pedrali Noy,G. and Weissbach,A. (1977) *J. Biol. Chem.*, **252**, 3351.
18. Tanaka,S. and Koike,K. (1977) *Biochim. Biophys. Acta,* **479**, 290.
19. Adams,W.J. and Kalf,G.F. (1980) *Biochem. Biophys. Res. Commun.*, **95**, 1875.
20. Bertazzoni,V., Scovassi,A.I. and Brun,G.M. (1977) *Eur. J. Biochem.*, **81**, 237.
21. Yamaguchi,M., Matsukage,A. and Takahashi,T. (1980) *J. Biol. Chem.*, **255**, 7002.
22. Wernette,P.M. and Kaguni,L.S. (1986) *J. Biol. Chem.*, **261**, 14764.
23. Wintersberger,U. and Blutsch,H. (1976) *Eur. J. Biochem.*, **68**, 199.
24. Christophe,L., Tarrago-Litvak,L., Castroviejo,M. and Litvak,S. (1981) *Plant Sci. Lett.*, **21**, 181.
25. Wong,T.W. and Clayton,D.A. (1986) *Cell,* **45**, 817.
26. Mignotte,B., Barat,M. and Mounolou,J.-C. (1985) *Nucleic Acids Res.*, **13**, 1703.
27. Pavco,P.A. and van Tuyle,G.C. (1985) *J. Cell Biol.*, **100**, 258.
28. Brun,G., Vannier,P., Scovassi,J. and Callen,J.-C. (1981) *Eur. J. Biochem,.* **118**, 407.
29. Castora,F.G. and Lazarus,C.M. (1984) *Biochem. Biophys. Res. Commun.*, **121**, 77.
30. Castora,F.G. and Kelly,W.G. (1986) *Proc. Natl. Acad. Sci. USA,* **83**, 1680.
31. Fairfield,F.R., Bauer,W.R. and Simpson,M.V. (1985) *Biochim. Biophys. Acta,* **824**, 45.
32. Cottrel,S.F. (1981) *Biochem. Biophys. Res. Commun.*, **98**, 1091.
33. Callen,J.-C., Dannebouy,N. and Mounolou,J.-C. (1980) *J. Cell. Sci.*, **41**, 307.
34. Charret,R. and Andre,J. (1968) *J. Cell Biol.*, **39**, 369.
35. Tourte,M., Mignotte,F. and Mounolou,J.-C. (1981) *Dev. Growth Differ.*, **23**, 9.
36. Gurdon,J.B. (1968) *J. Embryol. Exp. Morphol.*, **20**, 401.
37. Bouteille,M. (1976) *J. Microsc. Biol. Cell,* **27**, 121.
38. Berk,A.J. and Clayton,D.A. (1974) *J. Mol. Biol.*, **86**, 801.
39. Barat,M., Dufresne,C., Pinon,H., Tourte,M. and Mounolou,J.-C. (1977) *Dev. Biol.*, **55**, 59.
40. Sommerville,J. and Scheer,U., eds (1987) *Electron Microscopy in Molecular Biology – A Practical Approach*. IRL Press, Oxford.
41. Inman,R.B. and Schnös,M.J. (1970) *J. Mol. Biol.*, **49**, 93.
42. Kleinschmidt,A.K., Lang,D., Jackerts,D. and Zahn,R.K. (1962) *Biochim. Biophys. Acta,* **61**, 857.
43. Baldacci,G., Cherif-Zakar,B. and Bernardi,G. (1984) *EMBO J.*, **3**, 2115.
44. Bishop,M.J. and Rawlings,C., eds (1987) *Nucleic Acid and Protein Sequence Analysis – A Practical Approach*. IRL Press, Oxford.
45. Rickwood,D. and Hames,B.D., eds (1982) *Gel Electrophoresis of Nucleic Acids – A Practical Approach*. IRL Press, Oxford.
46. Winkley,C.S., Keller,M.J. and Jachning,J.A. (1985) *J. Biol. Chem.*, **260**, 14214.
47. Levens,D. and Howley,P.M. (1986) *Mol. Cell. Biol.*, **5**, 2307.
48. Kelly,J.L. and Lehman,R.I. (1986) *J. Biol. Chem.*, **261**, 10340.
49. Schinkel,A.H., Groot-Koerkamp,M.J.A., Van der Horst,G.T.J., Touw,E.P.W., Osinga,K.A., Van der Bliek,A.M., Veeneman,G.H., Van Boom,J.H. and Tabak,H.F. (1986) *EMBO J.*, **5**, 1041.
50. Groot,G.S.P., van Harten-Loosbroek,N., Van Ommen,G.J.B. and Pijst,H.L.A. (1981) *Nucleic Acids Res.*, **9**, 6369.
51. Docherty,R. and Rickwood,D. (1986) *Eur. J. Biochem.*, **156**, 185.
52. Boerner,P., Mason,T.L. and Fox,T.D. (1981) *Nucleic Acids Res.*, **9**, 6379.
53. Levens,D., Morimoto,R. and Rabinowitz,M. (1981) *J. Biol. Chem.*, **256**, 1466.
54. Edwards,J.C., Levens,D. and Rabinowitz,M. (1982) *Cell,* **31**, 337.
55. Wintersberger,E. (1972) *Biochem. Biophys. Res. Commun.*, **48**, 1287.
56. Kuntzel,H. and Schafer,K.P. (1971) *Nature, New Biol.*, **231**, 265.
57. Wu,G.J. and Dawid,I.B. (1972) *Biochemistry,* **11**, 3589.
58. Evans,I., Linstead,D., Rhodes,P.M. and Wilkie,D. (1973) *Biochim. Biophys. Acta,* **312**, 323.
59. Mukerjee,H. and Goldfedar,A. (1973) *Biochemistry,* **12**, 5097.
60. Kantharag,G.R., Bhat,K.S. and Avadhani,W.G. (1983) *Biochemistry,* **22**, 3151.
61. Goswami,B.B., Chakrabarti,S., Dube,D.K. and Roy,S.C. (1974) *Physiol. Plant,* **32**, 291.
62. Chang,D.D. and Clayton,D.A. (1984) *Cell,* **36**, 635.
63. Gaines,G. and Attardi,G. (1984) *Mol. Cell Biol.*, **4**, 1605.
64. Shuey,D.J. and Attardi,G. (1985) *J. Biol. Chem.*, **260**, 1952.
65. Gaines,G., Rossi,C. and Attardi,G. (1987) *J. Biol. Chem.*, **262**, 1907.
66. Goswami,B.B., Chakrabarti,S., Dube,D.K. and Roy,S.C. (1974) *Physiol. Plant,* **32**, 291.

67. Good,N.E., Winget,G.D., Winter,W., Connelly,T.N., Izawa,S. and Singh,R.M.M. (1966) *Biochemistry,* **5**, 467.
68. Newman,D. and Martin,N. (1982) *Plasmid,* **7**, 66.
69. Bhat,K.S., Kantharaj,G.R. and Avadhani,N.G. (1984) *Biochemistry,* **23**, 1695.
70. Levens,D., Ticho,B., Ackerman,E. and Rabinowitz,M. (1981) *J. Biol. Chem.,* **256**, 5226.
71. Monroy,G., Spencer,E. and Hurwitz,J. (1978) *J. Biol. Chem.,* **253**, 4481.
72. Hanausek-Walaszek,M., Walaszek,Z. and Chorazy,M. (1981) *Mol. Biol. Rep.,* **7**, 57.
73. Williams,J.G. and Mason,P.J. (1985) In *Nucleic Acid Hybridisation – A Practical Approach.* Hames,B.D. and Higgins,S.J. (eds), IRL Press, Oxford, p. 139.
74. Hames,B.D.H. and Higgins,S.J., eds (1984) *Nucleic Acid Hybridisation – A Practical Approach.* IRL Press, Oxford and Washington, DC.
75. Yoza,B.K. and Bogenhagen,D.F. (1984) *J. Biol. Chem.,* **259**, 3909.
76. Chang,D.D. and Clayton,D.A. (1985) *Proc. Natl. Acad. Sci. USA,* **82**, 351.
77. Nobrega,F.G., Dieckman,C.L. and Tzagoloff,A. (1983) *Anal. Biochem.,* **131**, 141.
78. Reid,D.B. and Parsons,P. (1971) *Proc. Natl. Acad. Sci. USA,* **68**, 2830.
79. Scragg,A.H. (1976) *Biochim. Biophys. Acta,* **442**, 331.
80. Kuntzel,H. and Schafer,K.P. (1971) *Nature, New Biol.,* **231**, 265.
81. Rogall,G. and Wintersberger,E. (1974) *FEBS Lett.,* **46**, 333.

CHAPTER 9

Synthesis of mitochondrial proteins

D.S.BEATTIE and K.SEN

1. INTRODUCTION

The mitochondrion is the only organelle of non-plant cells which has its own genetic system and machinery for protein synthesis. Less than 10% of the several hundred mitochondrial proteins, however, are synthesized on mitochondrial ribosomes; the remaining proteins are encoded by nuclear genes, translated on cytoplasmic ribosomes and imported into the mitochondria in a subsequent step. Indeed, all of the proteins of the outer membrane, the intermembrane space and the matrix are synthesized by the nucleocytoplasmic system. In addition, all of the proteins (except one ribosomal protein) of the mitochondrial protein synthesizing system and the vast majority of inner membrane proteins are synthesized extramitochondrially. The products of the mitochondrial protein synthesizing system include several hydrophobic components of the mitochondrial respiratory chain in the inner membrane, including, apo-cytochrome *b* of complex III (bc_1 complex), three subunits of complex IV (cytochrome *c* oxidase), two (or three depending on the species) subunits of the ATP synthetase complex (F_1-F_0) and one protein (Var 1), of the small subunit of the mitochondrial ribosome (1). Hence, the methods used in studying the synthesis of mitochondrial protein must involve a consideration of both protein synthesizing systems of the cell.

2. CYTOPLASMIC SYNTHESIS OF MITOCHONDRIAL PROTEINS

The transport of proteins synthesized in the cytoplasm into mitochondria has been shown to be a post-translational event (2). Mitochondria utilize at least three different mechanisms for importing cytoplasmically made proteins. These mechanisms vary according to the final destination of the protein.

(i) Proteins that have their final destination in the inner membrane, intermembrane or the matrix of the organelle are usually synthesized as longer precursors with an amino terminal extension (3−6); the precursors bind to a protease-sensitive component on the cytoplasmic side of the outer membrane (7,8); import of the precursors requires a membrane potential (9,10); and during import the precursors are cleaved to the mature form by a protease located in the matrix (11). The intermembrane polypeptides, for example cytochrome b_2 and cytochrome *c* peroxidase and some of the inner membrane proteins such as cytochrome c_1 (subunit IV of the bc_1 complex), and the iron-sulphur protein (subunit V of the bc_1 complex) undergo a two-step processing and are first cleaved into an intermediate form (12−15). The adenine nucleotide translocase is one protein of the inner mitochondrial membrane which is not synthesized as a larger precursor form.

(ii) Proteins destined for the outer membrane are not synthesized as larger precursors and do not require a protease-sensitive receptor to bind to the membrane or an energized inner membrane for import (16–18); however they are translocated into a space where they become resistant to digestion by exogenous proteases.

(iii) Cytochrome *c*, although a protein localized in the intermembrane space is not made as a larger precursor and does not require an energized state of the mitochondrial membrane for import. Interestingly, only the apoprotein of cytochrome *c* and not the holoenzyme can be imported into the mitochondria. The haem group is attached to the apoprotein once it is in the mitochondria to form the holoenzyme (19). The binding site for this protein on the outer mitochondrial membrane appears to be different from that of the proteins of the first category discussed above as these proteins do not compete with cytochrome *c* in binding to the membrane (20).

3. METHODS FOR STUDYING THE IMPORT OF PROTEINS INTO MITOCHONDRIA

Two main approaches have been used for studying the kinetics and mechanism of transport of newly-synthesized proteins from the cytoplasm into mitochondria: *in vivo* pulse labelling of the cells with radioisotopic tracers followed by a chase with the unlabelled compound; *in vitro* synthesis of the proteins in a cell-free protein translation system and their subsequent import into isolated mitochondria. More recently, coupled transcription–translation systems with isolated and cloned genes have been used successfully to define the sequences necessary for transport (21–24). Import studies *in vivo* have only been performed with inner membrane and matrix proteins. Firstly, the mature size of the imported protein is generally smaller than that of the precursor protein, and therefore a change in molecular weight clearly indicates that the protein has been imported into the mitochondrion. Secondly, the precursors of the inner membrane and matrix proteins can be easily accumulated in the cell after addition of uncouplers of oxidative phosphorylation such as carbonyl cyanide *m*-chlorophenylhydrazone (CCCP), since their import is dependent on a charged membrane (12,15). Import of proteins localized in the outer membrane has been studied with the help of *in vitro* translation systems (25).

Most of the research on the import process has been performed with the yeast *Saccharomyces cerevisiae*, the mould *Neurospora crassa* and rat-liver cells. In the following section the procedures that have been developed to study the import process in yeast will be described in detail. Studies using mammalian systems will be covered more briefly.

3.1 **Antibodies to membrane proteins**

The newly synthesized labelled polypeptides whose transport is being followed both *in vivo* and *in vitro* are present in extremely small amounts in the cell and consequently immunological procedures are needed to identify them. Antibodies are raised in rabbits by standard procedures (26–28). Many membrane proteins are insoluble in water and hence must be solubilized in suitable detergents for isolation, purification and subsequent injection into the animals. Proteins solubilized with either Triton X-100 or SDS have been used successfully to raise antibodies. Most of the antisera raised in our laboratory have been directed against SDS-denatured antigens. For example, complex III isolated from yeast mitochondria was subjected to electrophoresis on 12% preparative gels and the individual subunits excised from the stained gels, the proteins electroeluted from the gel and prepared for immunization (29). Before preparing the samples for immunization the eluted protein sample should be subjected to a second electrophoretic separation to ensure that a purified single band has been obtained for subsequent immunization. Antibodies to the entire complex III were also prepared by solubilizing the complex in 0.05% Triton X-100 before preparing it for immunization. About 500 μg of pure antigen is sufficient to raise the antibodies. Membrane proteins are usually poor immunogens and therefore for the initial immunization the proteins should be mixed with Freund's complete adjuvant or other adjuvant mixtures prior to subcutaneous injection into rabbits at multiple sites. Give booster injections of 100–200 μg of purified protein every 2 weeks. The antibodies thus prepared generally recognize not only the appropriate mature protein but also the precursor forms of the protein in the cell or the *in vitro* translation mixture. Sometimes, the antibodies recognize more than one antigen after immunoblotting, even when they were initially prepared against a single antigen. This result may have occurred because proteolytic fragments of higher molecular weight subunits co-migrated with the antigen against which the antibodies were raised. The antibodies can be analysed by simple immunodiffusion in agar. The presence of contaminating antigens can be avoided by isolation of the complex and the purification of the subunits as quickly as possible and in the presence of several protease inhibitors (29).

3.2 **Import of proteins into mitochondria *in vivo***

3.2.1 *Growth of yeast cells*

A wild-type strain of the yeast, *S. cerevisiae*, such as KL14-4A is generally grown in the semi-synthetic medium described in *Table 1*. If the cells are to be labelled with [^{35}S]methionine, they should be grown in a sulphate-free medium which contains all

Table 1. Growth medium for yeast.

In a final volume of 1 litre:	
Yeast extract	3 g
$CaCl_2 \cdot 2H_2O$	0.4 g
NaCl	0.5 g
$MgSO_4 \cdot 7H_2O$	0.7 g
KH_2PO_4	1.0 g
$(NH_4)_2SO_4$	1.2 g
$FeCl_3$	5 mg
Carbon sources:	
galactose	30 g
or	
glucose	50 g
Adjust to pH 5.0 with HCl before autoclaving	

the reagents listed in *Table 1* except that the $MgSO_4$ is replaced with 0.6 g of $MgCl_2.6H_2O$ and the $(NH_4)_2SO_4$ with 1.0 g of NH_4Cl.

The cells are grown at 29°C with vigorous shaking in 2 litre conical flasks containing at most 500 ml of culture medium, to mid-log phase (1.0–1.1 OD units at 650 nm). This optical density is usually reached after 14–16 h when the initial inoculum has an optical density of 1.8 and is added in a 1:50 dilution.

3.2.2 *Labelling of cells*

While [^{35}S]methionine is very well suited for fluorographic detection of labelled polypeptides, some of the polypeptides may not contain methionine. For example, core protein II of the cytochrome bc_1 complex does not label very well with [^{35}S]methionine or with [^{35}S]sulphate. For its detection cells must be labelled with [^{3}H]leucine (15,30). Similarly, subunit VI of cytochrome oxidase lacks methionine; however, its precursor contains methionine in the amino-terminal extension and therefore can be labelled with [^{35}S]methionine (30). A knowledge of the amino acid composition of the polypeptides is helpful in choosing the correct radioactive tracer and also in interpreting pulse–chase data (31).

Reagents.

Labelling medium:

The semi-synthetic medium described in *Table 1* is used for labelling, but the concentration of galactose and yeast extract are reduced to 0.3% and 0.5%, respectively. The low concentration of galactose minimizes any possible catabolite repression.

10 mM stock solution of CCCP in ethanol

[^{35}S]methionine: specific activity 800 Ci/mmol (30 TBq/mmol)

200 mM stock solution methionine

0.05% 2-mercaptoethanol

100% stock solution trichloroacetic acid (TCA)

Procedure.

(i) Harvest the cells by centrifugation at 3000 *g* for 5 min and wash them once with distilled water.

(ii) Resuspend the washed cells in the labelling medium at a concentration of 100 mg (wet wt)/ml.

(iii) Add CCCP to the cell suspension to a final concentration of 20 μM and pre-incubate the cells for 5 min, with shaking, at 29°C, prior to a 5 min pulse with [^{35}S]methionine (200 μCi/ml, 7.5 MBq/ml).

(iv) If the path of the precursor into the mitochondria is to be followed, then restore the membrane potential by the addition of 0.05% 2-mercaptoethanol (v/v) and chase the label with 10 mM unlabelled methionine for various periods of time.

(v) Take equal aliquots of cells from zero time to the end of the chase and stop the reaction by the addition of 20% TCA (w/v).

Figure 1 shows typical results obtained in our laboratory. Lane 1 shows the precursor form of the iron-sulphur protein which has accumulated in yeast cells in the presence

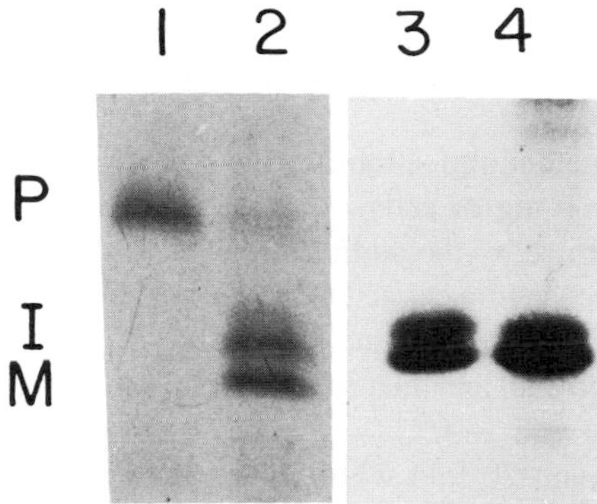

Figure 1. Comparison of precursor and mature forms of the iron-sulphur protein. Yeast cells (strain 777-3A) were pulsed with [^{35}S]methionine for 10 min at 28°C in the presence of 20 μM CCCP (**lane 1**) or in the absence of CCCP (**lane 3**). The membrane potential was re-established by the addition of 2-mercaptoethanol to the CCCP-treated cells and chased for 10 min. Aliquots were removed, mixed with 20% TCA and solubilized as described in Section 3.2.2. P, precursor; I. intermediate form; M, mature protein.

of CCCP. Lane 2 shows the intermediate and mature form of this protein after a 20 min chase with 2-mercaptoethanol. Lane 3 shows the intermediate and mature form of the iron-sulphur protein in cells labelled in the absence of CCCP.

3.2.3 *Immunoprecipitation of precursor and mature forms of proteins*

Reagents and solutions.

5% TCA
Glass beads (size 0.5 mm)
1.0 M Tris base
Sample buffer: 4% SDS, 0.15 M Tris–HCl, pH 7.4, 5 mM EDTA and 0.1% Bromophenol blue
Immunoprecipitation buffer: 0.1 M Tris–HCl, pH 7.4, 150 mM NaCl, 5 mM EDTA, 1% Triton X-100 and 1 mM phenylmethylsulphonyl fluoride (PMSF)
Specific antibody to a mitochondrial nuclear-encoded protein
Protein A-Sepharose: 2 mg/25 μl in 10 mM Tris–HCl, pH 7.4, 150 mM NaCl swollen for 60 min or overnight
Washing solution A: 50 mM Tris–HCl, pH 7.4, 150 mM NaCl and 0.2% Triton X-100
Washing solution B: 10 mM Tris–HCl, pH 7.4, 150 mM NaCl
Dissociation buffer: 5% SDS, 2 mM EDTA, 5% 2-mercaptoethanol, 100 mM Tris–HCl, pH 6.8.

Procedure.

(i) Add glass beads (3 g/ml of suspension) to TCA-denatured cells and mix the suspension vigorously on a vortex mixer at maximum speed for 3–4 min.
(ii) Carefully remove the suspension containing the denatured, lysed cells from the glass beads using a Pasteur pipette.

(iii) Rinse the glass beads several times with 5% TCA, pool the washings with the first decanted suspension and centrifuge the pooled supernatants at 1000 *g* for 10 min at 5°C.
(iv) Discard the supernatant and solubilize the protein pellet in 1.5 ml of 'sample buffer' for each 100 mg of cells.
(v) Neutralize the mixture by the addition of 1.0 M Tris base as indicated by the Bromophenol blue changing colour from yellow to blue, and then heat the mixture in a boiling water bath for 5 min.
(vi) Allow the mixture to cool and then remove the insoluble material by centrifugation at 23 000 *g* for 30 min at 4°C.
(vii) Dilute the supernatant 20-fold with the immunoprecipitation buffer, such that the final concentration of SDS is decreased to 0.2%.
(viii) Centrifuge the diluted supernatant at 100 000 *g* for 45 min and discard any remaining sediment.
(ix) Add the antibody against the protein to be detected to the supernatant and incubate the mixture with gentle shaking overnight at 4°C. The amount of antiserum to be added for a particular antigen should be first optimized by immunotitration which involves the addition of increasing amounts of antiserum to a fixed amount of labelled mitochondria or cell extract.

The immune complexes thus formed are then bound to protein A–Sepharose beads as follows:

(i) Incubate pre-swollen beads (12–15 mg swollen beads per 100 μl antiserum) with the mixture by end over end mixing for 60 min at room temperature.
(ii) Sediment the protein A–Sepharose-bound immunoprecipitates by centrifugation first in 15 ml glass Corex tubes, which can fit into adaptors for a 8 × 50 ml fixed-angle rotor, at 18 000 *g* for 10 min.
(iii) Remove the supernatant leaving about 1 ml in the tube.
(iv) Transfer the beads to 1.5 ml Eppendorf microcentrifuge tubes, and wash the sides of the Corex tube carefully so as to transfer all of the beads.
(v) Sediment the beads once more by centrifugation for 2 min in the microcentrifuge and remove the remaining supernatant.
(vi) Wash the immunoprecipitates thus obtained a total of five times.
(vii) Resuspend the pellet in 1 ml of washing solution A and mix vigorously using a vortex mixer then centrifuge in the microfuge. Remove the washing solution by suction. Using this technique wash the pellets three times with washing solution A and twice more with washing solution B. It is essential that the final wash is removed as completely as possible with a drawn out Pasteur pipette, taking care to see that no beads are withdrawn.
(viii) Release the immunoprecipitates from the Sepharose beads by heating them at 95°C for 5 min in 40 μl of dissociation buffer.
(ix) Remove the beads by centrifugation and save the supernatants at –20°C or run them directly on SDS–polyacrylamide gels.

After the protein A–Sepharose-bound immunoprecipitates have been removed by centrifugation at 18 000 *g*, the supernatant remaining can be used for immunoprecipitating another subunit of a multienzyme complex. Thus, we have successfully immunoprecipitated core protein I of complex III using its specific antiserum after first immunoprecipitating the iron-sulphur protein from the SDS- and Triton-solubilized mitochondrial extract.

3.2.4 *SDS–polyacrylamide gel electrophoresis and fluorography*

Reagents.

Acrylamide A: 22.2 g of acrylamide, 0.6 g of methylene bisacrylamide per 100 ml of water

Acrylamide B: 44.4 g of acrylamide, 1.2 g of methylene bisacrylamide per 100 ml of water

Ammonium persulphate: 100 mg/ml

Lower gel buffer: 1.5 M Tris–HCl, pH 8.8 with 6 M HCl (18 g of Tris base/ 100 ml)

Upper gel buffer: 0.5 M Tris–HCl, pH 6.8 with 6 M HCl (6 g of Tris base/100 ml)

Reservoir buffer: 6 g of Tris base; 28.8 g of glycine, 1 g of SDS, water to 1000 ml (pH 8.3)

Staining solution: Coomassie brilliant blue R 2.5 g; methanol 454 ml; glacial acetic acid 92 ml, water 454 ml.

De-staining solution: 300 ml of glacial acetic acid, 200 ml of methanol; 3.3 litres of water

Sodium salicylate 1.0 M

Procedure.

Slab-gels (14 cm × 15 cm × 1.2 mm thick) generally use the discontinuous SDS–Tris buffer system of Laemmli (32). The stacking gels are 5% acrylamide and the running gels are 10% or 12.5% acrylamide.

(i) Load the samples into their respective lanes and electrophorese them for 4–5 h at a constant current of 35 mA.

(ii) Stain the gels with Coomassie blue and destain overnight.

(iii) For fluorography soak the gel for 30 min in 1 M salicylate and dry onto Whatman 3MM paper.

(iv) The molecular weight standards can be marked on the dried gel with an ink pen which has a drop or two of [^{35}S]sulphate or methionine. Alternatively, radiolabelled standard proteins can be purchased and applied directly to the gel prior to running. Expose the dried gel to Kodak NS-5T X-ray film at −70°C for the appropriate time.

The amount of immunoprecipitated polypeptide can be quantitatively determined by scanning the autoradiogram with a densitometer. A quantitative determination can also be made by locating the band on the gel from the autoradiogram and excising the corresponding position. The piece can then be rehydrated by soaking it in distilled water for 30 min, incubated overnight at 60°C in 0.5–0.8 ml of NCS tissue solubilizer or 30% H_2O_2, cooled and counted in 10 ml of Liquiscint or any toluene-based water-miscible scintillation fluid.

3.3 ***In vitro* import of polypeptides into mitochondria**

Inner mitochondrial or matrix proteins synthesized in a cell-free protein synthesizing system can be successfully imported into isolated mitochondria in the presence of an energy source such as ATP plus a respiratory chain substrate. The mature form of the protein can be easily distinguished from the precursor because it is usually of lower molecular weight. The polypeptides are imported to their correct location within the mitochondria where they are protected against digestion by exogenous proteases. Under optimal conditions approximately 70% of a given protein can be successfully imported to the correct location; this can be assayed by testing whether the polypeptide is rendered resistant to protease digestion. Pre-treatment of mitochondria with trypsin allows the import of outer membrane proteins but not proteins destined for inner membrane or matrix fractions. The import system described below allows the study of the various steps involved in the import process.

3.3.1 *Isolation of yeast RNA*

Reagents.

Growth medium:

Yeast cells are grown in the medium described in *Table 1* with the exception that the galactose is replaced with 1 g of glucose and 22 ml of lactic acid per litre of water is added. The final pH is adjusted to pH 5.0 with 1 M NaOH.

Solutions for the isolation of RNA:

NETS buffer containing 150 mM NaCl, 50 mM Tris–HCl, pH 7.4, 5 mM EDTA, 5% SDS.

Phenol-chloroform-isoamyl alcohol mixed in the ratio 50:50:1.

2 M sodium acetate, pH 5.5

6 M LiCl

Procedure.

(i) Grow yeast cells to mid-log phase at 29°C under non-repressing conditions.

(ii) Cool the cultures rapidly by placing the flasks in an ice-salt bath for 15 min.

(iii) Harvest the cells by centrifugation for 5 min at 3000 *g* at 4°C, wash them once with cold water.

(iv) Resuspend the cells in cooled NETS buffer at 10 g wet wt/100 ml; do not overcool the buffer or the SDS will precipitate.

(v) Freeze the cells by mixing liquid nitrogen with the yeast suspension in a pre-cooled beaker. (CARE! use safety spectacles for this operation). Rotate the beaker every 15–20 sec so that there is uniform freezing of the yeast cells.

(vi) Transfer the frozen pellets rapidly to a cooled sterile blender and blend the cells at low speed for 15 sec followed by blending at high speed for 45 sec.

(vii) Thaw the white powder thus formed by the addition of 2 volumes of the phenol-chloroform-isoamyl alcohol mixture at 30°C, then transfer it to autoclaved Beckman polypropylene bottles and shake rapidly for 30 min. Siliconized Corex bottles can also be used in this step.

(viii) Centrifuge the mixture at 5000 *g* for 5 min at room temperature and transfer the upper aqueous phase to another bottle.
(ix) Extract the lower phenol layer by shaking it for 15 min with one half volume of NETS buffer and centrifuge at the same speed as before.
(x) Pool the upper aqueous layer with the first aqueous extract.
(xi) Extract the combined aqueous fractions three times with an equal volume of the phenol-chloroform-isoamyl alcohol mixture.
(xii) Shake each extract for 10−15 min and then centrifuge. It is essential that the interface remains with the lower layer at each separation.
(xiii) Precipitate the nucleic acids from the aqueous layer at −20°C overnight after the addition of 0.11 volumes of 2 M sodium acetate, pH 5.5 (so that the final acetate concentration is 0.2 M) and 3 volumes of ice-cold 100% ethanol.
(xiv) Collect the precipitated nucleic acids by centrifugation at 9000 *g* for 10 min at 4°C, dry completely *in vacuo*, and resuspend in 5 ml of water.
(xv) Add an equal amount of 6 M LiCl and allow the RNA to precipitate by incubating the solution at −20°C for 3−4 h.
(xvi) Collect the precipitate by centrifugation at 9000 *g* for 10 min, dry *in vacuo*, and resuspend in 5−6 ml of water.
(xvii) Subsequently, add 0.1 volume of 2 M sodium acetate and re-precipitate the nucleic acids by the addition of 3 volumes of cold ethanol as before. At this step, the ratio A_{260}/A_{280} should be 1.8−2. The concentration of RNA is calculated by assuming that a 1 mg/ml solution has an OD of 25 at 260 nm.
(xviii) Suitable small aliquots of RNA can be stored either in water or in sodium acetate, pH 5.5, and ethanol mixture at −80°C.

3.3.2 *In vitro translation*

Rabbit reticulocyte lysate for *in vitro* translation can be prepared from commercially available whole rabbit blood (Pel-Freeze Biologicals, Robers, Arizona) collected from rabbits made anaemic by injections with acetylphenylhydrazine (33).

(i) Store the lysate in small aliquots in sterilized Eppendorf tubes under liquid nitrogen.
(ii) Prior to nuclease digestion, thaw the lysate rapidly by warming by hand, place in an ice bucket, adjust to 20 μM haemin and 1 mM $CaCl_2$ and incubate at 20°C for 10 min in the presence of 4 μg of micrococcal nuclease per ml of lysate. The digestion conditions should be adjusted such that the incorporation of [^{35}S]-methionine is decreased to 5% of the incorporation achieved with the non-nuclease treated control.
(iii) Stop the nuclease digestion by addition of 2 mM EGTA and cooling the suspension in ice.
(iv) Alternatively, a nuclease-treated lysate can be obtained commercially in an *in vitro* translation kit from either New England Nuclear or Amersham. Store the nuclease-treated lysate in small aliquots at −80°C.

Optimal translation conditions. For each preparation of yeast RNA, the optimal concentration of RNA must be determined by titrating the nuclease-treated lysate with

Table 2. Incubation medium for cell-free translation.

In a final volume of 1ml
0.5 ml Reticulocyte lysate (nuclease-treated)
1 mM Hepes, adjust to pH 7.4 with KOH
100 mM KCl
1 mM $MgCl_2$
30 mg Creatine kinase
100 mM Creatine phosphate
400–1000 μCi (15–37 MBq) of [^{35}S]methionine
400–500 μg RNA (12–18 A_{260} units)

RNA. Usually 400–500 μg of RNA (12–18 A_{260} units) per ml of total reaction mixture is needed to give maximum incorporation. The composition of the reaction mixture is described in *Table 2*. The protocol supplied by New England Nuclear in their translation kit has also been used successfully to give high incorporation of [^{35}S]-methionine into proteins after optimizing the amount of RNA to be added. The reaction should be carried out at 30°C for 60 min.

Assay of the incorporation of radioactivity into proteins. The following protocol has been useful to optimize the RNA necessary for *in vitro* translation.

(i) Prepare an assay pre-mix containing all of the reagents except the RNA. The first tube contains 40 μl of the pre-mix, while the other tubes contain 20 μl.
(ii) Add RNA (4 μl) to the first tube and mix on a vortex mixer.
(iii) Remove 20 μl and transfer it to the second tube.
(iv) Continue this procedure taking care to mix well each time. In this way a series of assays containing 2-fold dilutions of RNA is established.
(v) In addition, keep one tube without any RNA as a control.

Incorporation of radioactivity into protein can be measured by two methods.

(a) *TCA precipitation*

(i) At fixed intervals throughout the incubation, remove 2 μl aliquots from each tube and add the sample to tubes containing 0.5 ml of 1 M NaOH and 5% (v/v) H_2O_2.
(ii) When all of the samples have been taken, place the tubes in a water bath at 37°C for 10 min to hydrolyse the aminoacyl-tRNA complexes and decolourize the samples.
(iii) Add 2 ml of ice-cold 25% TCA to each tube, mix the contents well and leave on ice for at least 30 min.
(iv) Filter the contents of each tube through a Whatman GF/C filter
(v) Wash the filter discs twice with 8% TCA (w/v) to remove unreacted labelled amino acids.
(vi) Dry the filters in an oven at 45°C overnight and measure the radioactivity of each in a liquid scintillation counter.

A 15- to 20-fold increase of incorporation of radioactivity from the basal level without added RNA should be observed for average preparations of RNA, while the best preparations obtained in our laboratory have resulted in as much as 100-fold stimulation. In addition, it is always useful to analyse the samples on a gel to ensure that high molecular weight proteins have been synthesized.

(b) *Slab-gel autoradiography*

(i) At fixed intervals during the incubations remove 5 μl aliquots and dissociate them immediately in 25 μl of dissociating buffer containing 2% SDS.

(ii) Separate the samples by electrophoresis on 10 or 12% polyacrylamide gels followed by autoradiography (see Section 3.2.4). Samples from later time points should reveal proteins of higher molecular weight relative to those observed at earlier time points.

For the studies of the import of proteins described in the next section the *in vitro* translations are generally performed in 1 ml aliquots for 60 min at 30°C. For immunoprecipitation of the precursors dissociate 100–125 μl of the translated lysate immediately with 3% SDS, 1 mM PMSF at 95°C for 3 min, dilute with immunoprecipitation buffer containing 1% Triton X-100 so that the SDS concentration is reduced to 0.2% and immunoprecipitate the proteins as described in Section 3.1.

3.3.3 *Import of labelled precursors into isolated mitochondria*

To study the import of precursors into mitochondria *in vitro*, chill the reticulocyte lysate after incubation with yeast RNA for 60 min at 4°C and then centrifuge it at 140 000 *g* for 45 min to sediment the polysomes. The post-ribosomal supernatant can then be frozen in liquid nitrogen and stored at −70°C prior to use or alternatively used directly for import studies.

Isolation of mitochondria from spheroplasts. To study the import of proteins into mitochondria, it is essential to isolate intact and well-coupled mitochondria from spheroplasts.

(a) *Reagents and solutions*

Growth medium: the semi-synthetic medium described above for the isolation of RNA.

Buffer A (washing buffer): 0.1 M Tris–HCl, pH 9, 0.35 ml of 2-mercaptoethanol/100 ml.

Buffer B (buffer to prepare spheroplasts): 1.2 M sorbitol, 20 mM KH_2PO_4, 0.35 ml of 2-mercaptoethanol/100 ml.

Buffer C (breaking buffer): 0.6 M mannitol; 20 mM Hepes–KOH, pH 7.4, 0.1% bovine serum albumin (BSA), 1 mM PMSF. (The BSA and PMSF are added fresh.)

Mitochondrial buffer; 0.6 M mannitol, 20 mM Hepes–KOH, pH 7.4.

(b) *Procedures*

(i) Grow up a wild-type yeast strain such as D273-10B or KL14-4A to mid-log phase, harvest the cells, wash them once with distilled water and pre-incubate them in Buffer A at a concentration of 1 g wet weight/3 ml at 29°C for 10–15 min.

(ii) Pellet the cells by centrifugation at 3000 *g* for 5 min, wash once with Buffer B and resuspend in this buffer at a concentration of 1 g wet wt/ml.

(iii) Add zymolyase 5000 to the suspension to a concentration of 4 mg/g of cells and incubate the cells at 30°C with shaking. Incubate the mixture for 40–60 min until the absorbance at 800 nm (of a 1:100 water-diluted cell suspension) is decreased to 10% of the initial zero time absorbance. Usually, the initial readings are 0.650 which decrease to 0.060–0.120 at 800 nm in about 45 min.

(iv) Pellet the spheroplasts by centrifugation at 3000 *g* for 5 min at room temperature, wash the cells twice with Buffer B and resuspend them in chilled Buffer C at a concentration of 0.3 g/ml.

(v) Transfer the suspension to a tight-fitting Dounce homogenizer kept in ice and homogenize with 10–12 strokes of the pestle.

(vi) Dilute the homogenate with an equal volume of Buffer C and centrifuge the suspension at 3000 *g* for 5 min at 4°C;

(vii) Save the supernatant in an ice bath and resuspend the pellet in Buffer C as before, homogenize and centrifuge again.

(viii) Combine the supernatant obtained from the wash with the first supernatant.

(ix) Obtain a crude mitochondrial pellet by centrifuging the supernatant at 18 000 *g* for 10 min at 5°C.

(x) Resuspend this pellet gently in Buffer C and centrifuge it at 3000 *g* for 5 min to remove the remaining cellular debris.

(xi) Centrifuge the supernatant from the low-speed spin at 18 000 *g* and wash the pellet containing the mitochondria four times with the mitochondrial buffer.

(xii) Suspend the final mitochondrial pellet in mitochondrial buffer at a concentration of 10 mg protein/ml.

(xiii) The approximate mitochondrial protein content can be determined by diluting 10 μl of the suspension with 1.0 ml of 0.6% SDS. A 10 mg/ml concentration of the original suspension gives an optical density of 0.2 at 280 nm after dilution.

(xiv) The structural integrity of the mitochondria can be determined by measuring the respiratory control ratio (rate of oxygen uptake in the presence of ADP divided by the rate of oxygen uptake in the absence of ADP) in an oxygraph using succinate or ethanol as a substrate (34). For a suitable experimental protocol see Chapter 1.

Precursor processing by isolated mitochondria.

(i) Thaw the centrifuged lysate containing the labelled proteins quickly and then pass through a Sephadex G-25 (medium) column equilibrated with 100 mM KCl, 20 mM Hepes–KOH, pH 7.4 prepared in a 5 ml syringe.

(ii) Incubate aliquots (0.1–0.15 ml, containing $4-6 \times 10^6$ c.p.m. of TCA-insoluble material) of the centrifuged and the gel-filtered lysate with 300 μg of mitochondria (~20 μl of suspension) in the presence of 1 mM ATP, 1 mM $MgCl_2$, 5 mM phosphoenolpyruvate, 4 units of pyruvate kinase in a final concentration of 0.6 M mannitol, 20 mM Hepes–KOH, pH 7.4.

(iii) To accomplish this, add an equal volume of 1.2 M mannitol, 20 mM

Hepes–KOH, pH 7.4 to the mixture. Also add GTP (5–6 mM) to inhibit any matrix-located protease which might have been released from the mitochondria; adjust the final volume to 0.4 ml.

(iv) Incubate the mixtures at 27°C for 30–45 min. If the entire processing of the precursor, from the intermediate (if there is any) to the mature form, is to be followed the mixtures should be incubated for different periods of time.

(v) Stop the reaction by the addition of 1 mM PMSF and cooling to 4°C.

(vi) Pellet the mitochondria by centrifugation for 10 min at 18 000 *g* at 4°C and wash them once with buffer containing 0.6 M mannitol, 20 mM Hepes–KOH, pH 7.4.

(vii) Dissociate both the supernatant and the mitochondrial pellet with 3% SDS and isolate the protein to be studied by immunoprecipitation (Section 3.2.3). The entire mixture can also be dissociated directly with 3% SDS without first isolating the mitochondria. For outer membrane proteins, however, it is necessary to separate the mitochondrial pellet from the supernatant. In our laboratory, we have obtained better results when the mitochondria are pelleted before use.

Assay of import. The import of proteins synthesized in the cytosol across the mitochondrial membrane can be assayed by two methods. For proteins localized in the inner membrane and matrix fractions, the appearance of the shorter mature form indicates that the protein has entered the mitochondria where the processing reaction occurs. However, it is essential that GTP is present in the incubation mixture to inhibit any protease that may have been released from the matrix during mitochondrial preparation. Another more powerful technique is based on the observation that once the processed polypeptides are within the mitochondria they become insensitive to digestion by exogenous proteases. Thus, the insensitivity of proteins to protease digestion can be used for proteins localized in the outer as well as in the inner membrane. For the protection assay, the transfer mixtures are chilled on ice following import for 30 min. Add proteinase K to a final concentration of 250 μg/ml and incubate the mixture for 30 min at 0°C. Stop the protease digestion by the addition of 1 mM PMSF, and dissociate the samples in 3% SDS and then immunoprecipitate as described in the following section. The supernatants will not contain any labelled proteins because they will have been digested by the added protease. Trypsin (final concentration: 120 μg/ml) can also be used in which case the digestion is stopped after 30 min by adding 1.2 mg of soybean trypsin inhibitor and 1 mM *N-p*-tosyl-L-lysine chloromethyl ketone (TLCK). A word of caution here: the concentration of proteinase K required to digest the precursor should be determined for each different polypeptide. For example, it has been reported that a concentration of 10 μg/ml is sufficient to digest the precursor of the adenine nucleotide translocase that is not transferred, while the form transported into mitochondria or the pre-existing mature form is resistant to 260 μg/ml proteinase K (35).

3.3.4 *Quantitative immunoprecipitation*

(i) After the samples have been dissociated in 3% SDS, add 300 μg of SDS-dissociated unlabelled mitochondria to all the supernatant fractions. This ensures that all of the samples, pellets as well as the supernatants, will have the same

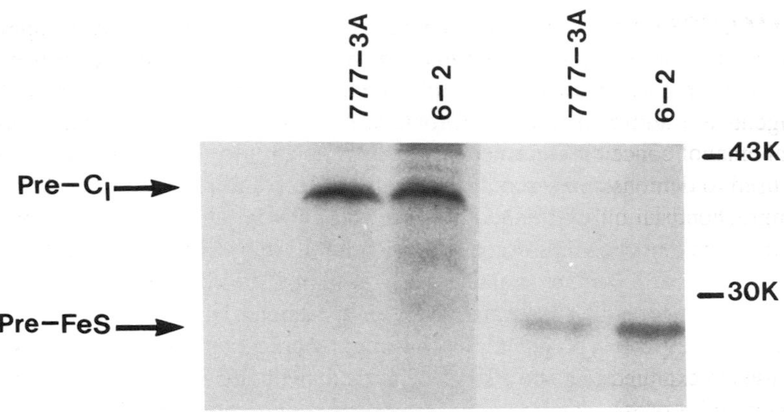

Figure 2. Precursors of iron-sulphur protein and cytochrome *c*, synthesized *in vitro*. Total yeast RNA was isolated from strains 777-3A and Box 6-2 and translated in a nuclease-treated reticulocyte lysate in the presence of [^{35}S]methionine. Each 125 μl of reaction mixture contained 50 μl of lysate, 1.25 A_{260} units of total yeast-RNA and 225 μCi (8 MBq) of L-[^{35}S]methionine (800 Ci/mol, 30 TBq/mol). The reaction mixture was incubated at 30°C for 60 min and the reaction terminated with 3% SDS at 95°C. Precursors of cytochrome c_1 and the iron-sulphur proteins were immunoprecipitated using the respective antiserum as described in Section 3.2.2.

amount of antigen to react with the immunoglobulins of the antiserum.

(ii) Carry out the immunoprecipitation with specific antibodies as described for the *in vivo* import process (Section 3.2.3) except that you should wash the protein A–Sepharose-bound immune complexes twice with buffer containing 50 mM Tris–HCl, pH 7.4, 150 mM NaCl, 0.2% Triton X-100 (Washing solution A), once more with the same buffer but with the salt concentration raised to 500 mM, and twice more with buffer containing no detergent (Washing solution B).

(iii) Perform slab-gel electrophoresis followed by autoradiography and quantitation of the radioactivity in the bands described in Section 3.2.4.

A typical result is shown in *Figure 2* which shows the precursor form of the iron-sulphur protein and cytochrome c_1 which have been synthesized in the reticulocyte lysate programmed with yeast RNA.

The uptake of different precursors into mitochondria varies from protein to protein. One reason for the differences observed may be that different polypeptides have different stabilities in the reticulocyte lysate. Furthermore, some polypeptides are imported more efficiently if they have not been stored at −70°C. The time required to detect an imported protein also varies for the different precursors. It should be noted, however, that mitochondria become sensitive to components in the reticulocyte lysate and begin to disrupt after incubation for 30–45 min. Consequently, processed mature forms of these proteins are then released where they become sensitive to digestion by exogenous proteases.

The *in vitro* import assay can be used to study the energy requirements for the import of proteins into the mitochondrion. Thus, pre-incubation of 200 μg of isolated mitochondria for 10 min with 2 mM KCN (to inhibit the respiratory chain), 20 μg of carboxyatractyloside (to inhibit the adenine nucleotide translocase) 1 μg of valinomycin or 2.5 nM CCCP (to discharge the membrane potential and pH gradient, respectively) prior to incubation with the filtered lysate under import conditions described above, will result in the accumulation of precursors in the supernatant. It should be noted that these agents are added as concentrated stock solutions prepared in ethanol such that the final ethanol concentration in the assay is 0.5% (v/v) or less. Similarly, the assay can be used to demonstrate that a trypsin-sensitive receptor component on the surface of the mitochondrial outer membrane is necessary for import of inner membrane and matrix proteins. Pre-treatment of 5 mg of mitochondria/ml with trypsin (10 μg/ml) at 0°C for 10 min will abolish the ability of the mitochondria to bind and import the precursor proteins destined for the inner membrane but has no effect on the import of outer membrane proteins and matrix proteins (7,8). The *in vitro* import system can also be used to examine whether mitochondria of species which are divergent in evolution can import and process precursors from other species. Information as to the universality of the mechanism, receptors and matrix proteases can thus be obtained.

3.4 *In vitro* transcription-translation systems

A major breakthrough in the methods to study the import of newly-synthesized proteins into mitochondria has been the development of efficient *in vitro* transcription-translation systems (21–25). These systems allow the accurate and efficient synthesis of single proteins from cloned DNA sequences. The high purity and specific activity of the [^{35}S]-methionine-labelled proteins produced in these systems has provided considerable knowledge of the protein sequences necessary for binding to the mitochondrial membrane as well as for the processing. A major feature in one system is the highly effective and efficient synthesis of mRNA using a promoter derived from coliphage T5 (21) which can produce specific mRNAs capped or uncapped. To do this insert the gene to be studied into a suitable plasmid containing the coliphage promoter and use purified RNA polymerase from *Escherichia coli* to transcribe the gene *in vitro*. Capping, necessary for the proper binding of eukaryotic mRNA to ribosomes, can be accomplished during transcription *in vitro* by including 7mGpppA in the transcription assay. The mRNAs produced in this way are subsequently translated in a cell-free system for translation.

The methods for preparation of the plasmids will not be covered in this chapter as these are covered in other volumes of the Practical Approach series (36). The reader should also consult the excellent articles by Hurt *et al.* (21) and Stueber *et al.* (23) for a detailed description of these methods as applied to genes for mitochondrial proteins.

Genes under the control of the coliphage T5 promoter are transcribed in a medium containing in a final volume 10 μl: 20 mM Hepes–KOH, pH 7.9, 10 mM magnesium acetate, 200 mM potassium acetate, 0.2 mM spermidine, 5 mM dithiothreitol, 0.5 mM each of GTP, CTP and UTP, 0.5 units of RNase inhibitor from human placenta, 6 μg of plasmid DNA (purified using a CsCl gradient) containing the gene to be expressed, 100 μM 7′-methylguanosine-adenosine (7mGpppA) and 1 unit of *E. coli* RNA polymerase.

(i) After a 3 min incubation at 37°C, add ATP to 1 mM and continue the incubation for 10–15 min at 37°C.
(ii) Stop the reaction by chilling and then add the reaction mixture directly to 30 μl

of nuclease-treated reticulocyte lysate in the medium described in Section 3.3.2 supplemented with 250 μg/ml of calf liver tRNA, 50 units of human placenta RNase inhibitor/ml and 1 mCi (37 MBq) of [^{35}S]methionine to a final volume of 50 μl.

(iii) After translation for 30–60 min at 30°C, spot aliquots of the reaction mixture on filter paper soaked in 5% TCA to measure incorporation of [^{35}S]methionine into protein or alternatively fractionate the entire mixture by SDS–polyacrylamide gel electrophoresis and fluorography as described in Section 3.3.2.

(iv) Then incubate the reticulocyte lysate containing the radiolabelled protein precursor with isolated yeast mitochondria as described in Section 3.3.2. The incubation medium generally contains 140 μg of mitochondrial protein, 40 mM KCl and 15 μl of reticulocyte lysate containing the radiolabelled polypeptide. After 30 min at 30°C, re-isolate the mitochondria by centrifugation through a 1 M sucrose cushion (30 000 *g* for 10 min at 4°C).

4. IMPORT STUDIES USING OTHER MITOCHONDRIAL SYSTEMS

The detailed descriptions of the synthesis of precursor proteins and their import into mitochondria both *in vivo* and *in vitro* presented in this chapter have involved the yeast *S. cerevisiae*. Similar studies have been performed in *N. crassa* and in various mammalian systems. The results obtained in these other systems have, in general, complemented those obtained with yeast.

The reader is referred to a series of articles published in a recent volume of *Methods in Enzymology* (37–39) which describe several aspects of studying mitochondrial biogenesis and assembly in *N. crassa*. As the articles originate from work in Neupert's laboratory a certain continuity of approach has been presented. The topics include the 'Biogenesis of Cytochrome *c* in *Neurospora crassa*' (37), the 'Biosynthesis and Assembly of Nuclear-Coded Mitochondrial Membrane Proteins in *Neurospora crassa*' (38), and the 'Isolation and Properties of the Porin of the Outer Mitochondrial Membrane from *Neurospora crassa*' (39). Thus, the biosynthesis and import of proteins localized in three different parts of the mitochondria are covered in sufficient experimental detail and will not be reiterated here.

4.1 **Import of proteins into mammalian mitochondria *in vitro***

The import of precursor proteins into mammalian mitochondria has also been studied successfully *in vitro*. The methods involve the synthesis of the precursor protein in a cell-free system programmed with RNA from the tissue of interest, followed by incubation of the newly-synthesized protein with mitochondria for various periods of time. In general, it has been necessary to isolate a mRNA fraction at an elevated pH and at a high ionic strength to avoid digestion with RNase by passage through an oligo(dT) column (40). For some other mitochondrial proteins, polysomes are prepared and then protein synthesis is initiated by read-out translation of free polysomes in a reticulocyte lysate in the presence of [^{35}S]methionine (41). Protein synthesis is stopped by chilling or by the addition of 0.2 mM cycloheximide (40). Post-ribosomal supernatants can then be obtained by centrifuging the chilled translation mixture at 190 000 *g* for 60 min.

For import studies, intact mitochondria with a good respiratory control ratio are necessary. Often the entire reticulocyte lysate is incubated with the mitochondria or in some cases the post-ribosomal supernatants are used. After a suitable period of time, the mitochondria are pelleted by centrifugation and the protein of interest immunoprecipitated and analysed on SDS−polyacrylamide gels. If the protein has been processed indicating its translocation into the proper site in the mitochondria, then the radiolabelled protein will have the same molecular weight as the mature form but will have a smaller size than the precursor form.

Now that the genes for several mitochondrial proteins have been isolated and cloned it has been possible to identify specific amino acid residues in the leader sequence of these proteins which are responsible for targeting these proteins to mitochondria (42). For example, Horwich *et al.* (42) recently reported that arginine in the leader peptide is required for both import and proteolytic cleavage of the precursor for ornithine transcarbamoylase (OTCase), a liver mitochondrial enzyme. The results were obtained using a coupled transcription-translation system similar to that described above for yeast. The plasmid $pSPOTC_2$ containing the entire human OTCase coding sequence plus plasmids containing an SP6 promoter sequence joined with a cloned coding sequence for a mutant OTC (pSP_mLOTC) are transcribed using SP6 polymerase. The transcription product can be isolated by ethanol precipitation and then incubated with a nuclease-treated reticulocyte lysate as described above. The products obtained are then either immunoprecipitated directly or incubated with freshly isolated mitochondria for 60 min at 27°C.

4.2 Import of proteins into mammalian mitochondria *in vivo*

Whole cell systems have also been used successfully *in vivo* for the study of protein import into mitochondria. These systems have included pulse-chased hepatocytes (43), chick embryo fibroblasts (43) or liver explants (40). Using these systems low levels of the precursors for ornithine transcarbamylase, carbamoyl-phosphate synthetase and aspartate aminotransferase were shown to accumulate during pulse-chase experiments. In these systems also CCCP prevented processing of the precursors which accumulated in the cytoplasm. The mature form again appeared after a chase with cysteamine to remove the uncoupler.

Kolarov and Nelson (44) have successfully used Zajdela hepatoma ascites tumour cells to study the import and processing of cytochrome bc_1 complex subunits. Both the iron-sulphur protein and cytochrome c_1 were shown to be synthesized as larger precursors of a similar size as those reported previously for yeast and *Neurospora*. Rhodamine 6B, a vital stain for mitochondria, was shown to be a potent inhibitor of processing in the isolated hepatoma cells. The authors suggest that Rhodamine 6B acts by affecting the matrix processing step rather than affecting the respiratory chain and consequently the membrane potential.

The reader is referred to the papers cited for specific details of these studies.

4.2.1 *Labelling of mammalian mitochondrial proteins in tissue culture*

Mitochondrial proteins of tissue culture cells can be simply labelled by incubation of cultured cells with [^{35}S]methionine in the presence of a drug that inhibits cytoplasmic protein synthesis such as emetine or cycloheximide (45). The following method is appropriate for cultured lymphocytes or fibroblasts. The specificity of the labelling can be assessed by using chloramphenicol with emetine which should result in the inhibition of all protein synthesis, both mitochondrial and nuclear.

Labelling of lymphoblast cells.

(i) Grow cells to a density of $10^6 - 5 \times 10^6$ per ml. An established cell line such as Raji cells is suitable.

(ii) Harvest the cells in 5 ml batches by centrifugation at 500 *g* for 5 min under sterile conditions.

(iii) Resuspend in 5 ml of methionine-deficient medium and 2% fetal calf serum in sterile flasks. Prepare three flasks.

(iv) Allow the cells to equilibrate by incubation at 37°C for 30 min in an incubator.

(v) To two flasks add emetine 100 μg/ml final concentration (from a stock solution of 10 mg/ml in sterile water, filtered prior to use). To one of these flasks also add chloramphenicol to a final concentration of 200 μg/ml (from a stock solution of 20 mg/ml in dimethyl sulphoxide, DMSO). To the final control flask add 5 μl of sterile water and 50 μl of DMSO.

(vi) Incubate for 10 min and then add 300 μCi (11 MBq) of [^{35}S]methionine (1 mCi/mmol, 37 MBq/mmol) and incubate for 2 h at 37°C.

(vii) Harvest the cells by centrifugation and wash four times with sterile phosphate-buffered saline (PBS) and store as pellets at −70°C.

(viii) Prepare a crude mitochondrial fraction as described in Section 2.1.3 of Chapter 1.

(ix) Dissolve the pellet in 2% SDS prior to electrophoresis or store at −70°C in pellet form.

Labelling of fibroblast proteins.

The fibroblasts adhere to the surface of the culture dish making manipulations easier.

(i) Grow three Petri dishes of cells to confluency (25 cm^2 surface area). Decant medium and wash cells twice with sterile PBS (Ca^{2+}/Mg^{2+}-free).

(ii) Add 4 ml of methionine-deficient medium and then 100 μg/ml emetine to two dishes. To one of these dishes add 200 μg/ml chloramphenicol (see previous section for details of stock drug solutions). The third dish is the control.

(iii) Add 80 μCi (3 MBq) of [^{35}S]methionine to each flask and incubate at 37°C for 2 h.

(iv) Wash the cells three times with PBS and harvest the cells by scraping. Trypsin is not recommended since it results in cleavage of some of the labelled proteins.

(v) Wash the cells three times with PBS by centrifugation at 500 *g* for 5 min. On the last centrifugation decant the buffer and centrifuge again. This allows a more complete removal of the buffer.

For the number of cells described above it is not practical to prepare mitochondria. Instead the whole cell pellet can be dissolved in 2% SDS and used for analysis by SDS−PAGE.

Figure 3 shows the result of labelling lymphoblast and fibroblast mitochondrial pro-

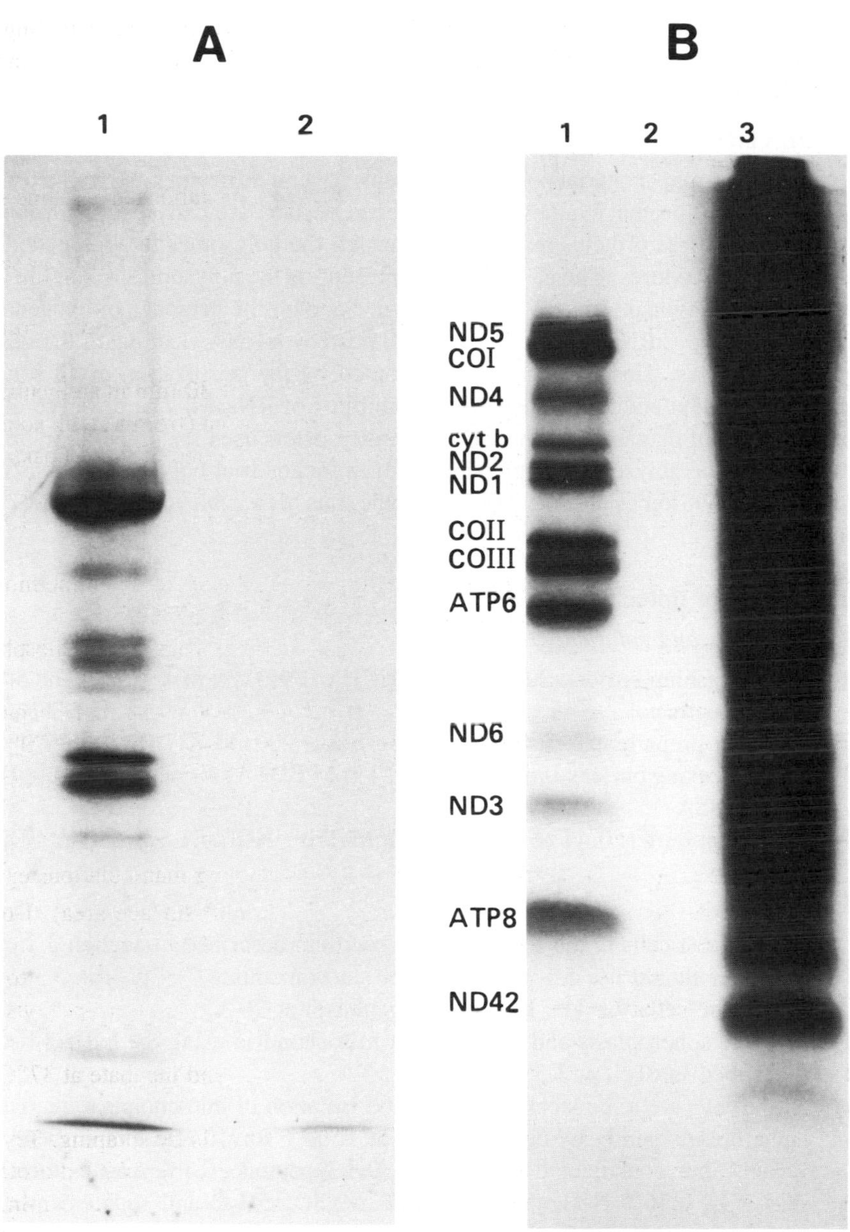

Figure 3. Labelling of mitochondrially-coded proteins in fibroblasts and lymphoblasts. (**A**) Shows the result of labelling fibroblast proteins and separating the products by SDS-PAGE in the presence of the cytoplasmic protein synthesis inhibitor emetine (**1A**) and emetine + chloramphenicol (**2A**). (**B**) Shows labelling of Raji lymphoblast cells in the presence of emetine (**1B**) chloramphenicol and emetine (**2B**) and no inhibitors (**3B**). The mitochondrial proteins (**1B**) are identified as described in (46,47). ND = components of NADH dehydrogenase, CO = cytochrome *c* oxidase subunits I, II and III, ATP = components of the ATP synthetase.

teins. The proteins are identified using the system of nomenclature described elsewhere (46,47).

5. ISOLATION OF MITOCHONDRIAL POLYSOMES

The isolation of mitochondrial polysomes presents several difficulties. First, the products of mitochondrial protein synthesis, as mentioned earlier, are extremely hydrophobic. Consequently, these proteins may interact causing the polysomes to aggregate during the isolation procedures. The nascent chains present on the polysomes would thus sediment during rate-zonal gradient centrifugation. Second, the presence of exogenous as well as endogenous RNases has a tendency to destroy mRNAs leading to the destruction of polysomes. This problem can be avoided by the sterilization of all solutions and the subsequent addition of heparin, an inhibitor of RNases. The presence of high concentrations of magnesium ions and KCl in the buffers used for preparation protects and stabilizes the polysomes. A preparation of mitochondrial polysomes consisting of two to eight monosomes plus the 74S monosome, has been achieved under these conditions (48).

5.1 **Isolation of mitochondria**

5.1.1 *Reagents and solutions*

Buffer A (washing buffer): 20 mM Tris–HCl, pH 9.0, 20 mM EDTA and 50 mM 2-mercaptoethanol

Buffer B (to prepare spheroplasts): 1 M sorbitol, 20 mM KH_2PO_4, pH 7.4

Buffer C (breaking buffer): 0.6 M mannitol, 1 mM EDTA, 10 mM KH_2PO_4, pH 6.8, 1 mg/ml BSA

Mitochondrial buffer: 0.44 M mannitol, 5 mM Tris–HCl, pH 6.8, 0.1 mM EDTA.

5.1.2 *Procedure*

(i) Grow yeast cells in the semi-synthetic medium described in Section 3.2 except that you should use 5% glucose as the carbon source.

(ii) Grow the cells for 11–13 h (mid-log phase) at 20°C.

(iii) Prepare spheroplasts and isolate intact mitochondria using the buffers listed as described in Section 3.3.3.

(iv) If the cells are to be labelled prior to the isolation of mitochondria, recover the spheroplasts gently by centrifugation at 1500 *g* for 10 min and suspend them in a medium containing 0.9 M sorbitol, 0.1% peptone, 0.1% yeast extract, 0.1% KH_2PO_4, 0.16% $(NH_4)_2SO_4$ and 0.1% $MgSO_4 \cdot 5H_2O$ and supplemented with 0.5% glucose and 1% ethanol as carbon sources.

(v) After incubation for 60 min at 30°C, add 80 μg/ml cycloheximide and continue the incubation for 15 min.

(vi) Add [^{14}C]formate (56 mCi/ml, 2 GBq/ml) to a final concentration of 20 μCi/ml (740 kBq/ml) and continue the incubation for 5 min at 30°C.

(vii) Terminate the reaction by the addition of 10 mM unlabelled formate and 4 mg/ml of chloramphenicol.

(viii) Isolate the mitochondria as described previously.

Polysomes isolated from labelled mitochondria can be used to assay whether the polysomes are of mitochondrial origin and are not contaminated by cytoplasmic polysomes.

5.2 **Isolation of polysomes**

Two methods can be followed for the isolation of polysomes from the mitochondria.

5.2.1 *Isolation of polysomes by sucrose density gradient centrifugation*

Reagents.

Buffer D: 50 mM Tris–HCl, pH 7.6, 14 mM $MgCl_2$, 300 mM KCl
Other reagents: Triton X-100

Procedure.

(i) Suspend the mitochondria by gentle homogenization in chilled Buffer D supplemented with heparin at 500 μg/ml at a concentration of 10–15 mg/ml and lyse the mitochondria by the addition of Triton X-100 to a final concentration of 1.7% for 5–10 min at 4°C.
(ii) Centrifuge the suspension twice at 27 000 *g* for 10 min at 5°C to remove mitochondrial membranes.
(iii) Layer the clear yellow supernatant, containing the crude mitochondrial ribosomes, onto a 35 ml 20–40% (w/w) linear sucrose gradient prepared in Buffer D supplemented with 100 μg/ml of heparin.
(iv) Centrifuge the gradient for 3.75 h at 95 000 *g* in a swing-out SW27 rotor or equivalent at 0–5°C.
(v) Determine the distribution of polysomes by puncturing the bottom of the tubes, collecting 1 ml fractions, and reading the absorbance at 260 nm for each fraction.

5.2.2 *Isolation of polysomes using a discontinuous sucrose density gradient*

(i) Layer the clear yellow supernatant, obtained by lysing the mitochondria with Triton X-100 and centrifuge as described above, on a discontinuous sucrose gradient. The gradient contains 1.5 ml of 2.5 M sucrose and 3 ml of 1.2 M sucrose prepared as described by Palacios *et al.* (49).
(ii) Centrifuge the gradient for 3.75 h at 170 000 *g* in a swing-out SW 50.1 rotor or equivalent at 4°C.
(iii) The opalescent polysomes which band at the boundary of the two sucrose layers in 0.5–0.7 ml of sucrose solution can be recovered by puncturing the side of the tube with a sterile syringe right below the band.
(iv) Prepare dialysis tubing by boiling in 5% $NaHCO_3$ containing 0.1 mM EDTA (to inactivate nuclease activity and chelate any metal ions) and then wash it thoroughly with sterilized distilled water.
(v) Place the polysomes in the tubing and dialyse for 4 h against Buffer D containing 40 μg/ml of heparin buffer placed in an ice-water bath.
(vi) After dialysis, centrifuge the preparation for 5 min at 37 000 *g* at 5°C.

The clear supernatant obtained contains polysomes which are free of tRNAs which may be present in polysomes obtained by the other method. These polysomes can be used for the study of *in vitro* protein synthesis. If the distribution of the polysomes is to be examined, the polysomes then can be layered a 20−40% continuous sucrose gradient onto and centrifuged as described above.

5.3 **Assay for presence of polysomes**

The presence of polysomes can be determined by measuring the absorbance at 260 nm of the different fractions which will reveal both the monosome plus polysomes consisting of 2−6 monosomes. Alternatively, polysomes can be prepared from ^{3}H-labelled cells and the radioactivity in the TCA-insoluble material in the fractions determined. Mix the mitochondrial polysome fractions at 1 ml/tube with 50 μg of BSA and 2 ml of 15% TCA and incubate at room temperature for about 30 min. Filter the contents through Millipore filter discs (0.45 μm Millipore filter) and the subsequent steps of washing the acid-insoluble material on the discs and the measurement of radioactivity in the scintillation counter are as directed earlier in Section 3.3.2.

6. IN VITRO SYNTHETIC SYSTEMS

Protein synthesis *in vitro* by mitochondria isolated from mammalian tissues, tumour cells, plants, *Neurospora* and yeast has confirmed that the proteins labelled after incubation *in vitro* are identical to those synthesized by the mitochondria *in vivo*. Thus isolated mitochondria have been shown to have the ability to synthesize the three polypeptides of cytochrome oxidase and cytochrome *b* (50,51). Protein synthesis *in vitro* has been studied also on isolated ribosomes and polyribosomes.

6.1 **Protein synthesis by isolated mitochondria**

For all studies with isolated mitochondria care has to be taken that the mitochondrial preparation is completely free of bacterial contamination, so that the possibility of bacteria contributing to the observed incorporation rate is completely eliminated. All solutions, especially the isolation medium, should be prepared using freshly distilled or autoclaved water. Solutions should be autoclaved and solutions which cannot be autoclaved should be passed through Millipore filters with a pore size of 0.45 μm prior to use.

Bacterial contamination of the incubation flasks can be determined at the end of each experiment by plating 0.1 ml of incubation medium on blood agar plates. The plates are incubated at 30°C for 72 h and then the colonies are counted. The bacterial contamination generally varies from 70 to 1400 bacteria/ml. Bacterial contamination of this level does not affect the results of mitochondrial protein synthesis *in vitro* (52).

6.1.1 *Reagents and solutions*

The protein synthesizing medium is described in *Table 3*.

Mitochondrial buffer: 0.6 M mannitol, 1 mM EDTA, 1 mg/ml BSA (0.1%), 10 mM KH_2PO_4, pH 6.8.

Table 3. Incubation medium for incorporating amino acids into isolated mitochondria.

In a final volume of 1.0 ml	A	B
Mannitol	60 mM	600 mM
KCl	50 mM	150 mM
KH_2PO_4	10 mM	15 mM
$MgCl_2$	10 mM	–
$MgSO_4$	–	12.5 mM
ATP	5 mM	4 mM
GTP	0.5 mM	0.5 mM
Phosphoenolpyruvate	5 mM	5 mM
Pyruvate kinase	6 units	10 units
α-Ketoglutarate	5 mM	5 mM
Tris–HCl, pH 7.2	20 mM	20 mM
L-leucine	20 μM	–
Amino acid mixture (minus leucine)	0.1 mM	0.1 mM
Bovine serum albumin	1 mg	3 mg
Cycloheximide	100 μg	50 μg
L-4,5-[^{3}H]leucine (55.9 Ci/mmol) (2 TBq/mmol)	0.08 mCi (3 MBq)	–

(A), from reference 52; (B), from reference 53.

6.1.2 *Procedure*

(i) Isolate yeast mitochondria carefully from spheroplasts as described in Section 3.3.3 and suspend them in the protein synthesizing medium at a concentration of 1 mg of protein/ml (100 μl of original suspension in mitochondrial buffer).

(ii) Add [^{3}H]leucine at a concentration of 10 μCi/ml (370 kBq/ml) for 5 min.

(iii) Carry out protein synthesis *in vitro* in batches of 0.5 ml for 40 min at 30°C. During incubation aerate the mixture by gentle shaking.

6.1.3 *Preparation of protein for radioactivity determination*

(i) Stop the reaction by placing the beakers or tubes in an ice bath and by adding 10 mM of unlabelled L-leucine to each vessel.

(ii) Precipitate the proteins by the addition of cold TCA to a final concentration of 5% and allow to stand at room temperature for 30 min.

(iii) The precipitated proteins are generally washed by two similar methods, one suitable for large amounts of proteins and one for small.

(iv) When several milligrams of protein have been precipitated, the pellet should be washed twice with 5% TCA at room temperature, once with 5% TCA at 70°C for 20 min to hydrolyse RNA, and twice with ethanol-ether (1:1, v/v) at 35°C.

(v) Wash the final pellet with water and then dissolve it in a known volume, generally 1.0 ml, of 0.4 M NaOH or a commercial solubilzer such as Soluene (Packard Company). Heating the samples to 60°C is sometimes necessary to solubilize all the proteins.

(vi) Remove aliquots (0.1 and 0.2 ml) for protein determination by the Lowry procedure and for measurement of radioactivity. Addition of 5–10 ml of a counting solution such as Bray's, a toluene-based solution containing Triton X-100, or a commercially available scintillation fluid for aqueous solutions gives comparable counting efficiencies.

When smaller amounts of protein are available, the precipitated protein in the tube should be first heated at 70°C for 15 min and then collected quantitatively by suction on Millipore filters (0.45 μm) which are then washed several times with cold 5% TCA. The filters are dried and then placed in 5.0 ml of scintillation fluids, such as toluene containing 4 g of 2,5-diphenyloxazole (PPO) and 0.2 g of dimethyl-1,4bis[2(5-phenyloxyazolyl)]benzene (POPOP) per litre.

6.3 **Protein synthesis on isolated ribosomes**

The sensitivity of protein synthesis to various antibiotics can be studied using ribosomes obtained from yeast and liver mitochondria (51).

6.3.1 *Preparation of ribosomes from yeast mitochondria*

Reagents and solutions.

Buffer E: 50 mM KCl, 10 mM $MgCl_2$, 10 mM Tris–HCl, pH 7.5.

Procedure.

(i) Convert yeast cells to spheroplasts and prepare mitochondria in the medium described above.

(ii) Suspend the mitochondrial pellet at a concentration of 3 mg/ml in Buffer E and lyse them by the addition of a tenth volume of 3% (w/v) sodium deoxycholate at 0°C.

Table 4. Incubation medium for protein synthesis on isolated mitochondrial ribosomes or polysomes.

In a final volume of 0.25 ml:	
10 mM Tris–HCl, pH 7.8	
50 mM KCl	
20 mM Magnesium acetate	
5 μM Tyrosine	
25 μM Amino acid mixture	
30 μM GTP	
1 mM ATP	
5 mM Phosphoenolpyruvate	
25 μg Pyruvate kinase	
6 mM 2-Mercaptoethanol	
100–250 μg S-100 from *E. coli*	
and either	or
Ribosomes (100 μg RNA)	Polysomes (1 A_{260} unit)
250 μg poly(U)	1.0 μCi (37 kBq) [^{3}H]leucine
50 μM [^{14}C]phenylalanine	(or [^{3}H]amino acid mixture)

(iii) Centrifuge the lysate at 26 000 *g* for 20 min and then centrifuge the supernatant at 160 000 *g* for 90 min.
(iv) Rinse the resulting pellet and resuspend it in the same medium.

6.3.2 *Protein synthesis on isolated ribosomes*

Reagents and solutions. Ribosomes obtained from either yeast or rat-liver mitochondria are active in protein synthesis when supplemented with poly(U) and supernatant factors from *E. coli* prepared as described previously (51). The medium used is described in *Table 4*. Generally, mitochondrial ribosomes should be added at a final concentration containing 100 μg of RNA. After incubation for 15 min at 30°C, precipitate the proteins and prepare the samples for counting by the Millipore filter method described above. Earlier studies had demonstrated that polyphenylalanine synthesis is dependent on the presence of ATP and the ATP-regenerating system, poly(U), supernatant factors, ribosomes and high concentrations of magnesium ions. A high-speed supernatant from liver mitochondria could replace the *E. coli* supernatant factors, in studies with liver mitoribosomes, but only one-half the activity was observed. A similar supernatant from yeast mitochondria would not support cell-free protein synthesis on yeast mitoribosomes to any significant extent. Furthermore, protein synthesis on both ribosomes was inhibited to the same extent by different concentrations of chloramphenicol, erythromycin and carbomycin but was unaffected by cycloheximide (54).

6.4 **Protein synthesis on isolated polysomes**

The isolation of polysomes from yeast mitochondria with a high activity of protein synthesis has been described in Section 5.2.

To study protein synthesis in isolated polysomes, 1 A_{260} unit is incubated in 0.25 ml of a medium similar to that used for cell-free protein synthesis on isolated ribosomes (*Table 4*). In some experiments the concentration of KCl was raised to 80 mM and that of magnesium acetate lowered to 10 mM. Using either [^{3}H]leucine or a [^{3}H]amino acid mixture results in almost identical results. In the latter case, the unlabelled amino acid mixture was eliminated from the incubation mixture. A high-speed supernatant fraction obtained from mitochondria could be substituted for the S-100 fraction from *E. coli*. Protein synthesis at 30°C under these conditions was linear for 40 min and was dependent on the ATP-generating system. The incorporation was inhibited by puromycin or chloramphenicol but unaffected by cycloheximide.

Polysomes with a similar activity in cell-free protein synthesis have also been isolated from Ehrlich ascites tumour mitochondria (55) by the Mg^{2+} precipitation procedure originally developed by Palmiter (56). The isolation of biologically active mitochondrial polysomes from two different organisms is highly significant, as mitochondrial polysomes may serve as a source for the isolation of mitochondrial mRNAs.

7. ACKNOWLEDGEMENTS

The studies from the authors' laboratory were supported, in part, by NIH grant HD-04007 and by NSF grant DMB83-20266.

8. REFERENCES

1. Schatz,G. and Mason,T. (1974) *Annu. Rev. Biochem.*, **43**, 51.
2. Gasser,S.M. and Hay,R. (1983) In *Methods in Enzymology.* Fleischer,S. and Fleischer,B.(eds), Academic Press, New York, Vol. 97, p. 245.
3. Viebrock,A. Perz,A. and Sebald,W. (1982) *EMBO J.*, **1**, 565.
4. Kaput,J., Goltz,S. and Blobel,G. (1982) *J. Biol. Chem.*, **257**, 15054.
5. Nagata,S., Tsunetsugu-Jokota,Y., Naito,A. and Kaziro,Y. (1983) *Proc. Natl. Acad. Sci. USA*, **80**, 6192.
6. Maarse,A.C., Van Loon,A.P.G.M., Riezman,H., Gregor,I., Schatz,G. and Grivell,L.A. (1984) *EMBO J.*, **3**, 2831.
7. Zwizinski,C., Schleyer,M. and Neupert,W. (1983) *J. Biol. Chem.*, **258**, 4071.
8. Riezman,H., Hay,R., Witte,C., Nelson,N. and Schatz,G. (1983) *EMBO J.*, **2**, 1113.
9. Scheleyer,J., Schmidt,B. and Neupert,W. (1982) *Eur. J. Biochem.*, **125**, 109.
10. Gasser,S.M., Daum,G. and Schatz,G. (1982) *J. Biol. Chem.*, **257**, 13034.
11. Bohni,P., Gasser,S., Leaver,C. and Schatz,G. (1980) In *The Organization and Expression of the Mitochondrial Genome*. Saccone,C. and Kroon,A.M. (eds), Amsterdam, North Holland, p. 423.
12. Reid,G.A., Yonetani,B. and Schatz,G. (1982) *J. Biol. Chem.*, **257**, 13068.
13. Daum,G., Gasser,S.M. and Schatz,G. (1982) *J. Biol. Chem.*, **257**, 13075.
14. Ohashi,A., Gibson,J., Gregor,J. and Schatz,G. (1982) *J. Biol. Chem.*, **257**, 13042.
15. Sidhu,A. and Beattie,D.S. (1983) *J. Biol. Chem.*, **258**, 10649.
16. Zimmermann,R., Paluch,U., Sprinzl,M. and Neupert,W. (1979) *Eur. J. Biochem.*, **99**, 247.
17. Freitag,H., Janes,M. and Neupert,W. (1982) *Eur. J. Biochem.*, **126**, 197.
18. Gasser,S.M. and Schatz,G. (1983) *J. Biol. Chem.*, **258**, 3427.
19. Zimmermann,R., Paluch,U. and Neupert,W. (1979) *FEBS Lett.*, **108**, 141.
20. Zimmermann,R., Hennig,B. and Neupert,W. (1981) *Eur. J. Biochem.*, **116**, 455.
21. Hurt,E.C.. Pesold-Hurt,B. and Schatz,G. (1985) *EMBO J.*, **3**, 3149.
22. Hurt,E.C., Pesold-Hurt,B., Suda,K., Oppliger,W. and Schatz,G. (1985) *EMBO J.*, **4**, 2061.
23. Stueber,D., Ibrahimi,I., Cutler,D., Dobberstein,B. and Bujard,H. (1984) *EMBO J.*, **3**, 3143.
24. Adrian,G.S., McCammon,M.T., Montgomery,D.L. and Douglas,M.G. (1986) *J. Mol. Cell Biol.*, **6**, 626.
25. Hase,T., Miller,V., Riezman,H. and Schatz,G. (1984) *EMBO J.*, **3**, 3157.
26. Williams,C.A. and Chase,M.W., eds (1967) *Methods in Immunology and Immunochemistry*. Vol. 1, Academic Press, New York.
27. Hurn,B.A.L., Chantler,S.M. (1980) In *Methods in Enzymology.* van Vanalicis,H. and Langone,J.J. (eds), Academic Press, New York, Vol. 70, p. 104.
28. Ouchterlony,O. (1953) *Acta Pathol. Microbiol. Scand.*, **32**, 231.
29. Sidhu,A. and Beattie,D.S. (1982) *J. Biol. Chem.*, **257**, 7879.
30. Teintze,M., Slaughter,M., Weiss,H. and Neupert,W. (1982) *J. Biol. Chem.*, **257**, 10364.
31. Gregor,I. and Tsugita,A. (1982) *J. Biol. Chem.*, **257**, 13081.
32. Laemmli,U.K. (1970) *Nature*, **227**, 680.
33. Pelham,H.R.B. and Jackson,R.J. (1976) *Eur. J. Biochem.*, **67**, 247.
34. Villalobo,A., Briquet,M. and Goffeau,A. (1981) *Biochim. Biophys. Acta*, **637**, 124.
35. Zimmermann,R., Paluch,V., Sprinzl,M. and Neupert,W. (1979) *Eur. J. Biochem.*, **99**, 247.
36. Glover,D. (ed.) (1985) *DNA Cloning: A Practical Approach, Vol. I and II.* IRL Press, Oxford.
37. Hennig,B. and Neupert,W. (1983) In *Methods in Enzymology*. Fleischer,S. and Fleischer,B. (eds), Academic Press, New York, Vol. 97, p. 261.
38. Zimmermann,R. and Neupert,W. (1983) In *Methods in Enzymology.* Fleischer,S. and Fleischer,B. (eds), Academic Press, New York, Vol. 97, p. 275.
39. Freitag,H., Benz,R. and Neupert,W. (1983) In *Methods in Enzymology.* Fleischer,S. and Fleischer,B. (eds), Academic Press, New York, Vol. 97, p. 287.
40. Shore,G.C., Rachubinski,R.A., Argan,C., Rozen,R., Pouchelet,M., Lusty,C.J. and Raymond,Y. (1983) In *Methods in Enzymology.* Fleischer,S. and Fleischer,B. (eds), Academic Press, New York, Vol. 97, p. 396.
41. Behra,R. and Christen,P. (1986) *J. Biol. Chem.*, **261**, 257.
42. Horwich,A.L., Kalousek,F. and Rosenberg,L.E. (1985) *Proc. Natl. Acad. Sci. USA*, **82**, 4930.
43. Doonan,S., Marra,E., Passarella,S., Saccone,C. and Quagliariello,E. (1984) *Int. Rev. Cytol.*, **91**, 141.
44. Kolarov,J. and Nelson,B.D. (1984) *Eur. J. Biochem.*, **144**, 387.
45. Attardi,G. and Ching,E. (1979) In *Methods in Enzymology.* Fleischer,S. and Packer,L. (eds), Academic Press, New York, Vol. 56, p. 66.
46. Chomyn,A., Mariottini,P., Cleeter,M.W.J., Ragan,C.I., Matsuno-Yagi,A., Hatefi,Y., Doolittle,R.F. and Attardi,G. (1985) *Nature*, **314**, 592.

47. Chomyn,A., Mariottini,P., Cleeter,M.W.J., Ragan,C.I., Doolittle,R.F., Matsuno-Yagi,A., Hatefi,Y. and Attardi,G. (1985) In *Achievements and Perspectives of Mitochondrial Research*. Quagliariello,E. *et al.* (eds), Elsevier, Vol. II.
48. Ibrahim,N.G. and Beattie,D.S. (1976) *J. Biol. Chem.*, **251**, 108.
49. Palacios,R., Palmiter,R.D. and Schimke,R.T. (1972) *J. Biol. Chem.*, **247**, 2316.
50. Poyton,R.G. and Groot,G.S.P. (1975) *Proc. Natl. Acad. Sci. USA*, **72**, 172.
51. Beattie,D.S. (1979) In *Methods in Enzymology*. Fleischer,S. and Packer,L. (eds), Academic Press, New York, Vol. 56, p. 17.
52. Everett,T.D., Fionzi,E. and Beattie,D.S. (1980) *Arch. Biochem. Biophys.*, **200**, 467.
53. McKee,E.E. and Poyton,R.O. (1984) *J. Biol. Chem.*, **259**, 9320.
54. Ibrahim,N.G., Burke,J.P. and Beattie,D.S. (1974) *J. Biol. Chem.*, **249**, 6806.
55. Lewis,F.S., Rutman,R.-J. and Avadhani,N.G. (1976) *Biochemistry*, **15**, 3362.
56. Palmiter,R.D. (1974) *Biochemistry*, **13**, 3606.

APPENDIX I

Single-letter code for amino acids

Amino acid	*One-letter symbol*
Alanine	A
Arginine	R
Asparagine (Asn)	N
Aspartic acid (Asp)	D
Asn and/or Asp	B
Cysteine	C
Glutamine (Gln)	Q
Glutamic acid (Glu)	E
Gln and/or Glu	Z
Glycine	G
Histidine	H
Isoleucine	I
Leucine	L
Lysine	K
Methionine	M
Phenylalanine	F
Proline	P
Serine	S
Threonine	T
Tryptophan	W
Tyrosine	Y
Valine	V

APPENDIX II

Phospholipid content of mitochondria

1. PHOSPHOLIPID ANALYSIS

Lipids were extracted from the samples in Triton X-100 (1). The extract [150–200 mg of total lipid in 2 ml of chloroform/methanol (1:1,v/v)] was separated on an activated silicic acid column (22 cm × 1 cm; Bio-Sil A; 200–400 mesh) eluted with methanol/chloroform mixtures in which the percentage of methanol was increased in steps. Phospholipid fractions were further separated by using two-dimensional t.l.c. on silica-gel G-60 plates (20 cm × 20 cm) with chloroform/methanol/NH_3 (260:19:9, by vol.) and chloroform/methanol/acetic acid/water (450:75:75:18, by vol.) as solvents I and II respectively. Spots were located with I_2 vapour and identified by comparison with authentic standards.

2. QUANTITATION OF PHOSPHOLIPIDS

The phospholipids were quantified after converstion to their fatty acid methyl esters (FAME). These were prepared by saponification followed by methylation with 14% boron trifluoride in methanol (2). The FAME were analysed quantitatively using a Perkin-Elmer gas–liquid chromatograph F17 equipped with a flame ionization detec-

Table 1. Weight of phospholipids (mg/g dry weight of sample).

	Tissue	Mitochondria	Associated with isolated complex IV
Beef	157	247	269
Dogfish	143	268	282
Cod	60[a]	170	94
	56[b]	168	130

Representative analyses are shown.
For cod analyses, results are shown for extractions performed in the presence ([a]) and absence ([b]) of 4-bromophenacyl bromide, a phospholipase inhibitor.

Table 2. Phospholipid composition (as percentage of total phospholipid) of mitochondria.

	PC	PI	PE	PS	CL
Beef	42	3	34	2.5	18.5
Dogfish	45	12.5	38	1.5	3
Cod	56	9	25	3	6

PC = phosphatidyl choline; PI = phosphatidyl inositol; PE = phosphatidyl ethanolamine; PS = phosphatidyl serine; CL = cardiolipin.

Table 3. %Composition

Fatty acid	PC			PI			PS			PE			CL		
	B	C	D	B	C	D	B	C	D	B	C	D	B	C	D
8.0	-	-	-	-	-	-	-	-	-	-	-	-	-	-	-
10:0	-	-	-	9.3	-	-	-	-	-	-	-	-	-	-	-
12:0	-	t	-	t	0.5	-	-	5.0	-	-	t	4	-	4.6	-
14.0	3.5	4.1	-	4	t	-	t	4.2	t	-	0.6	1.3	-	2.3	t
14:1	4.2	1.0	t	t	1.4	-	5.8	t	-	2.3	0.1	t	1.4	4.6	t
10:0	18.3	17.3	21.31	37.2	12.9	13.4	3.7	7.6	t	20.2	6.7	5.3	2.1	1.6	28
10:1	5.6	5.6	t	t	1.9	1.5	-	t	t	-	1.3	3.3	1.9	14.1	5.6
18.0	25.2	1.6	15.7	10.4	21.7	14.3	60.9	24.5	22.3	32	3.8	13.2	t	9.4	5.6
18:1	20.1	19.7	5.8	9.8	9.4	10.4	24.7	8	23.4	25	20.4	7.4	10.1	8:5	5.6
18:2	15.8	4.2	10.25	10.0	2.3	1.2	t	0.3	t	8.4	2.5	t	7.6	1.6	4.7
18:3	t	1.5	t	t	0.4	1.9	4.4	2.1	2.4	-	0.6	t	8.6	3.2	t
18:4	-	3.1	-	-	-	-	-	0.3	-	-	0.3	-	-	t	-
20:0	t	t	20.0	5.8	0.4	-	-	0.2	t	-	t	-	-	6.2	-
20:1	-	3.3	-	-	5.2	-	-	0.6	-	-	1.9	3.1	-	5.5	-
20:2	-	1.4	16.1	-	t	-	-	t	-	-	t	-	-	3.1	-
20:4	7.2	0.9	t	8.3	7.6	16.3	1.8	0.6	20.3	12.3	1.3	10.1	-	7.0	-
20:5	-	5.6	9.9	-	5.9	5.9	-	t	-	-	10.2	4.6	-	t	-
22:0	-	-	-	-	-	-	-	-	-	-	1.2	-	-	0.8	-
22:1	-	3.3	-	-	4.9	-	-	0.3	-	-	3.8	-	-	2.3	-
22:5	-	4.2	t	-	3.6	-	-	0.5	t	-	5.1	4.7	-	5.5	-
22:6	-	23.0	20.1	-	21.8	34.7	-	44.9	29.7	-	40.1	41.6	-	18.0	-

t = trace

Fatty acid composition of PC, PI, PS, PE and CL of Beef (B), Cod (C) and Dogfish (D) mitochondria. The fatty acids are named according to the convention that the first number represents the number of carbon atoms in the acyl chain while the second number is the degree of unsaturation. Thus 18:1 represents an 18-carbon mono-unsaturated fatty acid (oleic acid).

tor and fitted with a 185 × 0.6 cm glass column packed with 15% diethylene glycol succinate on 100–120 mesh gaschrom Q. The column was operated at 185°C.

Standard mixtures of FAME of known composition were used to both identify and quantitate the unknown FAME. Retention time was used for identification and there was a linear correlation between FAME concentration and peak area. Peak areas were calculated from the product of height and width at half peak height. The retention times and areas of unknown FAME were measured under identical operating conditions to those of FAME standards.

The percentage phospholipid composition of each extract may then be calculated from these data. The total peak area for each phospholipid was obtained by summing the areas of all its component FAME and could then be expressed as a percentage of the total phosophaolipid area.

For fuller details the reader is referred to standard texts (e.g. 3).

These tables illustrate the phospholipid composition of the mitochondria isolated from three species, i.e. a mammalian source and two aquatic species. These data are taken from refs 4 and 5.

3. REFERENCES

1. Bligh,E.G. and Dyer,W.J. (1959) *Can. J. Biochem. Physiol.*, **37**, 911.
2. Metcalfe,L.D. and Schmitz,A.A. (1961) *Anal. Chem.*, **33**, 363.
3. Kates,M. (1972) In *Laboratory Techniques in Biochemistry and Molecular Biology*. Work,T.S. and Work,E. (eds), North-Holland, Vol. 3, p. 269.
4. Al-Tai, W.F., Jones,M.G., Rashid,K. and Wilson,M.T. (1983) *Biochem. J.*, **209**, 901
5. Al-Tai,W.F., Jones,M.G. and Wilson,M.T. (1984) *Comp. Biochem. Physiol.*, **77B**, 609.

INDEX

Acetate in proton pumping, 41–51
Acetyl-Co-A chemical synthesis, 160
Aconitase assay, 161
ADP/ATP translocase, 38, 65–66
  preparation and reconstitution, 7
Amanitin, 280
Antibodies to mitochondrial proteins, 126, 284–285
Antimycin,
  mitochondrial spectrum, 14, 51
ARS sequences, 250, 262
Arsenazo III, 53–56
Asolectin,
  purification, 143
  proteoliposomes in, 143–151
Aspergillus nidulans, 227 ff (*see also* Mitochondrial mutants)
Aspartate/glutamate exchange, 65–66
ATP synthetase,
  assay, 109–110
  criteria of purity, 111
  electron microscopy, 31
  mitochondrial subunits, 301
  preparation, 108–109
  reconstitution, 147
  subunit composition, 110
ATPase, *see* ATP synthetase
Autonomous replication sequence, 250, 262

Bleomycin, 187

Calcium buffers, 57–59
Calcium/sodium exchange, 52
Calcium specific electrodes, 53, 57–61
Chloramphenicol, 124, 300, 302, 307
Cholic acid,
  in reconstitution, 144–147
  purification of inner membrane complexes, 85, 92, 97, 106
  recrystallization, 143
Citrate synthase,
  assay, 160
  mitochondrial integrity, 160
  preparation, 168
Collagenase, 5
Competent bacteria, 248
Complex I,
  assay, 89
  function in mitochondria, 9
  kinetic properties, 89
  mitochondrial subunits, 301
  preparation, 85–88
  purity, 89–91
  subunit composition, 90–91
Complex II,
  assay, 94
  criteria of purity, 95
  function in mitochondria, 9
  kinetic properties, 94
  polypeptide composition, 96
  preparation, 91–94
Complex III *see* Ubiquinol cytochrome *c* reductase
Complex IV, *see* Cytochrome *c* oxidase
Cycloheximide, 298, 305
Cytochromes–spectral characterization, 14–15
Cytochrome aa_3, *see* Cytochrome *c* oxidase
Cytochrome bc_1, *see* Ubiquinol cytochrome *c* reductase
Cytochrome *c*,
  absorption spectra in mitochondria, 137
  activity with complex III, 134
  activity with cytochrome oxidase, 137
    polarographic assay, 137
    proton pumping assay, 148–150
    spectrophotometric assay, 157, 148–150
  binding with complexes III and IV, 139
  function in mitochondria, 9
  modification of lysine groups, 130–134
    assay modified cytochrome *c*, 135–137
  synthesis and transport, 284
Cytochrome *c* oxidase,
  absorption spectrum, 102–105
    extinction coefficients, 105
  activity measurement, 107
    polarographic, 137, 138
    spectrophotometric, 137
  concentration in mitochondria, 14–15
  function in mitochondria, 9, 107
  labelling mitochondrial subunits in cells, 124, 283, 286, 301
  marker of inner membrane, 82
  preparation of, 101–107
    criteria of purity, 107
    subunit composition, 116
    subunit numbering, 116
    subunit purification, 117, 121
  proton pumping, 147–150
  reconstitution into liposomes, 145–146
  subunit II,
    amino acid sequence, 141

cytochrome *c* binding, 139–142
modification of carboxyl groups, 139
tryptic digestion, 141

Density gradients,
preparation mitochondria, 3–6
preparation DNA, 174, 248, 258
preparation RNA, 178
Deoxycholate, 85–86, 92, 93, 102, 154
Detergents, *see* also Digitonin, Deoxycholate, Cholic acid
subfractionation mitochondria, 80
membrane permeabilization, 153–158, 266
Digitonin, 80, 82
Dimethyl-oxazoline-2,4 dione, 41–51
DNA – mitochondrial (mt DNA), 245–246
associated proteins, 251
biotinylated, 186–187
cloning of, 247
detection biotin–avidin method, 184–187
DH–DNA (see D-loop)
polymerase (see Polymerases)
recombination, 178 (*see also* Recombination)
replication, 245–246
analysis by autoradiography
in vitro, 246
in vivo, 246
origin of, 247, 262
origin of by cloning, 262
origin of by electron microscopy, 262
origin of by sequence analysis, 263
purification associated proteins, 252–254
hybridization, 184–185, 276–278
isolation,
A. nidulans, 236–237
animal, 172–174, 184
CsCl–ethidium bromide method, 174
higher plant cells, 179–180
minilysate, 207
phenol chloroform extraction, 175–184
S. cerevisiae mutants, 223–225
tissue culture cells, 176
labelling *in vitro*,
blunt ends, 182
5′ single-stranded extension, 182
3′ single-stranded extension, 183
quantitation of, 256
Southern blotting, 184–186
structures, 171–172
synthesis,
liver mitochondria, 247
lysed mitochondria, 247
DNA-cellulose column, 253–254

Emetine, 124, 300
1-Ethyl-3-(3)trimethylaminopropylcarbodiimide protein modification, 139–141

Freeze etching, 21–24
Freeze fracture, 24–26
Fumarase
assay of preparation, 163
preparation, 168

H.p.l.c.-tryptic fragments, 141–142
Hybridization mt DNA, *see also* DNA, 276–278
Hydroxyethyl ethylenediaminetriacetic acid, 58

Immunoprecipitation, 287, 295–297
Inner membrane,
characteristics, 2, 9
labelling of proteins, 299–302
preparation, 81
Intramitochondrial volume, 43–44
Inulin, 42, 49, 64
Iodoacetamide–cytochrome *c* oxidase subunits, 123
Isocitrate dehydrogenase assay, 161–162
Isolation of mitochondria,
fungi, 7
heart muscle, 5, 84–85, 172–173
liver, 4, 184
plant tissues, 8, 179
tissue culture cells, 6, 177
yeast, 7, 302
Isopycnic centrifugation, 3
DNA purification, 174–257
gradient materials, 4

Ketoglutarate dehydrogenase assay, 162

Liposomes, *see* Reconstitution
Lysosomes–contaminating mitochondria, 159

Malate dehydrogenase,
assay, 81, 163–164
preparation, 168
Maternal inheritance, 172

Matrix,
composition, 2
enzyme assays, 159–164
enzyme preparation, 165–169
preparation, 81
volume, 43–44
Membrane permeabilization,
deoxycholate, 154
digitonin, 158
freeze-thawing, 157
lubrol, 81
octyl glucoside, 153
osmotic shock, 156, 157
sonication, 155, 156, 169
toluene, 157–158
Triton X-100, 153–154
Metabolon, 153
Metallochromic indicators, 53, 54–56
Methylamine–proton-motive force measurement, 41–51
Microscopic analysis,
electron microscopy,
freeze etching, 21–24
freeze fracture, 24–26
intramembraneous particles, 22, 25, 32, 33
negative staining reagents, 19–20
thin sectioning, 18–19
light microscopy,
analysis of mtDNA replication, 259
Mitcon proteins, 125
Mitochondrial coded proteins,
identification, 301
labelling, 124, 299–302
synthesis, 226, 304–306
Mitochondrial compartments–enzyme markers, 81–83
Mitochondrial DNA (mt DNA) *see* DNA
Mitochondrial genotype,
transfer cytoduction, 208
transfer sporulation, 207
Mitochondrial inheritance, 197
Mitochondrial integrity,
citrate synthase, 160
malate dehydrogenase, 80–81, 162–163
sonication, 14, *see also* RCR
Mitochondrial membranes,
freeze fracture, 29–32
Mitochondrial mutants,
Aspergillus nidulans, 222 ff
analysis of mitochondrial genome, 236
cleistothecia, 235
conidial suspension and treatment, 229
extracellular inheritance, 230
growth media, 228
heterokaryon, 230–231, 235, 241
isolation of mutants, 227
recombination, 233
restriction map, 238–239
sexual cross, 235
uniparental inheritance, 234
Saccharomyces cerevisiae, 188 ff
genome mapping, 211
growth media, 189–195
growth media antibiotics, 190
hybridization, 225
isolation of mutants, 195–202
minilysate analysis, 207
mutagenesis, 203
purity of strain, 205
recombination, 209–219
spontaneous mutation, 203
sporulation, 218
storage, 196
Mitochondrial polysomes, 302–304
Mitochondrial ribosomes, 306–307
Mitochondrial RNA (*see* RNA)
Monamine oxidase, 2, 82

Nagarse, 5
Negative staining, 19–21
Nick translation of DNA, 185, 255
Nonactin,
cell culture in, 123–125
Northern blotting, 272
Nuclear coded proteins,
import, 290, 293, 295, 298
nonactin, 123–125
precursor proteins, 287–289
processing, 294–297
transport, 283

Outer membrane, 2, 82, 156, 284
Oxygen electrode,
calibration, 12
H^+/O measurement, 38–40
measuring RCR, 12–14
preparation of proteoliposomes in, 151
respiration in mutants, 220

Phospholipids, *see* Appendix II
Phospholipid vesicles, *see* Proteoliposomes
Pi/OH antiporter, 38
Polymerases,
mitochondrial DNA, 251–254
mitochondrial RNA, 267–268, 280
Primase, 252

Primer extension analysis, 277–278
Protease inhibitors, 168, 295
Protein determination concentration of, 102, 168
Protein isolation–principles, 165–167
  affinity chromatography, 167
  ammonium sulphate, 103, 165
  dialysis, 167–168
  gel filtration, 167
  ion exchange, 166
  peg, 166
Protein synthesis, 220, 226
Proteoliposomes–principles, 74, 143–144
  ADP/ATP exchange carrier, 73–74
  cytochrome *c* oxidase, 145–146
  dialysis, 144
  F_0F_1 ATP synthetase, 147
  sonication, 144–146
  ubiquinol cytochrome *c* reductase, 147
Proton-motive force,
  definition, 36
  measurement, 43–51
Proton pumping,
  intact mitochondria by, 35–40
  potentiometric measurement, 148–150
  rapid kinetics of, 148–150
Proton/ATP ratio measurement, 38
Protoplasting,
  Aspergillus, 239–241
  enzymatic, 7, 9, 293–294
Pyruvate dehydrogenase assay, 16

Rampresa (Rapid Mixing and Sampling Pressure Filtration Apparatus), 72
Ramquesa (Rapid Mixing Quenching Sampling Apparatus), 71, 72
Rate-zonal centrifugation of DNA, 258
RCR (*see* Respiratory Control Ratio)
Recombination, 178, 209–219 (*see also* Mitochondrial mutants)
  analysis of genetic crosses, 216
  computation of distribution, 214
  genome mapping, 211
  isolation of recombinants, 217
    by sporulation, 218
    mating (MAT) system, 219
  linkage between markers, 213
  markers for rho mutants, 210
  quantitative random cross
Replicatiuon (*see* DNA)
Reconstruction (*see* Proteoliposomes)
Respiratory control ratio (RCR), 12–14, 148–151
Restriction analysis of DNA, 181–186
Rifampicin, 280
RNA, mitochondrial,
  denaturing formaldehyde agarose gels, 270
  isolation of,
    animal, 177, 268–270
    newly initiated transcripts, 273–275
    plant, 180
    yeast, 290–291
  isolation using,
    CsCl gradient, 178
    phenol, 178, 180, 269
  Northern blotting, 272
  synthesis in yeast, 269
  translation, 291–293
Rotonone, 43, 51, 79, 89
Rubidium uses in proton-motive force measurement, 41–51
Ruthenium red, as calcium blocker, 53

Saccharomyces cerevisiae, 188, *see also* Mitochondrial mutants
SDS–Page,
  autoradiography and fluorography, 122–125, 289, 293
  gel composition, 114
  hydrophobic proteins: anomolies, 113
  inner membrane proteins 113
  radioactive procedures, 122–125
  sample preparation, 114
  slicing and counting, 122
  staining, 116–117
  Western blotting, 123–129
Segregation tests, 203
  homoplasmic and heteroplasmic states, 188, 203–204
Sodium dithionite,
  oxygen electrode, 12
  special measurements use in, 14, 15, 102
Sonication,
  mitochondria, 14, 155–156, 169
  preparation of liposomes, 144–145
  sub-mitochondrial particles, 79
  TCA cycle enzymes, 156
Southern blotting, 184–186
Spin column chromatography, 185–186
Streptavidin, 186
Sub-fractionation of mitochondrial, 79–83
Sub-mitochondria particles, 38, 79
Succinate dehydrogenase, *see* Complex II
Succinate thiokinase assay, 162

Tetraphenylphosphonium,
TPP use in proton-motive force measurement, 41–51
Topoisomerase, 252
Transcription,
conditions, 266
end labelling, 272–273
general, 265–267
hybridization, 275–278
inhibitors, 279
initiation, 270–272, 275, 277
rate measurements, 270, 278
Transcripts, 272–275
Transformation of yeast, 249
Translation of RNA, 291
Triphenyl phosphonium (TPMP),
use in proton-motive force measurement, 41–51
Turbidity, 14–15, 101–192, 159

Ubiquinol cytochrome *c* reductase,
assay, 98–99
kinetic properties, 99
criteria of purity, 99–100
import of cytoplasmic subunits, 286, 299
mitochondrial subunits, 301
polypeptide composition, 100
preparation, 96–98
reconstitution, 147
S-Met labelling, 286
visible absorption spectra 14–15, 100

Western blotting, 123–129

Xenopus laevis mt-DNA cloning, 247

Yeast cells,
digitonin treatment, 158
growth, 285–286
labelling (S-MET), 286
transformation, 249
Yeast spheroplasts,
digitonin treatment, 158
mitochondria isolated from, 293

Zygotic suppressiveness,
tests for, 221–223
Zymolyase, 249